Bio-Privacy

Bio-Privacy: Privacy Regulations and the Challenge of Biometrics provides an in-depth consideration of the legal issues posed by the use of biometric technology. Focusing particularly on the relationship between the use of this technology and the protection of privacy, this book draws on material across a range of jurisdictions in order to explore several key questions. What are the privacy issues in the biometric context? How are these issues *currently* dealt with under the law? What principles are applied? Is the current regulation satisfactory? Is it applied consistently? And, more generally, what is the most appropriate way to deal with the legal implications of biometrics? Offering an analysis, and recommendations, with a view to securing adequate human rights and personal data protection, *Bio-Privacy: Privacy Regulations and the Challenge of Biometrics* will be an important reference point for those with interests in the tension between freedom and security.

Nancy Yue Liu is a research fellow at the Norwegian Research Center for Computers and Law and her main area of interests lie in legal issues about security and privacy. She is winner of the 2010 and 2008 Best Student Paper Award at the International Conference on Legal, Security and Privacy Issues in IT Law.

Bio-Privacy

Privacy Regulations and the Challenge of Biometrics

Nancy Yue Liu

Routledge
Taylor & Francis Group
a GlassHouse book

First published 2012
by Routledge
2 Park Square, Milton Park, Abingdon, Oxon OX14 4RN

Simultaneously published in the USA and Canada
by Routledge
711 Third Avenue, New York, NY 10017

A GlassHouse book

Routledge is an imprint of the Taylor & Francis Group, an informa business

British Library Cataloguing in Publication Data
A catalogue record for this book is available from the British Library

Library of Congress Cataloguing in Publication Data
Liu, Nancy Yue.
Bio-privacy: privacy regulations and the challenge of biometrics/Nancy Yue Liu.
p. cm.
"A GlassHouse Book."
Includes bibliographical references.
1. Privacy, Right of. 2. Biometric identification – Law and legislation.
3. Data protection – Law and legislation. 4. Biometry – Law and legislation.
5. Technological innovations – Law and legislation. I. Title.
K3263.L58 2012
342.08'58—dc22
2011010343

ISBN 13: 978–0–415–67790–5 (hbk)
ISBN 13: 978–0–203–80408–7 (ebk)

Typeset in Baskerville
by Swales & Willis Ltd, Exeter, Devon

Contents

Preface

This book is closely based on a thesis for which I was awarded the degree of Ph.D. at the University of Oslo in September 2010.

The research for this book has been a very pleasant and exciting adventure in my life. This is not only because the research topic is novel and challenging, it is also due to the compassionate people I encountered along the way and the fruitful research experience at the Norwegian Research Centre for Computers and Law (NRCCL), Faculty of law, University of Oslo. I feel especially thankful for all the help I have received during the past four years.

This doctoral research could not have been taken this far without the financial support from the Norwegian Research Council, in particular its research program titled 'Ethical, Legal and Social Aspects of Biotechnology, Nanotechnology and Neurotechnology (ELSA)'. For this I owe my thanks to the Council and to the senior advisor of the ELSA program, Helge Rynning, who has been very friendly and helpful from the beginning of the project and during the ensuing years.

I am extremely grateful to Professor Jon Bing of the NRCCL, and to Associate Professor Lee Bygrave of the Department of Private Law, Faculty of law, University of Oslo, for their contributions and helping hands. Professor Jon Bing with all his kindness and wisdom opened the door to this enriching research experience for me. His generous encouragement and gentle prodding for professional support and personal support during my sickness deserve my sincerest gratitude. Associate Professor Lee Bygrave, my supervisor, is the one who successfully applied for the funding of my research project by the Norwegian Research Council, and who has otherwise lit the way and persistently pointed out the right direction whenever I got lost. Our discussions have been of immense value for my work. His contribution is outstanding as he has not limited his input to the substantive content of the book, but also helped a lot with my language and format problems. He also encouraged me to build up contacts with many distinguished research organisations and professionals in the relevant field. This has greatly broadened my knowledge and contributed to the final research result.

I am also grateful for the contribution from the host institution representatives during my overseas research visits. Professor Graham Greenleaf at the Cyberspace Law and Policy Centre, University of New South Wales, Australia has read part of my book and provided valuable comments. He has been an inspiration through his writing on data protection issues in Australia and for his kind response to my e-mails. Professor Graeme Laurie facilitated my research visit to the Intellectual Property and Technology Law Centre, Edinburgh University, UK, and provided inspiring comments on my research plans and arguments. Professor Charles D. Raab, also of Edinburgh University, deserves special mention for his helpful comments on part of my work and for his assistance in making contacts. I also want to thank Professor Ian Walden of Queen Mary, University of London for his thoughtful comments.

My effort to analyse biometric technology and its impact on privacy would not have been possible without the information and inspiration from technical experts in biometrics and from 'policy practitioners' in the field. In both early and later stages of my work, many such experts and practitioners have donated time to patiently explain facts and practices on the ground, or comment on my arguments. Special thanks go to Professor Roger Clarke from the Australian National University, Dr. Ted Dunstone and Mr. Terry Aulich from the Australian Biometrics Institute, Mr. Stephen Wilson from Lockstep Consulting, previous Australian Privacy Commissioner Mr. Malcolm Crompton, Associate Professor Patrick Bours from Gjøvik University College and Professor Jan H.A.M. Grijpink from Utrecht University.

It has been said that the professional life of a research fellow consists of a complex legal problem and lots of deep water. However my pleasant working environment has made the efforts to stay afloat so much easier. Special thanks go to Eva Modvar, our Head of Administrative Service of the Department of Private Law, University of Oslo, and to the NRCCL Librarian Anne Gunn Bekken whose constant help and efficiency made my work much easier. Honourable mention goes to Professor Dag Wiese Schartum, Director of the NRCCL, for his valuable comments on my research plans and providing me inspiring resources for my research and to Dr. Peter Chukwuma Obutte and Dr. Jens Petter Berg, former doctoral research fellows at the NRCCL, who have given me a lot of mental support and encouragement both at early and later stages of my project. I also thank Professor Olav Torvund and all my other research colleagues at the NRCCL. Sincere thanks also go to Darren Read and Tim Challman who have helped with proof reading the final drafts of my book.

I am indebted to my parents, sister, brother and brother-in-law for being so kind and caring. I would like to thank my husband Sam for being there for me throughout the past years and my little son Jon for being so lovely and giving me so much courage to go on. I dedicate this work to them all for their good companionship, for the wonderful moments and all the sweet times we have together.

Nancy Yue Liu
12 December 2010

Table of cases

Court Decisions

European Court of Human Rights

Court of Justice of the European Communities

Swedish Supreme Administrative Court

United Kingdom, High Court

United States of America, Supreme Court

United States of America, Lower Courts

Decisions of Administrative Bodies

Norwegian Data Protection Tribunal (*Personvernnemnda*)

Swedish Data Protection Inspectorate (*Datainspektionen*)

Greek Data Protection Authority

Table of legislation and other legal texts

Australia (Commonwealth)

Council of Europe

European Union

Norway

Sweden

United States of America (Federal)

United States of America (State)

Part I

1 Introduction

1.1 Background

1.1.1 Development of biometrics

Heightened security concerns arising from the growth of various forms of crime, including identity theft and terrorism, have led to increased interest in the development and application of 'technologies of surveillance' (Lyon, 2001). These are technologies that – following the definition of 'surveillance' given by Lyon, a definition which is adopted for the purposes of this book – facilitate the 'collection and processing of personal data, whether identifiable or not, for the purposes of influencing or managing those whose data have been collected' (Lyon, 2001: 2). Technologies often have an inherent logic or bias which may strongly influence the way in which they are being used. The bias of technologies of surveillance is essential to augment surveillance capabilities. Thus, to some extent, the development of surveillance is also driven by new forms of technology (Bygrave, 2002). A prominent form of such technology is the rapidly expanding use of biometrics, that is, identification and authentication technologies based on unique characteristics of individual human bodies. Biometrics, though, is not exclusively a surveillance technology; beyond surveillance biometrics is also being developed and applied in a range of other contexts including personal computing, entrance security, and automated banking. Nonetheless, it is mainly in relation to their surveillance and control potential that biometric applications are attracting controversy. They are seen by some as part of an array of surveillance technologies that have advanced to the point where they now collectively possess the capability to threaten the most basic democratic notions of individual autonomy and privacy (Nuger and Wayman, 2004a).

The surveillance of cross-border traffic, especially through the use of biometric technology, has become more popular in the wake of the '9/11' terrorist attacks and the ensuing war on terrorism. As part of efforts to enhance security, governments around the world are investing large sums of money and human resources in biometric technology, which also act to accelerate the use of biometrics among private-sector entities. Technologies that previously had difficulty in surviving even their pilot stages (Australian Government, 2004; European Commission's

Joint Research Centre, 2005) were suddenly 'good enough' to be installed across vulnerable points of entry into a variety of countries, not only in the United States but also in many other Western and Asian countries (Ploeg, 2005).

Although biometric data have long been used in the realm of criminal proceedings such as in fingerprinting and DNA profiling, such data are increasingly being used by governments and private entities as a way of ascertaining individual identity in other contexts. Reference measures have expanded from the traditional types mentioned above to involve multiple human characteristics, such as the use of the facial image, iris, retina hand prints, and so on.[1] The use of biometrics is not restricted exclusively to travel and identity documents but extends also into commercial and other private sector domains, such as e-passports holding both the user's personal information and facial image;[2] fingerprints taken as a legal requirement for getting a visa, but not stored anywhere on the visa;[3] paying a chain store's bill by scanning the purchaser's fingers (Jain and Pankanti, 2006); access control of an oil depot using employee fingerprints[4] and checking in at the airport using a passenger's iris.[5] These systems seem different in their purpose and procedure, but each uses biometric technology in some way.

Biometric technology is not new. Quantitative measurement of biological human traits for the purpose of identification dates back to at least the 1880s, though the use of human characteristics to recognise people has arguably always been an integral part of human society. The first concrete record of the use of biometrics was a form of fingerprinting used in China in the 14th century, as reported by the Portuguese explorer Joao de Barros. Chinese merchants used fingerprints to settle business transactions, and Chinese parents used fingerprints and footprints to differentiate children from one another (Garfinkel, 2000). Despite its long history, biometrics never became a distinct field until a French police officer named Alphonse Bertillion sought to solve the problem of identifying convicted criminals by devising a system of physical body measurements (anthropometry) (Cole, 1999; Beavan, 2001). Bertillion's system included such measures as skull diameter and arm and foot length. 'Bertillion realised that even if names changed and even if a person cut his hair or put on weight, certain elements of the body remained fixed' (Garfinkel, 2000: 39). This system was called anthropometrical signalment and was very basic in its implementation. Nevertheless, it was a huge breakthrough and was adopted by prisons and police stations within national borders and across the world. Some drawbacks were found when this system was applied to a large population where no automatic means were applied, and anthropometrical signalment was found not to be totally unique. The true biometrics system[6] began to emerge in the 1960s, coinciding with the emergence of computer systems (Bolle et al., 1996). The nascent field experienced an explosion of activity in the 1990s, and it began to surface in everyday applications in the early 2000s (National Science and Technology Council, 2006b).

1.1.2 Public debate

The nature and diversity of subjects that can be explored in an analysis of biometrics and its social/legal implications makes it a particularly interesting field of study. Although much debate about the technology appeared before the dramatic changes brought about due to '9/11', the relevance of such debates appears only to have increased. Apart from the technological aspects, the public debate on biometrics focuses on privacy and body-integrity issues.[7] Contributors to the ongoing debate are drawn from the perspectives of politics, law, philosophy, and ethics (Woodward, 1997; Etzioni, 1999).

Radical, diverging positions are held on the issue of whether biometric technology is dangerous to individual privacy and democracy in general. On the one hand, biometrics is seen by Clarke as one of the most serious among the many technologies of surveillance that are threatening the freedom of individuals and of societies (Clarke, 2001). Chandrasekaran has also claimed that '[in] most cases the biometric technology is impersonal' (Chandrasekaran, 1997). It enables the bringing together of disparate pieces of personal information about an individual. If the technology meets with widespread success, individuals may find that they are required to provide a biometric identifier in unexpected and unwelcome future circumstances. In contrast, others maintain that biometrics constitutes a potential solution to privacy by safeguarding identity and integrity, and could even be privacy-enhancing when combined with cryptography techniques (Woodward, 1997). It has been further claimed that '[b]iometrics would put a sudden and complete stop to as much as 80% of all fraud activity' (Milbank, 1995: B1).

Beyond this debate, the pressing need for Western democracies to strengthen security and re-assess foreign policy, particularly in the aftermath of '9/11' and similar terrorist operations, such as those in Madrid (2004) and London (2005), gave rise to a host of proposals for using biometrics. Implicit in these proposals was that enhancing security unavoidably requires some 'trade-offs' with respect to civil liberties and privacy (Dersgowitz, 2002). Opponents to the 'trade-off' arguments respond in two ways. One opponent is Ann Cavoukian, Ontario's Information and Privacy Commissioner, who has called attention to the necessity of balancing the key factors involved in the debate. She argues that:

> At the heart of this debate is the balance between human rights and civil liberties, including privacy, and the intrusion on these rights in the name of public safety. We may have to rethink this balance. And that should come as no surprise. The balance between security and privacy has never been static, shifting in favour of security whenever we're faced with significant threats to public safety.
>
> (Cavoukian, 2001: 1)

Schneier and Sobel respond in another way: 'for any security proposal, regardless of its implications for privacy and other civil liberties, we must ask "is the security solution worth it?"' (Schneier, 2003: 14 and 15). Furthermore, '[l]iberties are far easier to give away than to get back' (Schneier, 2003: 243) and 'trade-offs made in the emotional aftermath of 9/11 will be with us for years, if not decades' (Sobel, 2002: 362). They go on arguing that privacy is diminished with little subsequent benefit, since very little has been put forward in terms of substantive security benefits.

On a more abstract level, there exists a political and philosophical discussion about the use of biometric technology which aims to understand how diverging assessments of biometrics, function in the public construction of the technology involved (Ploeg, 2005). This discussion does not seek to counter the constructions of technology by claiming that technology in itself is neutral, or that the degree of its threat to privacy depends on the specific use to which it is being put; instead, it is claimed that technology is highly political and normative (Winner, 1986; Adam, 1998). The main idea is that technology is the outcome of social processes and choices, and is continually in the process of being made (Ploeg, 2005).

It has been argued that ethical scrutiny begins with care for the other, and to relieve and to prevent suffering. The lesson taught by traditional bioethics should now apply to biometric technology (Mordini and Petrini, 2007). The effects of technology on the lives of people are studied, focusing not only on the immediate but also on the long-term social implications of biometric technology (Ashbourn, 2006).

1.1.3 State of legal research

The legal debate on biometric technology is generally restricted to the context of informational privacy, although occasionally physical privacy may also be relevant. Biometrics is treated as a technology that generates and processes a new type of personal data. In addition to general data privacy legislation, there are several legal initiatives to develop specific biometric regulations. For instance, there is 'soft law', exemplified by the Privacy Code of the Biometrics Institute in Australia. There are, for instance, legislative initiatives in the United States at the state level such as the Biometric Identifier Privacy Act passed by the State of New Jersey Senate.[8] Simultaneously, biometrics-related issues are beginning to emerge in concrete legal disputes that generate considerable regulatory controversy. Both in Europe and in the United States, there are now quite a large number of biometrics-related cases brought before the courts or the data protection authorities.[9]

Academic legal research on the use of biometric technology is relatively new and quite insufficient compared with the public debate. When the doctoral research project that this book represents began in early 2005, there had been (to my knowledge) little systematic and/or comprehensive study of biometrics from a legal perspective, although several articles, book chapters, and reports touched upon the topic.[10] In 1998, Prins, a renowned legal scholar, was one of the first to analyse the impact of biometric technology on the area of fundamental rights as laid down in the European legal context (Prins, 1998). She set down some tentative first steps towards insight into the legal consequences of, as well as conditions for, the application of biometric technologies. She discussed issues related to fundamental rights, legal requirements for security measures, and the status of evidence. However, her analysis left quite a lot of issues open to discussion, which underlined the need for further study.

Three years later, Grijpink made another attempt at addressing the issues raised by biometrics and privacy law (Grijpink, 2001). With a solid technical background, he attempted to clarify the concepts and practical applications of biometrics, and put forward a number of basic rules to accommodate the use of biometric technology in a legally acceptable manner. However, the issues of how the law should be interpreted or improved in response to the technology were largely left untouched.

Official documents subsequently appeared. In August 2003, the 'Article 29 Data Protection Working Party'[11] adopted a working document on biometrics. The working document reflects the view of the EU member states' data protection authorities on biometric systems and data protection, and it puts forward several recommendations for the development of such systems (Article 29 Data Protection Working Party, 2003). The Organisation for Economic Co-operation and Development (OECD) also focused on the matter and released in April 2004 a report on the use of biometric-based technologies (Working Party on Information Security and Privacy, 2004). In 2005, the Institute for Prospective Technology Studies at the European Commission's Joint Research Centre also published a report in which the impact of biometrics, including legal aspects, was described (European Commission's Joint Research Centre 2005). As part of that work, Hert wrote a background paper analysing the legal issues with respect to the application of biometrics in Europe. His paper concludes that although human rights and data protection laws establish a flexible framework for the assessment of the legal implications of biometrics, the framework is incomplete (Hert, 2005). In the same year, the Council of Europe issued a 'Progress report on the application of the principles of convention 108 to the collection and processing of biometric data'. This report was prepared by the Council of Europe's Consultative Committee of the Convention for the Protection of Individuals with regard to automatic processing of personal data, and it comments on various data protection issues resulting from biometric technology. Additionally, several national data protection authorities, including those of Norway (Datatilsynet, 2006), France, Spain, Sweden, Germany, and Italy, have issued opinions on specific legal issues pertaining to biometrics. While all of the above-mentioned literature provides concrete legal analyses concerning biometrics, there remain many legal issues on point which are not adequately clarified.[12]

In the United States, a number of legal scholars have commented on biometric technology with a special focus on facial recognition techniques used by government agencies and on the possible breach of the Fourth and Fifth Amendments to the Bill of Rights in the US Constitution that may result from the use of those techniques. For instance, it has been argued that the government should not be allowed to unreasonably intrude on individual privacy rights under the Fourth Amendment (Ng, 2006). However, others have taken a different position, arguing that the use of facial recognition technology as a form of public security is in harmony with both the Fourth and Fifth Amendments (McCormack, 2003).

In Australia, Malcolm Crompton, the previous Federal Privacy Commissioner in that country, initiated discussions on the implications of the technology for the Federal Privacy Act of 1988, in his paper 'Biometrics and Privacy' (Crompton, 2002). He points to a few areas in the interface between biometrics, privacy, and the Privacy Act of 1988 that raise legal issues or need to be carefully monitored. This chapter is of significant value as there has been, up until recently, a paucity of analysis concerning the legal implications of biometrics in Australia, despite the increasing use of such technology. The situation has changed to some extent in the last 2 years after the publication of a major report on reform of Australian privacy laws in 2008 by the Australian Law Reform Commission (ALRC). The report contains a fairly extensive discussion of biometrics regulation (Australian Law Reform Commission, 2008).

Relevant research also appeared in Canada as early as 1999, when Dr. Ann Cavoukian, the Information and Privacy Commissioner for the province of Ontario, expressed the idea that whether biometrics are privacy's friend or foe depends entirely upon how the systems are designed and how the information is managed (Cavoukian, 1999). She also explained how Canada has used both law and technology to safeguard privacy when applying biometrics-based identification systems.

1.2 Subject matter and aims of this book

1.2.1 Research objectives

As the above overview of research suggests, the introduction of biometric technology has been accompanied by enticing security arguments. It has been claimed that the measures are taken to ensure, *inter alia*, the legitimate rights of individuals and to combat terrorism and crime. However, the use of biometrics also gives rise to much concern that such measures may also pose a serious threat to the protection of individuals and their rights, in particular those rights related to privacy. The book does not aim to simply take sides in the earlier disputes, because I am convinced that from a practical point of view, such a general discussion on biometrics and privacy may be too narrow and abstract to be very fruitful. We need to ensure that the use of biometrics does not pose a danger of undermining or even destroying individual rights on the grounds of defending it. In the case of *Klass and Others v. Germany* (1978), the European Court of Human Rights affirmed that 'the contracting parties may not in the name of the struggle against espionage and terrorism adopt whatever measures they deem appropriate.'[13] As the Court further observed, 'the Court must be satisfied that whatever system of surveillance is adopted, there exist adequate and effective guarantees against abuse.'[14] Seen from this perspective, the central aim of this book is to answer the following question:

> *What are the privacy issues raised by the use of biometric technology and what might be the appropriate legal solutions?*

This book attempts to answer this question by analysing characteristics of biometric information and examining legal safeguards which exist on national and EU levels, with the aim of recommending appropriate safeguards in order to provide adequate human rights and personal data protection. However, the objective of the book is not to devise a comprehensive framework for reform but to facilitate the drafting of such an agenda by exploring possibilities and providing recommendations.

The fundamental idea on which the book is based is that an interaction exists between biometric technology and privacy protection. The consequences of biometric technology development will unavoidably affect the function and nature of privacy protection. The legal measures of privacy protection could, however, be adjusted according to the challenges posed by the technology and thereby ensure that the technology would be used in a privacy-friendly manner.

To achieve the central objectives of this book, four sets of sub-questions must be investigated:

1. What are the privacy issues in the biometric context?
2. How are these issues currently dealt with under the law? What principles are applied? What interests are served?
3. Is current regulation satisfactory? Does it facilitate a fair weighting of the diverse interests involved? Is it sufficiently clear? Is it applied consistently?
4. Generally, what is the most appropriate approach to deal with the legal challenges posed by the use of biometrics?

These sub-questions are dealt with in different parts of the book, and the answers to each provide the results that I delineate in the conclusion.

Treatment of each set of sub-questions falls within three general frameworks or pillars of analysis. The first such pillar is constituted by the discussion of the first-mentioned sub-question (see Part II). That discussion builds upon an analysis and evaluation of the various arguments and perspectives pertaining to biometric technology. Generally, the analysis seeks to explore the interactions or interplay of privacy and security concerns with respect to the use of biometrics, particularly arguments concerning the apparent 'trade-off' of privacy against security in this context.

The second and third sets of sub-questions (dealt within Part III) flow from the first. Discussion of them forms a legal framework or pillar to complement the more theoretical framework in Part II. The aim is to critically examine the role of law against the background of the observations made in Part II, focusing on the respective regulatory frameworks in Europe, United States, and Australia. While the first pillar provides a requisite rational groundwork for the study, the second pillar ('legal framework') provides not only insight into relevant legal instruments but also provides guidelines for their application and suggestions for their improvement.

The discussion in relation to sub-question 4 (dealt within Part IV) centres initially on the practicability of treating biometric information as a form of personal property. It is followed by a comparative study of large-scale biometric systems at national borders. The aim of the study is to canvass and test various potential regulatory approaches and thereby improve understanding of their respective weaknesses and merits *in practice*. This generates in turn a set of findings that enable the making of recommendations for regulatory strategies that are viable and effective, particularly in respect of providing appropriate privacy protection in the biometric context.

1.2.2 Research challenges

The research of biometrics has been challenging because this is an area that continuously evolves. Challenges exist on two fronts: legal and technological. The relevant legal initiatives, recommendations, and policies are in a state of flux, and it is therefore difficult to monitor and keep up to date with them. For instance, the recent emergence of a limited number of court cases on biometrics is generally of significant value for analysis. Yet attaining an accurate picture of the state of this case law is far from easy: a lower court decision for one case may be reversed in a higher court, and other cases may be pending for final decisions. Another difficulty concerns legislative initiatives. There are, for example, several such initiatives for regulating biometrics at the state level in the United States[15] but there may be a lengthy process involved before they are voted upon, enacted, and published. Some of them appear to not receive much public attention, and some may not be successfully passed by the legislature. To find and get hold of these legislative initiatives and keep up with their progress has therefore represented a quite challenging, yet critical task to fulfil.

Another challenge encountered during research for this book was the constantly evolving biometric technology. This book is a piece of legal scholarship that aims to address the legal issues raised by the technology. It is therefore essential to maintain a current and correct knowledge of the technology. The continued development of biometric technology may render legal issues, which used to be important, less valuable for discussion. To be able to discuss or evaluate the technology, formulations are required which stabilise biometrics into a well-defined subject, even though the borders of this field have not yet stabilised. Misconceptions about the technology may render a legal analysis based on them meaningless, while knowledge of the technology will help render correct legal questions. For example, one popular misconception about biometrics is that it cannot be stolen (NHS, 2008); in fact, biometrics (at least some forms) can be easily stolen, and spoofing has also been proven easy to do. So the question we need to pose is this: How might a biometricbe used if it is stolen? In the light of the answer to that question, what kind of legal measures should be adopted?

Legal research dealing specifically with biometrics is relatively new; hence, not much legal literature has been written – as the state of research discussed above indicates. The legal resources that exist for commentary are therefore quite limited, and most of them are focused on certain similar issues. Obtaining timely access to some of these resources has proven to be difficult. Exacerbating this problem of access is the language problem. For example, much literature in Europe about the use of biometrics tends to be written in Member States' languages, many of which I am not proficient in. The limitations in my language proficiency also significantly restrict the choice of research materials.

The lack of resources also stems from the reluctance of many organisations to disclose information about how biometric systems are being deployed and what kind of policy is at hand for the protection of privacy. At the outset of the research for this book, it was intended that personal investigation into practical biometric applications would be an important research method, but the lack of responses to my requests for interviews and information from various governmental and non-governmental entities meant that it was necessary to downplay this method. Therefore, it was difficult, if not impossible, to obtain first-hand information of biometric systems in practice. I have had to depend only on second-hand resources and published documents as the basis of my research, which to a large extent restricts the scope to public sectors where relatively more information is available.

1.3 Scope and limitations

1.3.1 Technology

Previously, the use of biometrics was mainly confined to the area of fingerprint testing. As the technology develops, greater numbers of human physical characteristics are being utilised as targets for biometric technology. DNA has recently sailed up as one particularly popular biometric;[16] DNA-based technologies are frequently thought of as the 'ultimate identifier'. Many jurisdictions now have forensic identification databases containing DNA samples, and many other DNA storage systems exist, including research databases, blood banks, and tissue storage facilities. Because of the extremely sensitive nature of DNA, it is not popularly utilised for identification and authentication biometric systems outside the context of criminal investigation. At the same time, it could be argued that DNA, strictly speaking, is not a biometric in the same way as other biometric features. In the OECD's report on biometric technology, it is pointed out that:

- DNA requires an actual physical sample as opposed to an image, photograph, or scan;
- DNA matching is not done in real time and for the most part, is not automated;
- DNA matching does not employ templates or feature extraction but rather represents the comparison of actual samples.

(Working Party on Information Security and Privacy, 2004: 11)

Despite this, DNA-based biometrics are included in the general meaning of biometric technologies in this book, as they share several common attributes with other biometrics. However, a proper treatment of DNA-based technologies is beyond the scope of this book. DNA issues are a topic of sufficient size to warrant another doctoral book altogether. They are not the focus of the discussion here. My focus of research has been on the more widely used categories of biometric technology such as finger scanning, hand geometry, facial recognition, iris scanning, retinal scanning, voice recognition and so on. In general, this book focuses on the use of automatically processed biometric data for the authentication or identification of persons in non-criminal scenarios.

Although each kind of biometric technology possesses some special characteristics, all biometric systems operate in essentially the same manner and they have much in common, from a legal perspective. However, from a technical point of view, it is undeniable that distinctions must be made between the various kinds of biometric technologies. It is indeed meaningful to mention these differences from a legal perspective at certain points. As our analysis continues, attention will be drawn to those areas. The trend of using combined multiple biometrics in a single biometric application also renders separate discussions of various biometric technologies less valuable. Therefore, it is not necessary to distinguish one biometric from another all the time. In this book, the analysis generally refers to all the biometric categories, unless specifically noted.

1.3.2 Jurisdiction

This book generally covers the following jurisdictions: Australia, the United States, and the European Union. Regarding the latter, the legal analysis targets the EU legal framework of privacy and data protection in general. However, case decisions and legal initiatives and regulations from Norway are used as examples for detailed analysis at a national level in Europe. While Norway is not a member of the EU, it is party to the 1992 Agreement on the European Economic Area and, as such, is required to adhere to the central piece of EU legislation on data protection – namely, Directive 95/46/EC.[17]

Focus on these jurisdictions is justifiable for several reasons. First, they represent in sum a reasonably broad field of coverage in terms of legal systems. Australia and the United States are common law states sharing origins in English law, whereas in Europe (apart from the United Kingdom and the Republic of Ireland), there is a long tradition of civil law. The jurisdictions also represent a reasonably broad field of coverage in terms of geographical locations. The jurisdictions chosen are spread among the Asia-Pacific region, North America, and Europe.

The United States is particularly interesting because of its political and economic influence in the development and application of biometric techniques and of biotechnology and information technology more generally. It also boasts a wealth of relevant case law and biometric-related legislative initiatives. While the vast majority of those cases and initiatives are related to the use of biometrics by government agencies, this fact also has relevance with respect to the characteristics of privacy protection in the United States. There, self-regulation is frequently promoted for privacy and data protection within the private sector. In Europe, by contrast, both public and private sectors are subject to top-down legislative regulations. In addition, taking into account the different cultural background and human rights' regimes of the two regions, it is interesting to note the apparently different mentality of US companies compared to their European counterparts.

Norway is chosen as an example of analysis with respect to European jurisdiction because of its tendency to rely on legislative solutions rather than soft law. Indeed, the central data protection legislation in Norway contains a provision that relates directly to biometric identification systems. The way in which that provision applies to such systems has been the subject of a series of case decisions. The provision is also the subject of ongoing legislative reform processes. In addition, Norway is one of the most technologically advanced countries in Europe, with a relatively high uptake of biometric technology. At the same time, although Norway is not a member state of the European Union, as part of the European Economic Area, its general legal regime for privacy and data protection conforms to the EU model. Thus, the study of Norway serves as an example of analysis of EU law at a national level. Finally, it is also interesting to contrast the legal regimes of the United States and Australia with that of the EU, which is generally regarded as the legal regime with the world's highest level of data protection.

In contrast to the United States and Norway, Australia's tendency to encourage 'soft law' initiatives may serve as a good example of alternative legal approaches. The Australian Biometric Institute has issued a privacy code on biometrics. The code is the first of its kind in the world and has thus gained widespread attention.

1.3.3 Privacy and data protection law

The use of biometric technology necessarily entails a discussion concerning personal integrity, autonomy, and identity. The rights to human dignity and self-determination are closely linked to privacy rights and are included within the objective of privacy protection in general. Personal identity, personal autonomy, privacy, and protection from the misuse of information are all hugely important within a stable and happy society. All of these rights are perceived by many persons as being seriously threatened by the current biometric aspirations, which are effectively changing the trust model between citizen and state (Ashbourn, 2006). Moreover, the deployment of biometrics may become a threat to procedural rights, such as the presumption of innocence. Collection of biometrical data samples from individuals may also violate their right against self-incrimination, as defined in European case law and the US Constitution. Further, the reliability of biometrics is also a central problem with regard to evidence law (Hert, 2005). However, it would be too ambitious to deal with all of these issues here. Instead, this book focuses basically on the privacy-relevant concerns of biometrics; other legal aspects of biometric technology are largely outside the scope of the book.

Privacy-relevant concerns refer here to the legal issues related to the use of biometrics which revolve primarily around the notion of privacy. Data protection is particularly central to these concerns. The erosion of personal control over the use of biometric information is at the centre of the public debate on biometrics. Informational privacy, which is defined as the ability to maintain control over the use and dissemination of one's personal information, is closely related (Electronic Privacy Information Centre and Privacy International, 2005). Informational privacy concerns freedom of choice and informational self-determination. Physical privacy, which is related to physical intrusion on one's personal body, is also involved in the debate. My justification for focusing on privacy-relevant legal issues is based on the emerging situation described in Section 1.1 and the existence of numerous privacy-specific regulations and instruments highlighting the significance and urgency of the topic. Moreover, the realisation of adequate privacy safeguards for the use of biometric data could also result in better protection of other related rights such as personal dignity, self-determination, and procedural justice.

1.4 Methodology

The methodology adopted in this book is informed by the nature of the biometric technology. It is somewhat different from more traditional legal analysis. This section describes the methodology necessary to answer the main book questions and their range of sub-questions.

1.4.1 A system-oriented analysis: allocation of risks

The focal point of my analysis is the various privacy and security risks posed by the technology. This is due to the following reasons: First, the allocation of risk is a prescient method of evaluating the efficacy of existing legislation with respect to the issues raised by the technology. This research method is especially prominent in Part III, where the Australian regulatory regime is discussed. Second, there is the issue of allocation of the categories of risks, namely informational privacy risks and physical privacy risks over biometric information. This allocation is followed throughout the whole of Parts II and III of the book.

The analysis in the book of the legal issues can be described as system-oriented. The primary legal focus is on the systematic interaction between the legal risks posed by the technology and the law. Risk can be viewed from both legal and social angles. In the legal setting, risk assessment results in a legal obligation to protect rights and ensures that losses are compensated for in the event of damage. In everyday life, risks signal that events will sometimes deviate from their predicted course. In this book, 'risks' usually refer to the possible threats to privacy rights that are cherished in a democratic society. This kind of risk does not necessarily stem from specific legal provisions but from the understanding of the right to privacy as a rational value, as highlighted in Part II of this book. On a *de lege ferenda* basis, the allocation of risk will likewise be the tool for suggesting regulatory amendments to the existing privacy regulations.

1.4.2 **De lege lata** *and* **de lege ferenda** *analyses*

Examination of the legal instruments will have two main prongs: analysis of what the law is (*de lege lata*) and what the law ought to be. In determining the current law of the selected jurisdictions studied in this book, I have employed the traditional legal dogmatic analysis for ascertaining the legally valid scope and content of the rules concerned.

The task of establishing with certainty the scope and content of valid law is not so easy to fulfil, especially when one is interpreting law from across jurisdictions. The difficulty is exacerbated by the fact that there would seem to be relatively few decisions on point from judicial or quasi-judicial bodies, and even fewer from high-ranking courts. Within the EU, no cases dealing directly with biometrics and privacy issues have been put before the Court of Justice (formerly European Court of Justice), and only one such case has been dealt with by the European Court of Human Rights (ECtHR) under the European Convention on Human Rights and Fundamental Freedoms (ECHR). In the apparently small number of instances where national courts in Europe have addressed such issues, their decisions may create binding precedents for the courts in their country (dependent, of course, on their place in the country's judicial hierarchy), and they may influence court decisions for similar cases in other jurisdictions. Yet, they will not be binding on either the Court of Justice or the European Court of Human Rights, and they cannot be automatically regarded as providing authoritative guidance on how, for example, the provisions of the Data Protection Directive or the ECHR are to be construed.

As for the non-European jurisdictions that are in focus in this book, biometric-relevant privacy case law is yet to emerge in Australia. The situation is somewhat different in the United States where there are a small number of biometric-related cases both at the federal and state level. However, some of these cases are now quite old, and their findings may be considered as not strictly fitting into the present biometric technology settings.

Fortunately though, there are sufficient sources of law available in all the jurisdictions in focus to enable reasonably solid conclusions to be drawn on the basic elements of privacy law pertaining to each jurisdiction. It is concomitantly possible to determine a minimum core of legal principles for the use of biometrics that are likely to be in accordance with what the judicial bodies in the respective jurisdictions would refer to as *de lege lata*.

The discussion about establishing the certainty of *de lege lata* is always closely linked with the discussion of *de lege ferenda*. This is especially the case when analysing the interaction of privacy laws and biometrics. Moreover, the book is, to a large extent, explicitly a *de lege ferenda* work. The assessment of the current legal initiatives and judicial and quasi-judicial decisions is followed up with suggestions of what the law(s) ought to be.

1.4.3 Legal methods

Although this book includes studies of the law of no fewer than three different jurisdictions, its primary aim is not to be a comparative study. The study intends to refrain from extensive and systematic comparisons of these jurisdictions. The account taken of various countries' regulatory initiatives aims primarily to illustrate the various possible strategies and to highlight possible conflicts and problems. With this aim in mind, the country-specific chapters are not written in a manner with systematic comparative elements. Instead, these country-specific chapters are based on the special characteristics of each jurisdiction. In Chapter 4, the intention in the analysis of the European Union's privacy and data protection laws is to avoid a general study of each relevant data protection principle, as many of the problems regarding how to construe these principles have already been extensively addressed by other commentators. Tiresome repetition is avoided; instead, the study focuses on controversial legal issues that arise specifically from the use of biometric technology. In Chapter 5, which examines US privacy law, the aim is to give a comprehensive study of *possible* privacy protection regarding the use of biometric technology in the United States, due to the fact that many of the elements touched upon remain a 'vacuum area' in and about which discussion is either non-existent or has not come very far. In Chapter 6, Australian privacy and data protection law is studied but, again, instead of embarking on a general study of those laws, the analysis attempts to go into more detail based on the risk allocation assessment. However, some limited normative comparisons are made of the rules in each jurisdiction, though these are tempered by awareness of the many potential methodological pitfalls of cross-jurisdictional comparisons.

The jurisdiction-specific chapters serve two very significant objectives. First, they provide an empirical base illustrating how different legal cultures and traditions have dealt with similar legal issues. Second, they provide an empirical base for the exploration of the suitability of existing privacy laws in the biometric context. Such background discussion in Part III is important for the subsequent discussion about finding the appropriate legal approach for dealing with the legal issues.

However, the book requires more than the mere application of legal methods. The nature of the issues it seeks to tackle demands the utilisation of a wide range of study methods and source materials. Specifically, the research has drawn on insights from overlapping fields including computer information technology, medical ethics, and bio-ethics.

The research has not attempted, however, to carry out an extensive empirical survey on the public perception of the use of biometrics. Rather, it has drawn on existing surveys and reports on such matters.

As part of the research, two field trips were carried out: one to Australia and the other to the United Kingdom. Both trips were very valuable for the research. Interviews conducted with the various technical experts and 'policy entrepreneurs' were fruitful and enriched the content of this book with practical and original views from different angles. The persons interviewed include the previous federal Privacy Commissioner of Australia, several professional strategic and security consultants with technical and/or political backgrounds, the CEO of the Biometrics Institute in Australia, as well as legal experts in academia. The interviews were informal and semi-structured; sometimes the interviewees were sent a list of general questions in advance based on their specific expertise. During interviews, some general questions were supplemented with more specific oral questions posed by the interviewer. These questions were randomly raised during the interviews when interesting discussions were generated. In general, the interviewees were informative and seemed candid in their responses. The findings and information gained from the interviews are used in this book where relevant.

The book also benefits from empirical research on the practical biometric application systems for border control, such as the systems of VIS and US-VISIT, elaborated in Part IV. The study of the VIS and US-VISIT systems involves analyses of legal instruments establishing the systems and official documents describing the systems. The study of changes in control techniques and places of control is based on the analysis of those legal instruments, along with national laws and practices. It establishes a concrete example of how biometric technology is applied in the real world, and of the policy and legal safeguards which are built into the system.

1.5 Resources

The Internet has proven to be an invaluable means of obtaining resources. The explosive use of the Internet means that it has become a major medium for fast dissemination of information. Generally speaking, the Internet has been used as often as possible for collecting research materials for this book. There is a large and growing range of materials available online; even traditional journals, books, law reports, and court decisions have become available in electronic form, though some require subscriptions, and in the case of some of the online materials it has not been possible to include page references. In addition, traditional paper resources in the form of books, journals, and other official documents have also been used extensively.

1.5.1 Sources of law

The term 'source of law' usually refers to both formal sources and material sources of law (Brownlie, 2008). According to Brownlie, formal legal sources are those legal procedures and methods for the creation of rules of general application that are legally binding on the addressees. Legislative texts, case law, international treaties and conventions, and general principles of law are examples of formal sources of law. The normative weight of each differs according to the context in which they are applied. For the purpose of this book, a traditional list of such sources is used and it includes the most important sources to enable the presentation of a comprehensive picture of the relevant biometric and privacy law regulations. These include general privacy statutes and court or tribunal decisions on those statutes within the EU, the United States, Australia, and Norway, along with biometric-specific privacy laws in those jurisdictions. International laws, such as the ECHR and the Data Protection Directive, are also used. This creates international obligations and rights for contracting parties, and it forms the legal basis for the general protection of privacy and personal information in the territories of each party. It also, thereby, forms the legal basis for the regulation of biometrics.

The weight, a particular source of law may be given by the courts when it comes to the use of biometrics, and regarding interpretation of privacy provisions more generally, deserves mention here. At present, court decisions on biometric-relevant privacy issues are scarce. Most decisions are found at the national level. Within the EU, there are countries whose courts have addressed quite a number of cases, whereas the courts of other states have seen no cases at all. The jurisprudence established by national courts or tribunals cannot give binding instructions to foreign authorities (judicial or otherwise), but it may at least be a good example of how the EU Data Protection Directive may be implemented at the national level, and be a good parameter for assessing the influence and value of the policy recommendations of, say, Article 29 Data Protection Working Party on biometric-related issues. Court decisions in the area of human rights are also important. Although privacy-related biometric cases are scarce at the European Court of Human Rights, the case law developed by that court in response to other technologies and to surveillance practices generally provides very useful legal resources for guidance and inspiration in relation to the regulation of biometrics. Hence, ECHR case law, especially pursuant to Article 8 of that Convention, is analysed extensively. So too is case law pursuant to the Bill of Rights in the US Constitution. Obviously not all such case law is dealt with, only those cases that seem particularly relevant to the central questions posed by this book.

Material sources of law are not laws in themselves. The material sources provide evidence of the existence of rules which, when proved, have the status of legally binding rules of general applications (Brownlie, 2008). They serve as evidence of law and aid in the interpretation of law. In general, material sources cannot be decisive in a case as they are not binding, but they can be given greater weight where the law is not clear, and when a clarification for a new legal issue is required.

Preparatory works and initiatives to draft and adopt biometric-related privacy codes are important points of reference for evaluation, especially when formal legislation is lacking. Although they may play, at the most, only a moderate role as a source of law, they can be a useful starting point to find out how the law is and, in some cases, ought to be. In countries where there is a tradition of generating and publishing preparatory works before the incorporation of international treaties into national law, the preparatory works can be an important source of law at the national level. In this book, the proposals for the revision of the federal Privacy Act (1998) in Australia, the initiative of the Norwegian Data Protection Authority (Datatilsynet) for revising paragraph 12 of the Norwegian Personal Data Act, and the initiatives to pass specific biometric legislation in California, all fall into this category and deserve relatively detailed study. Interpretative works such as the progress reports and working documents issued by the Article 29 Working Party and the Council of Europe's Consultative Committee of Convention no. 108[18] are part of the legal sources.

In the process of searching for the above sources, some web sites have significantly contributed to the research by providing reliable, authoritative, and easily accessible information, such as the EU databases, Human Rights Case Law Database 3030, the Electronic Frontier Foundation, the Australasian Legal Information Institute, and the Electronic Privacy Information Centre (EPIC). Apart from these web sites, the various privacy commissioner's web sites and government web sites, such as that of the Parliament of Australia, have also provided updated information about privacy- and biometric-relevant legislation and policy discussion.

1.5.2 Secondary sources: the commenting literature

Secondary sources mean any description or reference of the primary sources and any description of the biometric technology and the applications of such technology. As explained previously, commenting literature is an important component in the book's methodology. 'Commenting' literature includes books, legal journals, legal manuals, and articles in various newspapers and magazines (e.g. on information technology and forensic science). The secondary sources are widespread and found in many different arenas, though sources directly related to both biometric and privacy laws are limited.

First, there are bibliographies of articles pertaining to biometric and privacy issues which mainly comprise articles from technicians and policy commentators. Here, one can identify work written or edited by Clarke (2003) and EPIC,[19] which contains literature about the general knowledge of biometric technology, examples of various biometric applications, and some comments. These materials can be used as an introduction for getting a basic idea of the problem at stake. Second, there are single-standing articles on data protection. A number of privacy advocates have published articles commenting on surveillance technology in general, and some articles are on biometrics in particular. Although there are limited numbers of articles on biometrics, there are many interesting discussions being raised, and the general comments on surveillance technology are also of significant value for study. Early works by Clarke on 'Information Technology and Dataveillance' (Clarke, 1994), Davies on 'Touching Big Brother' (Davies, 1994), and Woodward on 'Biometric Scanning, Law & Policy: Identifying the Concerns – Drafting the Biometric Blueprint' (Woodward, 1997) serve as good examples here. Third, books on surveillance technology, biometric technology, privacy or data protection law, and genetic data are also of relevance. Since the discussion on surveillance technology relating to genetic information relies heavily on exchange of information, these books also comment on privacy and data protection issues. In addition, books solely on privacy laws serve as basic sources for understanding the status of privacy regulations. Some works could be mentioned here: Bygrave (2002), Rothstein (1997), Schwartz and Reidenberg (1996), and Wayman (2004). Fourth, there are dissertations and theses; theses on biometrics and privacy issues are available at the Masters' level or from a technical point of view. The author has not come across any other doctoral theses on biometrics and privacy regulations in the English language. Indeed, the lack of such doctoral dissertations prompted the choice of topic for this book. Finally, to keep updated with respect to recent developments in biometric technology, its applications, and legal discussions, the websites of the Statewatch database,[20] International Biometric Group (IBG),[21] International Biometric Industry Association,[22] BITE,[23] and Australian Privacy Foundation[24] are rather useful places to visit in order to keep updated with respect to current events concerning relevant legislation, decisions, policies, meetings, conferences, etc.

The commenting literature is used in three principal ways. First, the commenting literature contains assertions regarding the consequences of using biometric technology, and therefore also on the merits of the technology and its influence, as either friend or foe, on privacy. The book discusses some of these sources. Second, the commenting literature is related to empirical observations made by practitioners, for instance, the Roadmap from Biovision (2003). These observations relate both to conditions in the independent biometrics market and to the quality, cost, and human factor evaluations. Factual observations are a source for identifying the privacy and security effects of biometrics. In addition to my own observations through interviews with practitioners,[25] the main source of these factual observations is the commenting literature. Finally, the commenting literature also touches upon the legal aspects of surveillance technology and privacy law problems. This part of the literature can be used to assess whether the legal problems have been adequately and accurately dealt with, and may also assess what needs to be commented on in greater detail. The commenting literature is both a source of legal study and a target for assessment.

1.6 Terminology

Although readers of this book are presumed to have a certain degree of prior knowledge of law in general and privacy and data protection law more specifically, there are reasons for commenting briefly on the terminology used. Commonly encountered terminology in this book includes concepts of privacy, informational privacy, physical privacy, data protection, bio-privacy, personal information (data), biometrics, biometric technology, data subjects, data controllers, and security.

The term *privacy* is used in this book to 'denote a condition or state in which a person (or collective entity) is more or less inaccessible to others, either on the spatial, psychological or informational plane' (Bygrave, 2002: 23). This definition builds on and is quite similar to Gavison's definition of privacy, which expresses quite well the meaning I wish to underscore in this book. However, there is much controversy about the definition of privacy and it has been defined in many different ways. Bygrave has categorised the definitions into four major groups (Bygrave, 2002). The first group is represented by Warren and Brandeis, who argued that the right to privacy in US common law is a right 'to be let alone' (Warren and Brandeis, 1890). The second group conceives of privacy in terms of degree of access to a person. Gavison has defined it as the condition of limited accessibility consisting of three elements: secrecy, solitude, and anonymity (Gavison, 1980; Allen, 1988; Reiman, 1995). A third group of definitions denote privacy in terms of information control. The following definition by Westin is fairly representative of this group: '[t]he claim of individuals, groups, or institutions to determine for themselves when, how and to what extent information about themselves is communicated to others' (Westin, 1967: 7). A similar definition is also formulated by Miller: 'the basic attribute of an effective right to privacy is the individual's ability to control the flow of information concerning or describing him – a capability that often is essential to the establishment of social relationships and the maintenance of personal freedom' (Miller, 1971: 40). This could be regarded as the root of *informational privacy*. The fourth group relates privacy exclusively to those 'intimate' and 'sensitive' aspects of human life: 'the state of possessing control over a realm of intimate decisions, which includes decisions about intimate access, intimate information and intimate actions' (Inness, 1992: 140).

Which of the above-mentioned definitions is the most appropriate is a subject of substantial academic debate. That debate has neither reached a commonly accepted conclusion nor is it likely to do so in the near future; and it is not my intention to participate in it with this book. As it has been pointed out, law and policy do not always need to operate with precise, sharply defined definitions of values (Inness, 1992; DeCew, 1997). Furthermore, privacy in itself is not a homogenous right, but a bundle of very different underlying values. Although I agree that privacy in a strict sense is a condition of being, this book also uses the following categories of privacy built on Kang's analysis. Kang has identified three such categories: physical privacy, decisional privacy, and informational privacy.[26] *Physical privacy* is concerned with the protection of one's home and one's bodily integrity. Decisional privacy is about the right to render personal decisions without undue external pressure, while *informational privacy* is seen as the right to control one's personal information. Informational privacy concerns exist wherever personal information is collected, stored, transmitted, and used in digital form or otherwise. Informational privacy issues can arise in response to information from a wide range of sources, such as health care, ethical, criminal settings, residential, financial, and biological. These definitions are adopted in this book.

Data protection is used here to denote certain measures used for protecting personal data/information. These include legal measures aimed at ensuring that the processing of personal data is carried out in ways that ostensibly safeguard the privacy and related interests of the data subjects. Such measures are generally referred to in Europe as constituting 'data protection law'. Outside Europe (e.g. in the United States and Australia), such measures are more often referred to in terms of 'privacy law' or 'information privacy law'.[27] Given the fact that this book analyses such laws both within and outside Europe, the terms 'privacy law' and 'data protection' are sometimes used synonymously in the book.

Bio-privacy is a term used by the IBG[28] which refers to the privacy issues of biometric technology. A bio-privacy initiative was created by the IBG and defines best practices, as well as deployment and technology guidelines for maintenance of personal and informational privacy in biometric deployments (International Biometrics Group, 2002).

Personal information is a legal term used in Australian privacy law. According to the federal Privacy Act of 1988 Part II section 6, 'personal information' means 'information or an opinion (including information or an opinion forming part of a database), whether true or not, and whether recorded in a material form or not, about an individual whose identity is apparent, or can reasonably be ascertained, from the information or opinion.' In the US Federal Privacy Act of 1974 § 552a, personal information is referred to as 'records' maintained on individuals:

> The term 'record' means any item, collection, or grouping of information about an individual that is maintained by an agency, including, but not

> limited to, his education, financial transactions, medical history, and criminal or employment history and that contains his name, or the identifying number, symbol, or other identifying particular assigned to the individual, such as a finger or voice print or a photograph.

The notion of personal information is very close to that of personal data, which is the legal term mainly used in European privacy and data protection law, following the definition from the Data Protection Directive: '. . . data which relates to a living individual who can be identified from those data, or from those data and other information which is in the possession of, or likely to come into the possession of the data controller' (Article 2(a)). These terms are frequently used in this book according to the jurisdictions to which they belong, the meaning and scope of which may slightly differ.

The term biometrics is defined by the IBG as: the automated measurement of behavioural or physiological characteristics of a human being to determine or authenticate their identity (Dennis, 2002).

The term *biometrics* is used alternatively to describe two different aspects of the technology: characteristics and processes (National Science and Technology Council, 2006a).

- Biometrics as 'characteristics' refers to measurable biological or behavioural aspects of the person that can be used for automated recognition.
- Biometrics as 'process' refers to automated methods of recognizing an individual based on measurable biological and behavioural characteristics.

The emphasis here is on the 'automated' measurement and the utilisation of a 'human being's behaviour or physiological characteristics' (Nuger and Wayman, 2004b). According to Wayman, biometric techniques differ from forensic laboratory techniques such as latent fingerprinting, DNA, hair, and fibre analysis by complete automation (Nuger and Wayman, 2004b). The biometric authentication or identification is done by machines, generally computers. Hence, forensic detection which usually involves manual work is not biometrics. For instance, in general, DNA identification is not considered by many as a biometric recognition technology, mainly because it is not *yet* an automated process (it takes some hours to create a DNA fingerprint).[29] However, because of the accuracy level of the process and because many consider it as a possible future biometric trait, it is usually involved in biometric technologies as an exception (European Commission's Joint Research Centre 2005).

Another pair of terms used frequently in this book is data subject and data controller. A *data subject* is the person or organisation to whom the data relates. A *data controller* is a person or organisation who/which determines the purpose and means of data processing (Bygrave, 2002).[30]

As for the term information security, this is used in line with its definition as 'the process of protecting data from unauthorised access, use, disclosure, destruction, modification, or disruption'.[31] It is closely linked with *computer security*, which shares the goals of protecting the confidentiality, integrity, and availability of information.

Finally, some explanatory words on the term anonymity are provided. Anonymity means literally a state of namelessness. In this book, it is used generally to denote a condition in which it is not possible to identify a person. Thus, anonymous information is the information from which it is not possible, directly or indirectly, to correctly ascertain a given person's identity.[32] Anonymity is a relative concept in the sense that a state of anonymity will be context-dependant – i.e. relative to specific facts, specific subjects, and purposes. Anonymity is usually linked with the notion of privacy. However, it is not exactly the same. Generally, anonymity is regarded as a method to protect privacy. In some legal contexts, it is also recognised as a legal right.[33]

1.7 Progression

This book is divided into five parts, comprising nine chapters. Part I includes this introductory chapter and Chapter 2. The *Introduction* deals with the problem statement, literature choices, and methodology, and aims to give a general overview of the background and content of the book. Chapter 2 (*Biometric technology and its applications*) aims to present an advanced-level introduction to biometric technology. It draws attention to the different biometric technologies and applications, as well as the various factors that may affect the privacy and security level of a certain biometric application. The common misconceptions of biometric technology are taken into consideration. The chapter also attempts to give a concrete and clear illustration of the relevant terms, evaluation factors, applications, and other technologies being combined when biometrics are used, to the degree of definition that is necessary for the subsequent legal study.

Subsequent to this is Part II, Chapter 3 (*Rational concerns about biometric technology: security and privacy*). In this part, I consider my research in a social and ethical context, one which will contribute to answering the key book questions. By undertaking an analysis of the characteristics of the biometric information, I seek to elucidate the relevance of biometric technology with respect to the concepts of privacy and security. The purpose of this part is to establish a solid ethical and social basis for the further legal analysis and risk allocation study, and to explore the validity of the argument about trading off security by scarifying privacy.

Part III addresses the relevant legal regulations in different jurisdictions. In particular, an examination is made of two different mechanisms designed to tackle bio-privacy issues. These consist of the privacy regulations about personal information (data) in general, such as the EC Directive, OECD Guidelines, Australian Privacy Act of 1988, and the US Privacy Act of 1974, as well as other privacy norms in instruments for protecting basic human rights, such as the ECHR and US Constitution. Biometric-specific privacy regulations are also covered, such as the Biometrics Institute Privacy Code, biometric-specific case decisions, and biometric-specific legislation at the state level within the United States. The legal research also follows a risk allocation sequence in this part and concerns the informational privacy risks, physical privacy risks, and the proprietary privacy discussion on the biometric information.

Chapter 4 *Regulations of biometrics by privacy and data protection law in Europe*) explores the shortcomings and uncertainties of the existing privacy regulations and the bio-privacy initiatives in Europe. To a large extent, the discussion is based on the provisions and recommendations manifest in the written law and legal initiatives of biometrics, in order to analyse their applicability and adequacy.

Chapter 5 (*Biometrics and privacy protection in the United States*) analyses the major US privacy regulations and provisions that are relevant to biometrics generally at the federal level, and specific regulations at the state level. The analysis does not intend to give a very comprehensive study of all the possible relevant regulations, which is both impossible and unnecessary for the purpose of this book. Nevertheless, a general study of the major privacy regulations that may be relevant to biometrics is made. The purpose of this chapter differs slightly from that of Chapter 4. This chapter attempts to address the bio-privacy issues and to examine the possible remedies in the privacy framework in the United States, where a comprehensive study of bio-privacy issues in both the public and private domain is lacking. By contrast, some general recommendations and reports have already been put forward in the European context; hence, the emphasis in Chapter 4 is more on the issues that have not been addressed sufficiently by those recommendations/reports.

The last chapter of Part III, Chapter 6 (*Privacy regulations on biometrics in Australia*), is a shorter chapter focusing on two major pieces of privacy law at the federal level. The specific provisions of the legislation are considered, based on the risk allocation analysis of biometric technology.

Part IV of this book includes Chapter 7: *Towards an appropriate regulatory approach for bio-privacy* seeks to explore the best regulatory approach for resolving or reducing the potential privacy risks posed by biometrics through a comparative assessment.

Chapter 8 *Concluding remarks and future vision* summarises the findings of this book and indicate some substantial arguments in the debate concerning privacy laws and biometric technologies. Specific recommendations will be put forward to build a better legal infrastructure.

Notes

1 The specific analysis and overview of biometric technology occurs in Chapter 2.
2 See, for example, European biometric passports.
3 See, for example, US-VISIT Program, as elaborated in Chapter 8 (Section 8.2.2).
4 See, for example, Esso Norway's use of a fingerprint authentication system, elaborated in Chapter 4 (Section 4.3.2.2).
5 See, for example, the iris recognition system at Schiphol airport, Amsterdam, The Netherlands, elaborated in Chapter 2 (Section 2.9.1).
6 This refers to the kind of biometric system that is automatic. See further details in Section 1.6.
7 There are very interesting explorations of the normative and political implications of linking identification practices to the physical body, by Ploeg (1999), Ploeg (2003), and Lyon (2001).

8 For more discussion, see Chapter 5.
9 See, for example, the litigation in Sweden concerning the legality of using fingerprint authentication for pupils to enter and use a school canteen, as elaborated in Chapter 4 (Section 4.3.2.3).
10 This may be restricted by the author's knowledge of languages.
11 Under Article 29 of Directive 95/46/EC, a Working Party was established, made up of the Data Protection Commissioners from the EU Member States, together with representatives of the EU Commission. The Working Party is independent of, and acts in an advisory capacity for, the EU Commission. The Working Party seeks to harmonise the application of data protection rules throughout the EU, and publishes opinions and recommendations on various data protection topics. It also advises the EU Commission on the adequacy of data protection standards in non-EU countries.
12 The literature overview here is not exhaustive but it encompasses the most publicly accessible legal discussions about biometric technology in the European context.
13 *Klass and Others v. Germany* 6 September 1978, Series A no. 28 §49.
14 Ibid. §50.
15 See further discussion in Section 5.4.2.
16 DNA contains the genetic specifications for the biological development of all cellular forms of life. DNA is a long polymer of nucleotides and encodes the sequence of the amino acid residues in proteins using the genetic code, a triplet code of nucleotides. See, for example, Watson and Berry (2003).
17 Directive of 24.10.1995 on the Protection of Individuals with Regard to the Processing of Personal Data and on the Free Movement of Such Data (O.J. L 281, 23.11.1995: 31) – hereinafter often termed 'Data Protection Directive'.
18 Convention no. 108 is shorthand for the Convention for the Protection of Individuals with Regards to Automatic Processing of Personal Data (ETS no. 108), adopted 28.01.1981; in force 01.10.1985 – hereinafter also termed simply 'Convention108'.
19 EPIC http://www.epic.org/.
20 See further http://www.statewatch.org.
21 See further http://www.biometricgroup.com/index.html.
22 See further http://www.ibia.org/.
23 'BITE' stands for Biometric Identification Technology Ethics; see further http://www.biteproject.org/.
24 See further http://www.privacy.org.au/.
25 Cf. Section 1.4.
26 Kang (1998) has added a fourth category which he terms 'formational privacy' and envisions it as providing a common foundation for the other three.
27 For further analysis of the concepts of data protection and data protection law and their relationship with privacy and privacy law, see Bygrave (2002).
28 International Biometrics Group (2002). The International Biometrics Group describes itself as a 'consulting and technology services firm (which) has provided technology-neutral and vendor-independent biometric services and solutions to financial institutions, government agencies, systems integrators, and high-tech firms since 1996.' See International Biometric Group homepage at http://www.biometricgroup.com/.
29 Beyond the definition, to most observers DNA looks, acts, and may be used like other standard biometrics. In practice, it is used more often in forensic and medical labs than for simple authentication and identification. Hence, although it is still included in biometric technology, it will not be central to this book.
30 See also the definition of 'personal data' in Article 2(a) of the Data Protection Directive, along with Article 2(d), which defines a controller as a 'natural or legal person, public authority, agency, or any other body which alone or jointly with others determines the purposes and means of the processing of personal data.'
31 44 U.S.C§3542 (b) (1) (2006).
32 See further the definition of 'anonymous transaction' in Clarke (1996).

33 For example, in France, the Law on the freedom of communication of 30 September 1986 (modified by Law of 16 June 2000) recognises a general right of anonymity on the Internet, formulated as the right to have access to the Internet anonymously (Article 43–6–4.II).

References

Adam, A. (1998) *Artificial Knowing, Gender and the Thinking Machine*, London, New York: Routledge.

Allen, A. (1988) *Uneasy Access: Privacy for Women in a Free Society*, Totoma, NJ: Rowman & Littlefield.

Article 29 Data Protection Working Party (2003) *Working Documents On Biometrics (12168/02/EN WP 80)*. Available at http://ec.europa.eu/justice_home/fsj/privacy/docs/wpdocs/2003/wp80_en.pdf.

Ashbourn, J. (2006) 'The Social Implications of the Wide Scale Implementation of Biometric and Related Technologies', Background paper for the Institute of Prospective Technological Studies, DG JRC – Sevilla, European Commission, January 2005.

Australian Government (2004) *Overview of Smart Gate Trial*, Online. Available at http://www.customs.gov.au/webdata/resources/files/medBackSmartgate0401.pdf.

Australian Law Reform Commission (2008) *ALRC Report 108: For Your Information: Australian Privacy Law and Practice*. Online. Available at http://www.austlii.edu.au/au/other/alrc/publications/reports/108/.

Beavan, C. (2001) *Fingerprints: The Origins of Crime Detection and the Murder Case That Launched Forensic Science*, New York: Hyperion.

Biovision (2003) *Roadmap for Biometrics in Europe to 2010*. Retrieved on 3 January 2005 from http://ftp.cwi.nl/CWIreports/PNA/PNA-E0303.pdf. (no longer accessible)

Bolle, R.M. et al. (1996) 'Veggie Vision: A Produce Recognition System', Paper Presented at the Third IEEE Workshop on Automatic Identification Advanced Technologies, New York.

Brownlie, I. (2008) *Principles of Public International Law*, Oxford: Clarendon.

Bygrave, L.A. (2002) *Data Protection Law: Approaching Its Rationale, Logic and Limits*, The Hague: Kluwer Law International.

Cavoukian, A. (1999) 'Privacy and Biometrics – A "Wake Up Call"', Paper Presented at the 21st International Conference on Privacy and Personal Data Protection. Available at http://www.austlii.edu.au/au/journals/PLPR/1999/62.html.

—— (2001) *Public Safety Is Paramount – But Balanced Against Privacy*. Available at http://ipc.on.ca/images/Resources/pub01-e.pdf.

Chandrasekaran, R. (1997) Brave new whorl: ID systems using the human body are here, but privacy issues persist, *Washington Post*. p. HO 1.

Clarke, R. (1994) Human identification in information systems: Management challenges and public policy issues, *Information Technology & People*, 7(4): 6–37.

—— (2001) *Biometrics and Privacy*. Available at http://www.rogerclarke.com/DV/Biometrics.html.

—— (2003) *Biometrics Bibliography*. Available at http://www.rogerclarke.com/DV/BiomBibl.html.

Cole, S. (1999). What counts for identity? The historical origins of the methodology of latent fingerprint identification. *Science in Context*, 12(1): 139–72.

Crompton, M. (2002) Biometrics and Privacy. *Privacy Law and Policy Reporter*, 9(3): 53–58.

Datatilsynet (2006) *Notat Fra Datatilsynet – Forslag Til Revisjon Av Personopplysningslovens § 12. og Ny Bestemmelse Om Bruk Av Biometriske Data.* Online. Available at http://www.datatilsynet.no/upload/Dokumenter/saker/2006/Revisjon_12_biometri.pdf.

Davies, S.G. (1994) Touching Big Brother: How Biometric Technology Will Fuse Flesh And Machine. *Information Technology & People,* 7(4): 38–47.

Dennis, C. (2002) *Biometrics and the Prevention Of Identity Theft.* Available http://www.biometricgroup.com/in_the_news/senate.html.

Dersgowitz, A.M. (2002) *Why Terrorism Works: Understanding the Threat, Responding to the Challenge,* New Haven, CT, Yale University Press.

DeCew, J.W. (1997) *In Pursuit of Privacy: Law, Ethics, and the Rise of Technology,* Ithaca, NY: Connell University Press.

Electronic Privacy Information Centre and Privacy International (2005) *Comments of the Electronic Privacy Information Center.* Available at http://www.epic.org/privacy/us-visit/comments080405.html.

Etzioni, A. (1999) *The Limits of Privacy,* New York: Basic Books.

European Commission's Joint Research Centre (2005) *Biometrics at the Frontiers: Assessing the Impact on Society (EUR21585 EN).* Available at ftp://ftp.jrc.es/pub/EURdoc/eur21585en.pdf.

Garfinkel, S. (2000) *Database Nation: The Death of Privacy in the 21st Century,* California: O'Rielly & Associates.

Gavison, R. (1980) Privacy and the limits of law, *Yale Law Journal,* 89(3): 421–28.

Grijpink, J. (2001) Privacy law: Biometric and privacy, *Computer Law and Security Report,* 17(3): 154–60.

Hert, P.J.A. de (2005) 'Biometrics: Legal Issues and Implications', Background paper for the Institute of Prospective Technological Studies, DG JRC – Sevilla). Available at http://www.statewatch.org/news/2005/apr/jrc-biometrics-paul-de-hert.pdf.

Inness, J.C. (1992) *Privacy, Intimacy and Isolation,* Oxford, Oxford University Press.

International Biometrics Group (2002) *How Is Biometrics "Defined"?* Available at http://www.biometricgroup.com/reports/public/reports/biometric_definition.html.

Jain, A.K. and Pankanti, S. (2006) A touch of money. *IEEE Spectrum,* 43(7): 22–7.

Kang, J. (1998) Information privacy in cyberspace transactions, *Standard Law Review,* 150, 1193–1294.

Lyon, D. (2001) Surveillance Society: Monitoring Everyday Life, *Issues in Society,* Buckingham: Open University.

McCormack, D. (2003) Can corporate America secure our nation? An analysis of identix framework for the regulation and use of facial recognition technology, *Boston University Journal of Science & Technology Law,* 9: 128–55.

Milbank, D. (1995, Dec.4) Measuring and cataloguing body parts may help weed out welfare cheats, *The Wall Street Journal,* B1.

Miller, A. (1971) *The Assault on Privacy: Computers, Data Banks and Dossiers,* Ann Arbor, Michigan: University of Michigan Press.

Mordini, E. and Petrini C. (2007). Ethical and social implications of biometric identification technology, *Ann Ist Super Sanità,* 43(1): 5–11.

National Science and Technology Council (NSTC) (2006a) *Privacy Overview.* Available at http://www.biometrics.gov/Documents/BioOverview.pdf.

National Science and Technology Council (2006b) *Biometrics History.* Available at http://www.biometrics.gov/Documents/BioHistory.pdf.

NHS (2008, 29 *February) Sidewinder Makes the Case for Biometrics in the NHS.* Available at http://www.microsoft.com/uk/nhs/content/articles/sidewinder-makes-the-case-for-biometrics-in-the-nhs.aspx.

Ng, R. (2006) Catching up to our biometric future: Fourth amendment privacy rights and biometric identification technology. Hastings Communications and Entertainment Law Journal, 28: 425–43.

Nuger, K. P. and Wayman, J. L. (2004a). Biometrics and the US Constitution, in J. L. Wayman, J. Anil, M. Davide, and M. Dario (eds.), *Biometric Systems: Technology, Design and Performance Evaluation*, Nottingham: Springer, 311–33.

Nuger, K.P. and Wayman, J.L. (2004b) An introduction to biometric authentication systems, in J, L. Wayman, et al. (eds.), *Biometric Systems: Technology, Design and Performance Evaluation*, Nottingham: Springer, 1–20.

Ploeg, I.V.D. (2005) *The Machine-Readable Body*, Maastricht, Netherlands: Shaker Publishing.

Prins, C. (1998) Biometric technology law, making our body identify for us: legal implications of biometric technologies', *Computer Law & Security Report,* 14(3): 159–65.

Reiman, J.H. (1995) Driving to The Panopticon: a philosophical exploration of the risks to privacy posed by the highway technology of the future, *Santa Clara Computer and High Technology Law Journal,* 11(1): 27–44.

Rothstein, M.A. (1997) *Genetic Secrets: Protection of Privacy and Confidentiality in the Genetic Era*, New Haven: Yale University Press.

Schneier, B. (2003) *Beyond Fear: Thinking Sensibly About Security in an Uncertain World*, New York: Copernicus Books.

Schwartz, P.M. and Reidenberg, J.R. (1996) *Data Privacy Law*, Charlottesville, VA: Michie Law Publishers.

Sobel, R. (2002) The demanding of identity and person hood in national identification systems, *Harvard Journal of Law and Technology,* 15(2): 320–86.

Warren, S.D. and Brandeis, L.D. (1890) The right to privacy, *Harvard Law Review,* 4: 19–211.

Watson, J. D. and Berry, A. (2003) *DNA: The Secret of Life*, New York: Random House.

Wayman, J. L., et al. (2004) *Biometric Systems: Technology, Design and Performance Evaluation*, London: Springer.

Westin, A.F. (1967) *Privacy and Freedom*, New York: Atheneum.

Winner, L. (1986) Do artifacts have politics?, In L. Winner (ed.), *The Whale and the Reactor: a Search for Limits in an Age of High Technology*, Chicago: University of Chicago Press, 19–39.

Woodward, J.D. (1997) *Biometrics: Identifying law & Policy concerns*. Available at http://www.cse.msu.edu/~cse891/Sect601/textbook/19.pdf.

Working Party on Information Security and Privacy (2004) *Biometric-based Technologies* (DSTI/ICCP/REG (2003)2/FINAL), Report for Directorate for Science, Technology and Industry Committee for Information, Computer and Communications Policy. Available http://www.olis.oecd.org/olis/2003doc.nsf/LinkTo/NT000070D6/$FILE/JT00166988.PDF.

2 Biometric technology and its applications

2.1 Introduction

The objective of this chapter is to present a high-level introduction to biometrics. Discussions range from basic processes to technical factors and highlight the characteristics of certain biometric technologies. Some common misconceptions are also discussed. The discussions are essential for grasping the possible legal implications of using biometrics and, concomitantly, for finding solutions for attendant legal problems. To establish a solid technical background for further study, some other closely related technologies that are commonly used in combination with biometrics are also explained. Although this is a rather technical chapter, an attempt has been made to make it comprehensible to legal professionals. Hence, the most technical aspects (such as mathematical parameters) have been purposefully omitted.

The chapter begins with an explanation of the relevant terminologies pertaining to biometrics. It continues with an explanation of the two basic biometric application types and the technology process. Section 2.6 describes the principal categories of biometric technologies and their advantages and disadvantages. Thereafter, the chapter highlights some common myths surrounding biometric technologies as well as some other factors that may affect their efficiency. The subsequent section illustrates some common technologies that are being used together with biometrics; namely, RFID, smart cards, and databases. Finally, some specific biometric applications are presented to better inform the reader of the ways in which biometric technology is being deployed.

2.2 Characteristics of biometrics

Biometrics is a technique that uses biometric features to verify the identity of, or to identify, human beings. Examples of biometric characteristics that have been used include physical characteristics (e.g., finger geometry, fingerprint, hand geometry, retinal pattern, DNA, iris, facial geometry (2D/3D), and voice print) and behavioural characteristics (e.g., handwriting, keystroke dynamics, and gait). The features are usually closely linked to the physical manifestation of a human being, and compared with a password or tokens, biometric characteristics are more difficult to lose or forget. It should be emphasised that some human characteristics can be used to derive knowledge, with potentially far-reaching consequences. For example, DNA can reveal not only the identity of a person but also his or her genetic properties, such as the likelihood that they will develop a particular disease.[1]

Various biometric features have characteristics that make them more or less useful for particular applications. Biological features that qualify as 'biometric' generally have five desirable properties: robustness, distinctiveness, availability, accessibility, and acceptability (Wayman, 1999, 2001). These properties are elaborated as follows.

Robustness: this refers to the extent to which the characteristic or trait is subject to significant changes over time (Woodward et al., 2003). These changes can occur as a result of age, injury, illness, occupational use, or chemical exposure. A highly robust biometric does not change significantly over time, whereas a less robust biometric will change. For instance, the iris changes very little over a person's lifetime and is usually regarded as more robust than a face or voice. It is important to note, however, that there is no absolutely robust biometric feature.

Distinctiveness: this refers to the degree to which a biometric shows a great degree of variation across a population. The higher the degree of distinctiveness, the more individual the identifier is (Woodward et al., 2003). A low degree of distinctiveness indicates the biometric feature can be frequently found throughout a general population. The retina and iris possess a higher degree of distinctiveness compared to hand or finger geometry (Woodward et al., 2003). The purpose of the biometric application determines the degree of robustness and distinctiveness required.

Availability: this denotes the extent to which a biometric is available to all persons. Ideally, a biometric measure should be capable of being derived from all persons in a given population (Nuger and Wayman, 2004b). However, it has been claimed that the only universally available biometric feature is DNA. Hence, DNA is an example of a biometric with a high degree of availability, whereas other biometric features can be missed for various reasons. For instance, less than a minute of exposing a finger to certain household cleaning chemicals can damage the fingerprints, and many people have chronically dry skin and cannot present clear prints.

Accessibility: this refers to the degree to which a biometric is easily presented to an imaging sensor.

Acceptability: high acceptability indicates that the process of collecting a feature is not regarded as intrusive and is accordingly considered to be 'acceptable' by most people. For example, facial recognition is typically considered as more acceptable than using other kinds of biometric features, as it is quite similar to traditional and widely used photo identification.

Dr. Wayman discusses these qualities by comparing fingerprints to hand geometry. Fingerprints are very distinctive, but are not robust. As mentioned previously, there are a number of ways to easily damage one's fingerprints. Hands are believed to be much more robust than fingerprints; to change hand geometry, a person would have to hit their hand very hard with a hammer or mangle it in a machine. However, a hand is not as distinctive as a fingerprint, as many people may have quite similar hands. Both fingerprints and hands are approximately accessible to the same degree, and can easily be 'presented'. They are probably also similar in terms of 'acceptability'. Yet with respect to availability, a person can present at least six nearly independent fingerprints, but only one hand geometry (assuming one's left hand is a near mirror image of the right) (Wayman, n.d.).

It is the combination of all these properties that determines the effectiveness of employing a biometric in a particular application (Jain et al., 1999). There is no biometric that satisfies any of these properties absolutely or simultaneously to the same degree, especially if acceptability is taken into account (Jain et al., 1999). Thus, any kind of biometric application system implies certain compromises.

2.3 Biometric authentication vs. identification

Regardless of specific uses and deployment scenarios, most biometric systems will serve one of two fundamental purposes: identification or authentication/verification.[2] Biometric authentication or verification involves a 'one-to-one' (1:1) search in which a live biometric sample presented by a person is compared to a stored sample collected previously from that individual, and the match is subsequently confirmed (Cavoukian, 2007). This answers the question of whether the person is who he claims to be. There are two possible database configurations in the context of authentication: storage of biometric information on the enrolled subjects in a centralised database or a distributed database in which the biometric information is stored in a distributed fashion (Bolle et al., 2003). For authentication purposes, use of a distributed database (e.g., smart cards) will be adequate. However, a central database can still be used, provided that some other data, such as serial numbers, are used to 'look up' an individual in the database, so as to find a particular biometric template therein and conduct a one-to-one match.

As Ford and Baum point out, authentication generally 'is the most essential of all the security services because reliable authentication is needed to enforce access control, to determine who is authorised to receive or modify information, to enforce accountability, and to achieve non-repudiation'[3] (Ford and Baum, 1997 126). Criteria for authenticating human identity have traditionally been placed in three categories (Grosche, n.d.). One category comprises things a person 'knows' (e.g., a password or PIN). Another comprises things a person possesses (e.g., ATM card or tokens). A third comprises, essentially, what a person *is* in the form of their biological features, in other words biometrics. The latter category is generally considered as the most convenient and reliable as they are more difficult to lose or forget than the other categories of criteria. However, it should be noted that biometric authentication criteria, though traditionally considered as 'irreproducible or non-transferable', are sometimes possible to 'reproduce' and 'transfer'.[4] In general, any combination of these categories of authentication criteria heightens security. Requiring all three for an application provides the highest form of security.

Biometric identification refers to the ability of a computer system to uniquely distinguish an individual from a larger set of individual biometric records on file (Cavoukian, 2007). This is also known as a one-to-many (1: *n*) match, in which searches are designed to determine identity based solely on biometric information. This match intends to answer the question: who is the person concerned? To support identification, a central database must be built containing a large set of individual biometric records. Theoretically, a central database of biometric records could allow the system controller to discover who a person is provided that he is already registered in the central database. During the process, the live biometric sample will be compared with all the registered biometric samples in the central database. Upon the discovery of a successful match, the person's identity is released from the central database.

In an identification system, biometric information is always stored in a database controlled by a custodian, whereas in an authentication system, the biometric information can be stored either in a database, in a portable medium or both. The storage of biometric information solely on a portable token held by the data subject (e.g. a smart card) is generally considered to be more privacy-friendly, as the data subjects have greater control over their own biometric information. In practice, however, such storage is very rare, since for back-up reasons, a central database is generally used even when the portable media also store the biometric information.

Some current biometric systems can be used for both identification and verification. A typical example is when a person applies for a biometric passport. In this scenario, the person's biometric sample is first collected and compared against a watch-list of criminals and terrorists in a central database via a one-to-many match. This is to ensure that this person has a 'clean' background and is not registered in the database under a different identity. Once the person is cleared, they are issued the passport, which will in most cases be used only for a one-to-one match at a later time. In this scenario, the biometric information is stored both in a central database and on the biometric passport.

2.4 Biometric system operation

All biometric technology systems have certain aspects in common Biometric applications depend on comparing a new measure against a previously captured measure (Clarke, 2001). A system for using biometrics for identification or authentication consists of two phases, enrolment and testing, which can be depicted as follows:

During enrolment/registration, the data subject's biometric sample is collected by the 'measuring device' (usually a biometric sensor). The biometric system then extracts feature information from the biometric sample (e.g. a fingerprint image) and creates a template (i.e., data representing the biometric measurement). This template is usually called a master template, which is then stored (e.g. in a database, token, or distributed system).

During the testing process, the data subject uses a biometric measuring device to provide a biometric sample, and a new biometric template is then created (this is usually called a live template). This template is then compared against one or more master templates. This process is commonly called matching. Matching can involve authentication or identification.

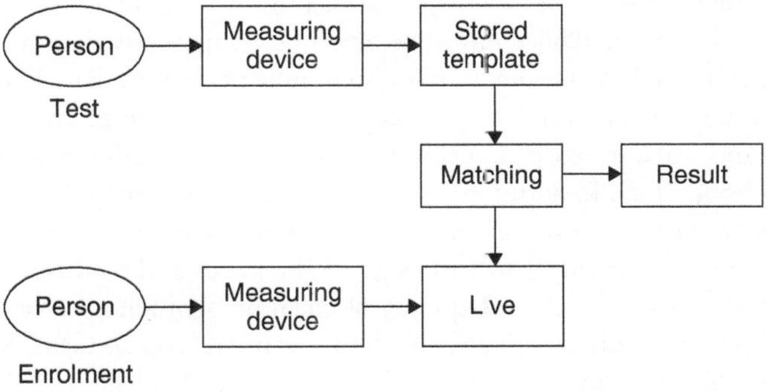

Figure 2.1 Biometric system design.[5]

The result might be an estimate of the likelihood of the two templates being from the same person. Although such a process commonly results in the generation of a Yes/No result, such Yes/No decisions can by no means be absolute. It is actually a decision based on some more or less arbitrary possibility threshold which may be fixed, computed by some algorithm, or manipulated by the system operator (Clarke, 2003). Each biometric system has a different method of assigning a score to the biometric correlation (the degree to which the candidate biometric sample matches a previously enrolled biometric sample) (Working Party on Information Security and Privacy, 2004). A threshold is a score which determines when a match is declared. Scores above the threshold generate a 'Yes' result in the matching process and scores below the threshold generate a 'No' result. As shown by Wilson, biometric authentication entails a long chain of processing steps, all of which may suffer from imperfections (Figure 2.2) (Wilson, 2005).

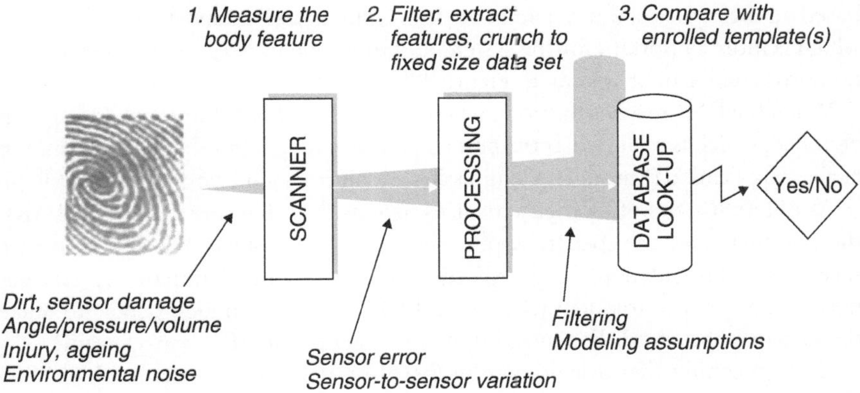

Figure 2.2 How uncertainty accumulates during biometric processing (Wilson 2005: 15).

Imperfection occurs for various reasons and is inherent to biometrics, as the body parts being measured can never be measured to precisely the same value each time. A biometric device has to be able to tolerate the variations of input to some extent; inevitably, this means the system may make mistakes. The biometric correlation may or may not be 'accurate' depending on the biometric type, the hardware and software used, and the algorithm and operational settings of the system (Wilson, 2005). In addition, biometric samples may be gathered from people in environmental conditions that are uncontrollable, employing equipment that may slowly be wearing out, and using technologies and methods that vary in their precision. For example, the quality of facial recognition may be affected by different lighting conditions, angle, subject movement, and so forth. Simply put, biometrics is 'fuzzy' – no two samples will be perfectly identical.

No single metric is sufficiently adequate to give a reliable and convincing indication of the identification accuracy of a biometric system (Jain et al., 2004). Consequently, the performance of a system in the context of a given population is a random variable and, strictly speaking, cannot be computed or measured. However, it can be estimated from empirical data, but estimates of performance are highly data dependent. Hence, they are meaningful only for a specific database in a specific testing environment (Jain et al., 2004). The accuracy of the biometric system is commonly assessed in light of these confounding variables via its tendency to exhibit a false acceptance rate (FAR) (which is equivalent to a false match rate (FMR) or false rejection rate (FRR) (which is equivalent to a false non-atch rate (FNMR). These error rates are often estimated for the portion of the user population that is not subject to exception handling.[6] FAR and FRR vary inversely, so for one technology under fixed operating conditions, lowering one error rate will necessarily raise the other.

The decision threshold must be adjusted according to the desired characteristics for the application considered. Usually, a high-security application requires a very low FAR which unavoidably results in a high FRR. In low-security applications, where a too-high FRR may frustrate users, FAR can thus be higher and accordingly decrease the FRR. Hence, decisions have to be made to adjust the accuracy based on the nature of the system in question. This must be done in order to decide which is more important for the system and to decide the degree to which exception processing can be tolerated (Figure 2.3).

When the FAR reaches the point where it equals the FRR, the equal error rate is obtained, which is regarded as the best choice for setting a threshold for a common civilian application (European Commission's Joint research Centre (JRC), 2005: 8).

As different types of biometric devices have widely differing FRRs and FARs, the parameters of the specific application must be considered. Where a device serves a small population, a higher FRR may not make much difference. In larger populations or with frequent use, a high FRR can affect enough people to make the system impractical. For most biometric systems, the FRR ranges from 0.1% to 20%, meaning that a legitimate user will be rejected at a rate of 1 of 1000 times up to 1 of 5 times on average. FAR ranges from 1 in 100 to 1 in 10,000,000 (Cavoukian, 2007). Provided we have a system that is average, with a FRR of 0.1%

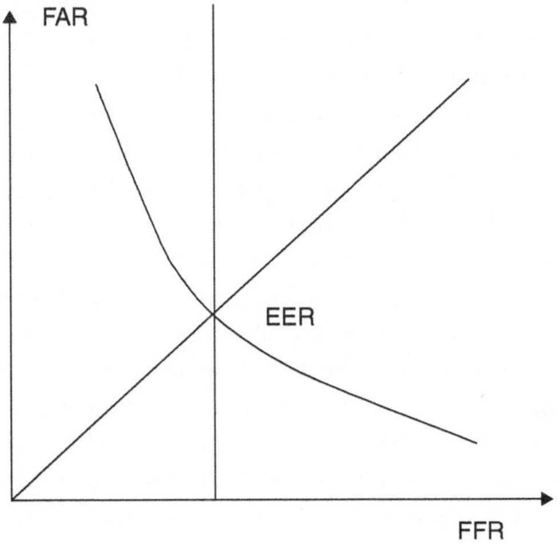

Figure 2.3 Biometric system error trade-off curve.

and a FAR of 1 in 10,000,000, and we assume a database of 10,000,000 entries, the implication is that a single biometric sample will generate 10,000 false rejections and 1 false acceptance. Clearly this will still be unacceptable for any real-time automatic identification and would require significant human intervention in order to function. In order to overcome this limitation of present technology, biometric designers have to resort to other techniques. One possible solution is to collect and compare multiple biometric samples; for example, two fingerprints could be collected instead of one, which would diminish the false match/non-match rate considerably. This is also the primary reason behind emerging international requirements for the inclusion of two separate biometrics (face and finger) in machine-readable travel documents such as passports.

In most cases, biometric systems must be evaluated as a part of a larger and more complex application. Even an ideal system can be easily defeated if it is incorporated into an insecure or poorly designed overall system architecture. When assessing the utility or the cost of a biometric system, it is important to consider some common questions for all such systems, such as How will the biometric samples be gathered and catalogued? Can a live sample be captured without the subject's knowledge? What are the possible implications of such outcomes? Is the system secure enough to prevent the interception of user biometrics?

It is very important to provide a fallback procedure[7] to deal with errors. Fallback procedures are also necessary for assisting people who experience difficulty in providing a sample. For example, some people with chronic skin disease cannot provide a clear fingerprint. The percentage of the population which exhibits such problems may be small but significant. Fallback procedures will require flexible human involvement to handle a variety of potential problems.

There is an oft made point that biometric authentication, as opposed to biometric identification, represents a lesser threat to privacy, as it does not need to store the biometric in a central database. In theory this is correct. In practice, as mentioned before it is usually not, given that a back-up system is typically required. Without back-up storage, a lost, stolen, or damaged smart card will require that an individual visiting an enrolment centre present his ID and have his biometric data processed again. This so-called 'privacy-enhancing' processing is actually considered impractical and inconvenient for users. An interview was conducted in November 1999 with an INS inspector[8] at Los Angeles Airport. In this interview, the configuration of a hand geometry-based system was discussed as it related to INSPASS, a program allowing enrolled frequent travellers quick passage through immigration. It was claimed by the INS inspector that the functionality of the system did not require central storage of biometric data after issuance of the card – whenever someone lost or damaged their card, a new one could be sent by mail without the person having to come to an INSPASS enrolment centre to register again. The implication, though, is that the system maintains relevant biometric data somewhere else as well (Ploeg, 2003).

2.5 Privacy and security risks of a biometric system

Like all computer systems, a biometric system may be vulnerable to attack (Ratha and Connell, 2001). Biometrics consists of a secure identification process only in that it can provide a strong link between a physical person and their identity data (European Commission's Joint Research Centre (JRC) 2005). It has also been claimed that biometrics' most notorious liability is its vulnerability to impersonation, which is perceived as a security flaw of biometric technology (Bolle et al., 2003). A biometric may be presented when the 'owner' is not present, which is a form of 'spoofing'. It has been noted that lifting a latent fingerprint and using the fake print to impersonate its owner is functionally identical to the covert observation of password input (Bolle et al., 2003). However, in gaining unauthorised access to a system, there is substantial difficulty in applying knowledge pertaining to a biometric, as opposed to a similar situation involving the use of a password.

The security of a biometric system depends on the secure operation of each step of the process (enrolment, storage, collection, and matching). In addition, the biometric system can be compromised when any other part of the system is open to circumvention. The way in which biometric data are created, stored, compared, linked, and shared may also raise a set of privacy concerns. Identity theft may be difficult to tackle when biometric data are compromised, and switching to another identity will be more difficult, if not impossible, when biometrics is used as a representation of people's identity.

The security of biometric systems is dependent on a variety of factors. Even with protections built into biometric systems, some biometric technologies may require live human or video supervision of the sampling process in order to protect against certain types of security attacks on the biometric system. Examples of such situations include use of an artificial mould of a finger on a finger scanner or the presentation of a printed picture of an iris. Dr. Cavoukian illustrates the privacy and security issues of a biometric system in her article (Cavoukian, 2007), accordingly the security and privacy concerns of a biometric system might involve:

Enrolment:

- Have the data subjects been clearly informed about the purpose of the collection of data and their subsequent use? This is a concern pertaining to the transparency level of the biometric system.
- Is the system compulsory or voluntary?
- Are they presented with an equivalent alternative if they choose not to use the biometric system?
- With respect to the quality of the sensor, are the collected biometric data sufficiently accurate? Insufficient accuracy may result in high FAR/FRR.

Storage:

- What kind of biometric information is being stored: a biometric template, biometric image or both?
- What kind of security measures are taken with respect to prohibiting unauthorised access to storage facilities/media either locally or remotely? How significant is the risk of a substitution attack?[9]
- Are the biometric data stored together with other personal information or are they linked with other personal information or databases?
- Are the biometric data revocable?[10] That is, is there a possibility of generating the original biometric image from the biometric template? This has been proven to be possible under certain conditions.
- Can the biometric data disclose health information? What kind of health information can be disclosed?[11]
- How can the system prevent tampering (template)?[12]

Matching and transmission:

- Can the system detect a 'replay attack'? Professor Clarke has demonstrated that 'masquerade'[13] is possible with the use of biometrics. As biometric data are not secret, it is tenable for an artefact to be constructed that makes it appear as if a particular person is presenting to the measuring device. Such an artefact may be a synthesised physical 'thing' or in some cases a synthesised set of signals. It does not even have to be a close copy of the original, but merely a near-equivalent with respect to the aspects/traits upon which the technology relies (Hill, 2001; Clarke, 2002).
- How can a system tackle a virus? Like other computer systems, the biometric system is vulnerable to viruses. Trojan horse programs can replace a matcher in the biometric system and compel the system to always output high-matching scores.

- How does one prevent a decision from being overridden? Despite the actual result produced by the system, an attacker may be able to interject a false 'Yes' response at a proper point in the transmission and thereby pose as a legitimate user in any system.
- The accuracy of the biometric system can also be of great concern. Biometric technologies are subject to a range of errors. These may be due to inaccurate reference measures, inaccurate test measures, an incorrect result, or incorrect manipulations.

(Hill, 2001; Clarke, 2002)

Collection:

- Can the sensor detect spoofing? Many biometric devices can be fooled by relatively simple mechanisms in a process known as 'spoofing'. As a result, these devices still require human supervision when used in high-security applications to ensure that this weakness is not exploited.[14]
- Applications: the application of the biometric system will also tend to involve various privacy threats such as function creep, covert surveillance, misuse of the biometric data, and loss of anonymity.[15]

2.6 Types of biometrics

A proper assessment of biometric technology is built not only upon a general understanding of biometrics but also upon an understanding of specific technologies. In order to better understand the challenges posed by biometric technologies, this section provides an in-depth illustration of the various biometric technologies that are currently being used or being researched. According to Maltoni et al., biometric artefacts that have received attention in the research and development industry are as follows:

> Physical characteristic: finger geometry, finger surface (3D), finger wrinkles, fingerprint, palm print, odour, pores, reflection of acoustic waves in the head, bite marks, hand geometry, hand pressure profile, hand thermogram, hand vein pattern, retinal pattern, dental geometry, ear, DNA, facial geometry (2D/3D), iris, skin spectrum, skin pattern, smile, facial thermo-gram, lips, voice print, bone sound transmission.
>
> Behavioural characteristic: dynamic grip recognition, handwriting, tapping, eye movement tracking, keystroke dynamics, gait, mouse dynamics.
>
> (Maltoni et al., 2003)

In this book, I focus on physical characteristics that are currently most commonly used as biometric identifiers in biometric devices (as well as some that show promise). The *In 2007 International Biometrics Group* reveals the recent comparative market share of technologies being implemented. Among which fingerprint occupied 25.3%, face recognition occupied 12.9%, AFIS live scan occupied 33.6%,[16] iris recognition occupied 5.1% and some other biometric systems occupied the rest of market.

2.6.1 Fingerprint recognition

The fingerprint is a highly familiar and well-established metric in biometric science. It is one of the most popular biometric technologies used in forensic divisions worldwide in criminal investigations and therefore has a stigma of criminality associated with it. The Department of Immigration and Citizenship in Australia considers fingerprint technology to be the most mature. It has the highest number of implementations compared with other types of biometric technologies (Department of Immigration and Citizenship, 2008).

It is necessary to explain some of the confusing terminology in this area. 'Fingerprint' refers to the unique patterns that exist on the underside of every human finger. 'Fingerprinting' refers to an impression or image of a fingerprint, which usually involves the creation of an impression on paper with ink when created intentionally, or in oil or dirt on a rigid surface when created unintentionally. 'Finger scanning' is the process of capturing an image of a fingerprint with a digital sensor for use as a biometric identifier (Fingerman, 2003).

Police forces have used fingerprints for decades. The positions of details which are supposed to be unique[17] for every person are used for identification or authentication. There are three major fingerprint features: the arch, loop, and whorl. Each finger has at least one major feature. These details, known as minutiae, can be endpoints, junctions, arches, loops, and whorls in the fingerprint pattern. The systems which do not utilise minutiae use pattern-matching, which extrapolates data from a particular series of ridges (Working Party on Information Security and Privacy, 2004). The series of ridges used at enrolment are the basis for future comparison (Working Party on Information Security and Privacy, 2004). The templates created in pattern-matching are generally, but not always, two to three times larger than that in minutiae (Working Party on Information Security and Privacy, 2004).

Despite extensive research and development, fingerprint recognition is still a challenging and important pattern-recognition problem (Maltoni et al., 2003). An essential goal in the field of fingerprint recognition is the development of technical standards. Interoperability is related to technology standards and is a crucial aspect of product implementation. Templates obtained by one system must be capable of being interpreted by a computer using another system (National Science and Technology Council (NSTC) 2006b). Moreover, although fingerprints exhibit a good balance with respect to the five desirable properties,[18] when only one finger is used, availability and robustness may be problematic.[19] It is estimated that about 5% of people would not be able to register and deliver a readable fingerprint (Sasse 2004). Many biometric systems can be spoofed using fake prints; some require the cooperation of the authorised user, whereas others only require a digital image of a fingerprint.[20]

Fingerprint identification is widely used in law enforcement agencies to identify criminals. In Europe, the most famous central fingerprint database is Eurodac, which is a large-scale on-line system. It stores the fingerprint of asylum seekers.[21] Fingerprints have also been selected as one of the biometrics recommended by

ICAO for standard machine-readable documents such as a passport (International Civil Aviation Organisation, 2004a).

2.6.2 Iris recognition

The iris is the coloured portion of an individual's eye. It is 'a thin diaphragm stretching across the anterior portion of the eye and supported by the lens' (Wayman, 2005: 63). The iris is located behind the cornea and the aqueous humour but in front of the lens (which is the only internal organ of the body that is normally visible externally) (Wayman, 2005). The iris' distinctive characteristics lie in the trabecular meshwork – a web of fibrous tissue that is permanently fixed by the eighth month of gestation and is stable throughout a person's life (International Biometrics Group, 2002). This meshwork gives the iris the appearance of being divided in a radical fashion. Other visible characteristics include rings, furrows, freckles, and the corona (International Biometrics Group, 2002). Iris patterns are both highly complex and robust – even a person's right and left iris patterns are completely different, making them very well-suited for biometric identification purposes (Daugman and Malhas, 2004). It is claimed that the iris recognition system offers very good performance even in identification modes with huge databases of enrolled users, though the necessary infrastructure is still costly (European Commission's Joint Research Centre (JRC), 2005). Of particular significance for the purpose of recognition is that following adolescence, the healthy iris varies little for the rest of a person's life. However, a variety of diseases of the eye can drastically alter the appearance of the iris (Adler, 1965). Also, certain drug treatments for eye disease and intensive exposure to certain environmental contaminants (e.g. metals) may alter iris pigmentation (Newell, 1991).

The first step in iris recognition is to locate the iris using landmark features. These landmark features and the distinct shape of the iris itself allow for imaging, feature isolation, and image extraction (National Science and Technology Council (NSTC) 2006a). The process involves the use of a high-resolution camera and a light source to capture a digital image of the iris. The resulting photography is analyzed using algorithms to locate the iris and extract feature information to create a biometric template. Since about 85% of the iris is normally visible, no invasive measures are needed to acquire an image of the iris (Roethenbaugh, 1998). However, the enrolment process necessarily requires the awareness of the user, as it cannot at present be done without the cooperation of the user.

Until 2004, the largest deployment of iris recognition was in all 17 border entry points (air, land, and sea ports) of the United Arab Emirates. Immigration control checks all incoming passengers against an enrolled database of around 420,000 iris codes of persons who were expelled from the United Arab Emirates (Daugman and Malhas, 2004). In Europe, the iris recognition system has also been deployed in Schiphol Airport (NL),[22] Frankfurt airport (DE),[23] and some UK airports for certain group of passengers.[24] Canada has also initiated a voluntary registration program that utilises iris scanning to confirm passengers' identity.

2.6.3 Facial recognition

The face is an obvious choice for a biometric because it is a physiological characteristic used every day by humans to identify each other. Facial recognition systems depend on a recognition algorithm, such as eigenface[25] and feature analysis. Eigenfaces are a set of eigenvectors used in the computer vision problem of human face recognition. Variations of eigenface are frequently used as the basis for other face-recognition methods. Feature analysis is a very widely utilised implementation of facial recognition technology related to eigenface; however, feature analysis is more capable of accommodating changes in appearance or other facial aspects. It may use dozens of features from different regions of the face and also incorporates the relative locations of these features (Working Party on Information Security and Privacy, 2004). The term facial recognition actually encompasses several technologies including 2D, 3D, and infra-red (IR) facial scans, with 2D face recognition being the most common type by far. Facial recognition differs from other biometric modalities in that the cooperation of the subject is not necessary, and an image of the face can be captured covertly without notice. However, the current performance levels of the technology limit the capabilities of a large-scale surveillance system. Nevertheless, better performance coupled with advances in computer vision could potentially enable an automated system to identify people in a crowd using images captured at long distances in the near future.

It is a challenge to develop face recognition techniques which can tolerate the effects of aging, facial expression, slight variations in the imaging environment and variations in the position of the face with respect to the camera (2D and 3D rotations). There is also the problem of distinctiveness, as when there are many similar faces among a large population, the patterns of faces show less variation compared to fingerprints or irises. It has been pointed out that face recognition is clearly not yet a mature technology. Its performance ranks far below iris and fingerprint systems (European Commission's Joint Research Centre (JRC), 2005). The low accuracy rates of face recognition make it easier for impostors to be falsely accepted. In 2002, an experiment by the local police department in Tampa, Florida, had disappointing results (Greene, 2002). Although the best-performing systems are not significantly affected by normal changes in indoor lighting conditions, face recognition is not yet ready for outdoor use. It is unsuitable for large databases and large watch-lists, and even for moderate-sized lists, it exhibits mediocre performance (European Commission's Joint Research Centre (JRC), 2005).

Nevertheless, having been chosen by the ICAO as the primary biometric identifier for travel documents, face recognition still shows great promise for implementation in the future. Facial recognition technology has been introduced by Australian Customs to coincide with the release of the Australian e-passport.[26]

2.6.4 Hand geometry

In recent years, hand geometry has become a very popular access control biometric and has captured a large share of the access control market.[27] Typically, devices using this feature map the size and shape of a person's hand, compute a template from it and verify this with a previously enrolled template (FIDIS, 2006). Hand geometry systems use an optical camera to capture two orthogonal two-dimensional images of the palm and sides of the hand (see Figure 12) offering a balance of reliability and relative ease of use. The most widely used cameras are charge-coupled device cameras and infrared light-emitting diodes (LEDs) with mirrors and reflectors to capture black and white images of the human hand silhouetted against a 32,000 pixel field. They typically collect more than 90-dimensional measurements, including finger width, height, and length, distances between joints and knuckle shapes (Global Security, n.d.). The scanner records no surface details, ignoring fingerprints, lines, scars, and colour. The process is much like placing a hand on a beaded projector screen. The hand scanner reads hand shape by recording the silhouette of the hand. In combination with a side mirror and reflector, the optics produces two distinct images: one from the top and one from the side. This method is known as orthographic scanning (Zunkel, n.d.). It is said that hand geometry readers can function in extreme temperatures and are not impacted by dirty hands (as fingerprint sensors can be). They are commonly used for access control to facilities, time clocks, or controlled areas (Security, 2007).

The EFF states that perhaps the most ubiquitous electronic biometric systems are based on hand geometry (EFF, 2003). Hand geometry is popular because of its ease of use, public acceptance, and integration capabilities. However, because of its insufficient distinctiveness and robustness, when stakes are high, these systems are not relied on exclusively to confirm identity; rather, they are used to provide an additional layer of security above and beyond that provided by an existing security system.

Hand geometry recognition is utilised in the US Immigration and Naturalisation Service Passenger Accelerated Service System (INSPASS). With this system, frequent international business travellers can simply use their hand geometry to enter the United States, rather than waiting in long immigration lines at the airport (Department of Homeland Security, 2007). The technology is also making its presence known in the financial sector, with a number of banks (e.g. Bank of Hawaii, Zion First National Bank) adopting it (Stakelbeck, 2004).

2.6.5 Voice recognition

Voice is a behavioural characteristic of an individual and voice capture is acceptable to most people. Voice recognition is probably the most natural and economical method for solving the problem of unauthorised use of computer and communications systems and multi-level access control (Campbell, 1997). Accurate voice verification or speaker recognition offers one great advantage, which is that it can allow remote verification using the phone system, an infrastructure that has already been built, and thus, no special reader needs to be installed at home (EFF, 2003). However, like hand geometry, it is not expected to be sufficiently distinctive to permit identification of an individual from a large database of identities. While the accuracy of voice recognition has improved over the last few years, some users still experience problems with accuracy either because of the way they speak or because of the nature of their voice. Moreover, the quality of a voice signal for authentication is typically degraded by use of a microphone, communication channel, and as a result of digitiser characteristics (Pankanti et al., 2000). Voice verification systems have to account for a greater number of variables than do other systems, and even the same person's voice can vary tremendously due to colds, aging, and fatigue.

A usual approach to cope with the voice recognition problem consists of classifying acoustic parameters derived from the input speech signal by short-time spectral analysis. These parameters contain both phonetic information related to the uttered text and individual information related to the speaker (Falavigna, 2000). As the task of separating the phonetic information from the individual one is not yet solved, many voice recognition systems behave in a text-dependent way (i.e. the user must utter a predefined key sentence) (Falavigna, 2000). Text-dependent voice verification authenticates the identity of a subject based on a fixed pre-determined phrase. Text-independent voice verification is more difficult and verifies speaker identity independent of the phrase. Language-independent voice verification verifies speaker identity irrespective of the language of the uttered phrase and is even more challenging (Pankanti et al., 2000).

Voice recognition technology is being deployed in, *inter alia*, the financial services industry (Simons, 2002) and in the food industry. Pizza Hut, for example, implements a voice recognition solution in its call centres in Australia to handle the 7.8 million calls that are put through its centres annually (Creed, 2002).

2.6.6 Vein recognition

Vein recognition works by identifying the subcutaneous (beneath the skin) vein patterns in the hand of a person. The scanner uses a special light to map the location of the veins in the hand. The red blood cells present in the veins absorb the rays and show up on the map as black lines, whereas the remaining hand structure shows up as white. After the vein template is extracted, it is compared with previously stored patterns and a match is made (Khan, 2006). There are different types of vein recognition technology which include finger vein, wrist vein, palm, and backhand vein recognition.

The vein patterns are very distinctive to each individual. Apart from size, the pattern is also robust; it does not change over time. Because veins are located underneath the skin surface, they are not prone to external distortion the way fingerprints are. And the extraction process can be contactless. This feature makes it suitable for one-to-many matching, for which hand geometry and face recognition may not be suitable. Compared with fingerprinting, it is more acceptable for most people as it does not have a link with traditional forms of criminality. However, its accuracy level is still doubtful.

Vein recognition technology has been deployed predominantly in the Asia-Pacific region. In countries such as Japan (Human Rights in Japan, 2006), where there is strong opposition to fingerprinting, vein recognition has become the biometric technology of choice. Vein recognition technology can be found in thousands of ATMs in Japan. In Europe, a Scottish school was the first to use palm-vein biometrics (Williams, 2006).[28]

2.7 Summary of biometrics

A review of the types of biometric technologies that comprise the largest proportion of market share leaves us at present to conclude that the current state of biometric devices leaves much to be desired. Although vendors claim that biometrics have become very accurate, secure, reliable, and convenient, most biometric technology is not yet sufficiently refined for very large population identification purposes.

Each biometric technology has its strengths and limitations. No single biometric is expected to effectively fulfil the needs of all applications. The applicability of a specific biometric technique depends heavily on the requirements of the application domain. Table 2.1 shows a brief comparison of 14 different biometric techniques that are based on the perceptions of three biometric experts:

Despite being claimed as a very secure authentication method, biometrics cannot yet deal with 'spoofing' very efficiently. According to FIDIS's report, most biometric technologies can be spoofed in one way or another. Even biometric devices with 'liveness' detection do not really live up to the claims of manufacturers. Cold, static, and inanimate objects can be enrolled and verified (FIDIS, 2006). Hence, biometrics can actually be 'stolen', and the theft of one's biometric characteristic can occur in a manner similar to the theft of information – the owner may still possess the biometric and does not even notice its theft. All that the imposter needs to do is to make a copy which is up to the standard necessary for cheating the machine.

A popular misconception held by the general public is that an individual can be positively identified using a biometric device. However, the fact is that all a biometric system can produce is a degree of correlation. Match scores that exceed the match threshold are considered to be matches, and such a process will inevitably involve some error.

Table 2.1 Comparison of various biometric technologies[29]

Biometric identifier	*Universality*	*Distinctiveness*	*Collectability*	*Acceptability*	*Performance*
DNA	H	H	L	L	H
Face	H	L	H	H	L
Fingerprint	M	H	M	M	H
Iris	H	H	M	L	H
Hand geometry	M	M	H	M	M

Contemporary database and transmission technology are not inherently secure. The security of a biometric system depends on many factors. The technical infrastructure of a biometric system is not necessarily secure. In addition to human factors, it may be rendered notably insecure by a combination of poor controls, ill-considered links with other data sources and poor policy.

It is difficult to conclude whether biometrics is actually privacy-enhancing or security-enhancing. A low-accuracy biometric device may actually be even less secure than some traditional authentication device. The extent to which biometric technology can enhance privacy also depends on how it is being deployed within a defined technical architecture and operational process.

From a technical point of view, the advantages and disadvantages of biometric technology can be examined in two ways: (1) as positive recognition applications that may work either in verification or in identification modes, and (2) as negative recognition applications that require identification (Jain et al., 2004).

Positive recognition

- As mentioned in a previous section, biometrics is more difficult to copy, share, and distribute compared with passwords and tokens. A biometric pattern cannot be lost and forgotten, which renders it convenient to use.
- To attack a biometric-based system, one needs to generate (or acquire) a large number of samples of that biometric, which is much more difficult than generating a large number of PINs/passwords (Jain et al., 2004).
- In identification mode, wherein the user does not need to claim an identity, speed is considered as a problem in scaling-up an identification application. However, even more problematic is that accuracy scales much more poorly than speed, with the implication that an impostor has a good chance of gaining access to the system.[30]

Negative identification

Although biometrics is not yet accurate enough to support large-scale identification systems, traditional personal recognition tools such as passwords and PINs cannot do the job. In negative recognition, it is required that personal recognition is performed in the identification mode. However, achieving the same level of accuracy in an identification system as in a verification system is a much more difficult problem due to a large number of comparisons.

Moreover, biometric systems also have some general limitations as highlighted in the previous sections:

- The sensed data may be noisy and distorted (for example, a fingerprint with a scar).
- The biometric data acquired from an individual during authentication may be very different from the data generated in the enrolment. The variation can be caused by a user who incorrectly interacts with the sensor.

- While a biometric trait is expected to be distinctive, there may be large inter-class similarities in the feature sets used to represent these traits.
- As mentioned, most biometrics are not universal.
- Spoofing is possible.

Because of a variety of advantages and disadvantages, a multi-biometric approach has been suggested by most technical experts as a good solution. For instance, Brunelli described a multi-modal biometric system that uses facial and voice traits of an individual for identification. Their system combines the matching scores of five different matchers operating on voice and facial features to generate a single matching score that is used for identification (Brunelli and Falavigna, 1995). Hong combined face and fingerprints for personal identification. The system consolidates multiple cues by associating different confidence measures with the individual biometric matchers and achieved a significant improvement in retrieval time and accuracy (Hong and Jain, 1998).

A carefully designed combination system is expected to perform better than the best of the individual ingredient modalities. A combination of uncorrelated modalities is expected to result in a better improvement in performance compared to a combination of correlated modalities. It has been noted that a combination of uncorrelated modalities can significantly reduce the failure to enrol rate and provide greater security against 'spoofing' (Jain et al., 2004). However, convenience and cost factors remain the largest barriers in the use of such multi-modal biometric systems in civilian applications. Jain has predicted that high-security, large-scale identification applications will increasingly use multi-modal biometric systems, whereas small-scale commercial applications are more likely to continue striving to improve unimodal biometric systems (Jain et al., 2004).

2.8 Other related technologies

Biometric technology is seldom used alone. The combination of several different technologies in combination with biometrics appears to play a crucial role. The integration of these technologies may offer increased/decreased levels of security and higher/lower degrees of user control over personal data depending on how they are applied. Section 2.8.1 briefly illustrates the major technologies frequently used in combination with biometrics.

2.8.1 Smart cards

A smart card is a device that includes an embedded integrated circuit chip that can take the form of either a secure microcontroller or equivalent intelligence with internal memory, or a memory chip alone (Smart Card Alliance, 2007). The microcontroller card can add, delete, and otherwise manipulate information on the card, while a memory chip card (for example, pre-paid phone cards) can only undertake a pre-defined operation. Unlike magnetic stripe cards,[31] a smart card can carry all necessary functions and information on the card. Therefore, it is not necessary that remote databases be accessed during a transaction. Smart card technology is available in a variety of forms including plastic cards, fobs,[32] subscriber identification modules (SIMs) used in GSM mobile phones, and USB-based tokens (Smart Card Alliance, 2007).

The smart card can be classified on the basis of a variety of parameters. One popular classification is based on card interface: contact card, contactless card, and combi card. A contact smart card needs to be inserted into a card reader. Transmission of commands, data, and card status takes place over these physical contact points. The limitation of the contact card is that with age, these contact points get worn out. Rough handling and stresses occurring during card insertion contribute to card damage. A contactless smart card requires only close proximity to a reader. The reading distances vary from a few centimetres to 50 cm. They merely have to be passed near an antenna for the transaction to be carried out. Since there is no insertion, this contactless card solves most of the limitations inherent to contact cards. Contactless cards are more expensive than contact cards, but they have a greater life span and are more reliable. A Combi or hybrid card has two chips, one with a contact interface and one with a contactless interface, thus facilitating its use for both methods. The two chips are not interconnected. With dual-interface cards, it is possible to access the same chip using either a contact or contactless interface with a very high level of security (Wolfgang and Wolfgang, 2003).

When classified on the basis of the components they contain, a smart card can be categorised as having microcontroller chips and memory chips. A memory chip is like a small floppy disk with optional security. Memory chip cards depend on the security of the card reader for processing and are ideal for situations that require low or medium security. Microcontroller chip cards, as the name implies, incorporate a microcontroller. A microcontroller chip can add, delete, and otherwise manipulate information in its memory. The microcontroller acts like a miniature computer, with an input/output port, operating system, and hard disk. Microcontroller smart cards have the unique ability to store large amounts of data, carry out their own on-card functions (e.g., encryption and digital signatures), and interact intelligently with a smart card reader.

Smart cards are used in a variety of applications including credit or ATM cards, fuel cards, SIMs for mobile phones, high-security identification and access-control cards, and public transport and public phone payment cards. A rapidly developing application of smart cards is in digital identification. In this application, the cards are used for authentication of identity. The most common example is in conjunction with a public key infrastructure.[33]

When smart cards are combined with biometrics, they usually provide a portable storage mechanism for the biometric template, which means that template management is eliminated across the biometric reader network. The card users present their smart card to the biometric reader at any location where the card is valid. The biometric template stored on the card will be compared to the live biometric. If the two match, the system grants the user access to the system.

This design requires that the smart card have sufficient memory to store the proper biometric data. As the biometric data are stored on the smart card, the number of users for any particular reader is unlimited. An ID system based on a smart card can support a private and secure biometric comparison process: the live biometric template on the reader is extracted and is transferred to the trusted smart card for comparison. Thus, the original stored template never leaves the card and matching is done within the secure card environment. Hence, this application is considered to be more secure and privacy-friendly. Alternatively, matching can also occur on a reader or a central server if the smart card has insufficient processing capacity (Smart Card Alliance, 2002).

2.8.2 Radio frequency identification (RFID)

Radio frequency identification (RFID) is a type of automatic identification system (Electronic Privacy Information Centre, 2007). RFID tags consist of a small microchip attached to an antenna or other coupling element. The tag communicates via radio frequencies with a transceiver or tag reader. The tag ID may be read automatically. The technology can be used to identify, track, sort, or detect a wide variety of objects. Communication takes place between a reader (interrogator) and a transponder (silicon chip connected to an antenna), which is often called a tag. Tags can either be active (powered by battery) or passive (powered by the reader field), and come in various forms such as smart cards, tags, labels, and watches. They may even be embedded in mobile phones (LARAN, 2004). The communication frequencies used depend to a large extent on the application and range from 125 kHz to 2.45 GHz (LARAN, 2004: 5).

The system must be supported by a sophisticated software architecture that enables the collection and distribution of location-based information. The complete RFID system combines the technology of the tags and readers with access to global standardised databases, ensuring real-time access to relevant information.

Tags contain a unique identification number called an electronic product code and other potential information of interest. Communication of data between tags and a reader is by wireless communication. All information on RFID tags such as product attributes, physical dimensions, prices, or laundering requirements can be scanned wirelessly by a reader at high speed and from a distance of several metres (Holloway, 2006).

The use of RFID in tracking and access applications first appeared during the 1980s (Electronic Privacy Information Centre, 2007). RFID has quickly become a popular technology of choice because of its ability to track moving objects, such as vehicles, pets, products, and even people. Compared with the bar code, an RFID system has several advantages. The read and write capability of an active RFID system enables the use of interactive applications. It enables distance reading and can be passed through a variety of substances such as snow, fog, or ice. RFID technology has been combined with biometrics in the European biometric passport. The passport's biometric information is stored on a tiny RFID computer chip, much like in a smart card.

2.9 Biometric applications

The applications of biometric technology can generally be divided into three main groups (Bateman, 1998):

- Commercial applications such as physical access control, electronic data security, e-commerce, internet access, financial services, personal electronic device security (e.g., mobile and PDA), and computer network login.
- Government applications such as border control systems, biometric passports, social security, drivers' licenses, national ID cards, and mass surveillance in a public area.
- Forensic applications such as criminal investigation and terrorist identification.

Biometric technology is increasingly deployed in various governmental or commercial uses. Wayman (n.d.) suggests characterizing biometric applications in terms of seven variables:

Cooperative vs. non-cooperative user: this refers to the behaviour of the user. Is the user trying to cooperate with the system? In verifying a positive claim of identity such as access control, a user might cooperate with the system in an attempt to be recognised as someone he or she is not. A user in such a situation may be asked to identify himself or herself by providing an ID card or a PIN code, which will limit the database search of stored templates to that of a single claimed identity. This is called a 'cooperative' application.

In applications verifying a negative claim to identity, a user might employ deception by attempting to not cooperate with the system with the objective of not being identified, which is a 'non-cooperative' application. In non-cooperative applications, users cannot be trusted to identify themselves correctly, so the entire database may need to be searched.

Overt vs. covert biometric system: this refers to whether the user is aware of the biometric sampling and identification. Most access control systems are overt, as most of the biometric systems require that a user 'does' something. In a covert situation, a deceptive user cannot be 'cooperative' or 'non-cooperative'.

Habituated vs. non-habituated user: is the intended user expected to be experienced in the use of the system? In applications where all intended users are 'habituated', the application is considered a 'habituated application'; otherwise, it is considered to be a 'non-habituated application'. In general, all applications in use will be considered 'non-habituated' during the first period of application; after several days, it may have a mixture of habituated and non-habituated users. For instance, with respect to access control, employees may be habituated, whereas in the case of access control to an airport, persons may be non-habituated.

Supervised vs. unsupervised user: will the user be supervised while using the biometric system? Generally, most biometric applications will be under supervision or guidance during their initial use. Non-habituated applications may require supervision or guidance throughout the whole application period, in order to assist initial users and prevent spoofing.

Standard Environment vs. Non-Standard Environment: Are environmental conditions under control throughout the operation of the biometric application? When indoors, it can generally be considered that temperature and lighting are controlled and we can therefore consider this to represent a standard environment. Most outdoor systems and other in-door systems are usually considered as non-standard environments.

Public vs. Private: This issue concerns the identity of system users. Generally, a public system is concerned with large-scale non-habitual situations in which the users are customers. Private use involves certain groups of users such as employees.

Open vs. Closed: Will the application be required to exchange biometric data with other systems?

This list is intended to be 'open' which means that additional partitions might also be appropriate. Every application can be classified according to the above partitions.

2.9.1 Examples of biometric applications

2.9.1.1 Biometric border passage system: Amsterdam Schiphol Airport

One of the most famous examples of a biometric application is the automated border passage system at Schiphol Airport, Amsterdam. In late 2001, Schiphol Airport officially implemented an iris recognition system for rapid checking-in of registered frequent flyers. The system is now owned by the Schiphol Group, the public company that owns Schiphol Airport (Figure 2.4).[34]

The biometric system at Schiphol Airport uses a smart card called Privium, in which identification is performed by analyzing iris data on the card.[35] At present, the smart card consists of three different sets of data: the iris-scan template, membership data, and the traveller's personal data such as place and date of birth. To

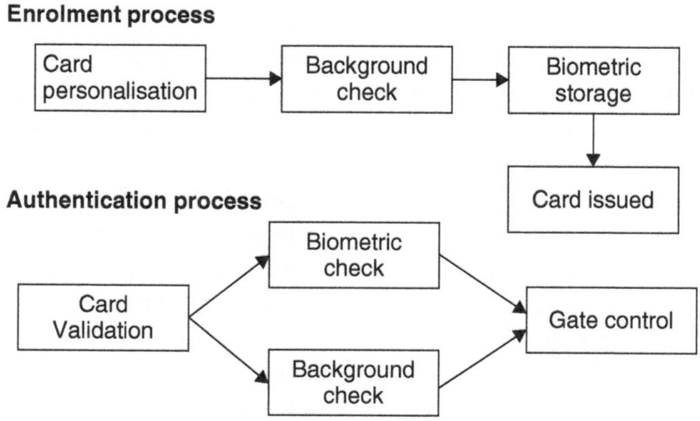

Figure 2.4 Customer processing at the biometric border control system at Schiphol Airport.[38]

use the system, the passenger is required to have their iris template[36] registered on the card.[37] According to the Schiphol Group, the storage of the biometric data is only on the smart card and no central database is used. The matching process involves one-to-one authentication.

When a passenger arrives at the system, they need to first enrol the card (on which three sets of data are stored). After the card is inserted (the first step), three categories of information stored on the card are read:

- Data used to identify the user in the Access and Border passage facility;
- A unique personal key used for governmental verification purpose;
- Biometric information saved on the card.

On the basis of the first two categories of data on the chip, the system will determine whether the passenger is permitted access to the Access and Border passage facility. If the answer is positive, the first access facility will be released. The passenger will then be permitted to proceed to the second step in which they enter the verification zone where the iris photograph is taken and directly matched to the biometric template on the card. A background check is carried out simultaneously. If both verifications show a positive result, the passenger is permitted to leave the verification zone. Immediately after the verification process, the stored data, including biometric information, will be automatically deleted from the passage and verification facility. The Privium smart card has a security authorisation module for encrypted data communication. Schiphol Group is the owner of both the iris recognition system and the Privium smart card. Consequently, Schiphol Group controls user data management. It is an opt-in program available to citizens of 18 European countries; Schiphol complies with statutory regulations such as the Dutch Personal Data Protection Act (Lips et al., 2004).

2.9.1.2 Eurodac

Another example is the EU-wide database, 'Eurodac', launched on 15 January 2003 and set up pursuant to the Eurodac Regulation.[39] It compares the fingerprints of asylum seekers and illegal entrants. It is the first common EU computerised central database equipped with a fingerprint recognition system (AFIS). All EU countries plus three additional European countries, Norway, Iceland, and Switzerland participate in Eurodac.

The system consists of a Central Unit, equipped with a computerised fingerprint database and an electronic data transmission application allowing member states to exchange information pertaining to asylum seekers and illegal entrants. All EU member states have in place automated fingerprinting systems for asylum seekers. The procedure for taking the fingerprint is governed by national law, but the applicant for asylum must be informed as to the purpose of taking the fingerprints, as provided in Article 5 of the Eurodac Regulation. The Central Unit processes the fingerprints along with the following types of data on individuals above the age of 14 years:

- Category 1: data on asylum applications. Fingerprints (full 10 fingerprints and 4 control images) of asylum applicants are sent for comparison against fingerprints of other asylum applicants who have previously lodged their application in another member state (European Commission, 2004).
- Category 2: data on aliens apprehended in connection with the irregular crossing of an external border and who were not turned back. These data (full 10 fingerprints and 4 control images) are sent for storage only, in order to be compared against data of asylum applicants submitted subsequently to the Central Unit (European Commission, 2004).
- Category 3: data relating to aliens found to be illegally present in a member state. These data, which are not stored, are searched against the data of asylum applicants stored in the central database (European Commission, 2004).

Once fingerprint images are transmitted by the member states to the Central Unit, they are recorded in the central database. Category 1 data are then compared against the transmitted data; if a 'hit' is returned, it indicates that the asylum seeker's fingerprints have been recognised by the Central Unit as a match against the stored fingerprints of an existing asylum supplicant (European Commission, 2004). The unit then compares the category 1 data against category 2 data and if a hit is returned, it means that the fingerprints of an asylum seeker match the stored fingerprints of an alien who has illegally crossed the border and who could not be turned back (European Commission, 2004). Finally, the Central Unit compares the category 3 data against category 1, and if a hit is returned, it means that the fingerprints of an alien found to be illegally present within a member state are recognised by the Central Unit as a match against the stored fingerprints of an asylum seeker (European Commission, 2004).

In addition to the fingerprints 'images', data sent by member states include the place and date of the asylum application (if applicable), as well as sex, reference number, date on which the fingerprints were taken, details with respect to the recipients of the transmitted data, and the dates on which the data were transmitted to the Central Unit and entered in the central database.[40] The collection is mandatory for people above 14 years of age and is encoded directly into the database by the Central Unit or the member state of origin. For asylum seekers, data are kept for 10 years until they obtain the citizenship of member states. Data relating to foreigners who attempted to cross the border irregularly and were apprehended are kept for 2 years, until they receive the residence permit of a member state or leave the territory of the member state. Data in category 3 are to be erased and the transmission media is destroyed once the results of the comparison have been transmitted to the member state of origin.

The EU Commission is responsible for operation of the central database, and national authorities are responsible for the technical aspects of data transmission. Access to data in the central database is limited to the member state which transmitted and recorded the data in the central database (European Commission, 2004). The Regulation also prohibits searches by a member state of data transmitted by another member state. It further prohibits the receiving of data by a member state with the exception of data resulting from a comparison carried out in accordance with the provisions in the Regulation (Article 15(1) of Council Regulation (EC) No.2725/2000). The exchange of data recoded in the central database with authorities of any third country is prohibited, unless it is specifically authorised within the framework of a community agreement on the criteria determining the state responsible for examining an application for asylum (Article 15(5)).The European Data Protection Supervisor commented in its audit report of 2007 that the security measures initially implemented with respect to the Eurodac system, and the way that they have been maintained, have provided a fair level of protection to date. The European Data Protection Supervisor further commented that only a few aspects of the system and its organisational security apparatus present weaknesses that require attention (European Data Protection Supervisor, 2007).

2.9.2 Analysis of the examples in light of Wayman's classification criteria

The Privium scheme can be classified as a cooperative, overt, unsupervised, habituated, standard environment, public, closed application. The system is cooperative because those wishing to defeat the system will attempt to be identified as someone already holding a card, and they will have to provide a smart card for the authentication. It is overt because all participants voluntarily join the program, and they are aware of the requirement of providing the iris as a condition of enrolment into this system. It is unsupervised and in a standard environment because collection of the biometric occurs near the passport inspection counter inside the airports, but not under the direct observation of the Schiphol employees. It is habituated because the users are intended to be frequent international travellers. The system is public because enrolment is open to any frequent travellers. It is closed because Schiphol Group does not exchange biometric information with any other system.

Fingerprint identification by Eurodac for the purpose of preventing illegal immigration to Europe can be classified as a non-cooperative, overt, supervised, non-habituated, standard environment, public, open application. It is non-cooperative because those wishing to defeat the system attempt to not be identified as someone already registered as an asylum seeker or illegal immigrant. It is overt because all participants are aware of the requirement to give a biometric measure as a condition of receiving a permit. It is supervised and in a standard environment because collection of the biometric occurs under police supervision. It is non-habituated because applicants are only required to give a biometric identifier once upon application. It is public because the system will be used by all EU countries in addition to Demark, Norway, and Iceland. Current systems are open, as countries using it are exchanging biometric information from time to time.

2.10 Summary

This chapter has provided an overview of biometric technology and has established what biometrics is, as well as the criteria it needs to satisfy for certain functional and practical purposes. The chapter introduced some of the major issues surrounding the application of biometric technology and the deficiencies and benefits of biometric technology with respect to meeting the needs of business and government agencies.

Distinguishing different types of biometric technology may be helpful in determining the possibilities and risks entailed by the use of a particular biometric technology. A closer look at how each of the main types of biometric technology work is necessary for understanding how their technical differences impact society. When more than one kind of biometric technology is being used, the risks and vulnerability of certain biometric types may decrease dramatically. When biometric information is supposed to be used as evidence in court (such as in forensic science), it is important to be aware of the limitations of the technology and understand how strong or weak the evidence is. For example, it is important to note that a biometric system can be spoofed, even when there is a declaration of liveness detection. To that end, the effectiveness of liveness detection needs to be investigated.

Notes

1 I return to this in Chapter 3.
2 According to the European Commission's report, there is a third type of purpose, screening, which makes use of a watch-list. A watch-list is a database containing data on individuals who are to be apprehended or excluded (e.g., from a service). The key feature of a watch-list is that people are not on the whole identified; they will only be identified if they appear on the list. However, according to my understanding, this is only a kind of negative way of using identification, which does not fundamentally differ from identification. Identification still happens when the match is found. See European Commission's Joint Research Centre (2005: 9).
3 Non-repudiation basically refers to a situation in which person/organisation who has done something cannot deny/repudiate the existence of that action.
4 I return to this in later sections.
5 Based on Clarke (2003).
6 Any biometric system will need an 'exception handling' procedure that involves a manual matching process, see Bolle et al. (2003).
7 A fallback procedure is a measure intended to address failures and accidents which occur as a result of the use of biometric technology. Examples of failures/accidents include matching failures, enrolment failures, compromised biometric data, etc.
8 INS stands for Immigration and Naturalisation Service Institutional Removal Program.
9 This occurs when attackers gain access to the template and overwrite the legitimate user's template with that of an impostor.
10 Several authors have shown that when given access to match-score data, it is possible to reconstruct a good estimate of an unknown enrolled image from a fingerprint or face recognition template. See Hill (2001: 469–472), Adler (2003), and Adler (2004).
11 Further discussion on this topic occurs in the following text.
12 To modify the features set upon verification or in the templates in order to obtain a high verification score.
13 In the biometric context, this could happen when another party obtains a copy of the stored biometric data. For example, by obtaining another person's fingerprint, one could *masquerade* as that person in various circumstances.

14 For a further demonstration of the various methods of spoofing, see FIDIS (2006).
15 I return to this issue in Chapter 3.
16 AFIS refers to automated fingerprint identification system. Therefore, systems using fingerprints account for 58.9% of biometric technologies.
17 This 'uniqueness' has not been proven by forensic science. I return to this issue in Chapter 3.
18 See Section 2.2.
19 Examples are discussed in Section 2.2.
20 For further illustration concerning the various methods of fingerprint spoofing, see FIDIS (2006, pp.45–61).
21 For more details, see European Commission (2000).
22 Amsterdam Airport Schiphol began using its automated border passage system in October 2001. That system combines iris recognition technology and an advanced smart card. It is the first of its kind to use iris recognition to secure border control.
23 The ABG pilot program at the Frankfurt/Main Airport involves 17 EU nations and Switzerland. Frequent passengers of Deutsche Lufthansa AG who are interested in the program and reside in EU countries or Switzerland can register their iris data. Seven iris recognition systems in total were installed at the immigration and emigration control gates of the airport.
24 As reported in 2007, the United Kingdom is testing several technologies as part of its 'e-Borders' program, which aims for more thorough oversight over travellers coming to the United Kingdom. See Cargo Security International (2007).
25 For more illustration, see Answers.Com (2010).
26 Australia's first e-Passport was introduced in October 2005. The first version of the e-passport, identified as the M series, was replaced by the N series in May 2009. The chip embedded in the centre page stores the holder's digitised photograph, name, gender, date of birth, nationality, passport number, and the passport expiry date. This is exactly the same information that appears on the printed data page of the passport. See further Department of Foreign Affairs and Trade (2009).
27 See Figure 7.
28 The school is Todholm Primary School, in Paisley, Scotland. Its cafeteria has gone cashless and students buy lunches by holding their hands over a palm-vein recognition unit produced by Glasgow-based Yarg Biometrics.
29 H: high level; M: middle level; L: low level. This table is according to Jain et al. (2004).
30 The accuracy level of biometric technology has raised much concern in response to success rates of 96% for iris, 81% for fingerprints, and 69% for face Wilson (2006).
31 That is, cards with a black stripe. The black stripe consists of three tracks of magnetic parties bonded to the card substrate and is the core of a magnetic stripe card. The magnetic stripe cards were introduced to: (1) store data in a machine-readable form, (2) minimise paper utilisation in financial transaction, and (3) allow for automation. The main problem with a magnetic stripe card is that data can be easily read and altered by anyone with access to the right kind of equipment. See further Dhar (n.d.).
32 Noun 1 a chain attached to a watch for carrying in a waistcoat or waistband pocket. 2, (also fob pocket) a small pocket for carrying a watch. 3, a tab on a key ring. In Compact Oxford English Dictionary. Retrieved from Oxford Reference Online Premium database.
33 Essentially, public key infrastructure refers to a collection of security services, security component, and organisations that enable large-scale deployment of digital signatures (Riisnæs, 2007).
34 Shares in the Schiphol Group are held by the Kingdom of the Netherlands, the city of Amsterdam, and the city of Rotterdam.
35 The iris system was launched at Schiphol Airport in October 2001. By the end of 2003, the system had about 10,000 users (mainly business travellers). In case of failure, a fall-back manual check is provided (Lips et al., 2004).

36 This is not a biometric image – it is instead a few extracted features of the biometric image.
37 Though not in any kind of database, as it is claimed.
38 Based on Blauuw (2002).
39 Council Regulation (EC) No. 2725/2000 of 11 December 2000 concerning the establishment of 'Eurodac' for the comparison of fingerprints for the effective application of the Dublin Convention (Official Journal L 316, 15/12/2000: 1–10); in force 15/12/2000. The implementation rules for *Eurodac* are set out in *Council Regulation* (EC) No 407/2002 of 28 February 2002. The European commission presented an amended proposal for the revision of Eurodac Regulation (No. 2725/2000) on 10 September 2009 (COM2009(342)).
40 For more information of Eurodac system, see European Council (2000).

References

Adler, F.H. (1965). *Physiology of the Eye: Clinical Application*, London, The C.V. Mosby Company.

Answers. Com (2010). *Eigenface.* Available at http://www.answers.com/topic/eigenface?method=5&linktext=eigenface.

Adler, A. (2003). Sample images can be independently restored from face recognition templates, *Electrical and Computer Engineering,* 2: 1163–66.

—— (2004). 'Images can be regenerated from quantised biometric match score data', in IEEE (ed.) *Proceedings of Canadian conference of Electrical and Computer Engineering*, Niagara Falls: IEEE, 469–72.

Bateman, S. (1998). Biometrics initiatives signal need for digital identification, *Computer Shopper.* Available at http://web3.searchbank.com/infotrac/session/547/703/2396287w1/5!xrn_1&bkm.

Blauuw, A.de. (2002). *Automatic Border Passage at Amsterdam Airport Schiphol.* Available at http://www.icpc.tudelft.nl/ibm/Automatic%20Border%20Passage%20for%20ACM%20ICPC%2016112002.ppt.

Bolle, R.M., Connell, J.H., Pankanti, S., Ratha, N., and Senior, A.W. (2003). *Guide to Biometrics,* New York: Springer.

Brunelli, R. and Falavigna, D. (1995). Person identification using multiple cues. *IEEE Transactions on Pattern Analysis and Machine Intelligence,* 12, 955–66.

Campbell, J. P. (1997). Speaker recognition: a tutorial. *Proceedings of the IEEE.* 85(9): 1437–62.

Cargo Security International (2007). *Border Security: Surveillance and Protection.* Available at http://www.cargosecurityinternational.com/_dataimages/CSIAM07UKTI.pdf.

Cavoukian, A. (2007). *Biometric Encryption: A Postive-Sum Technology that Achieves Strong Authentication, Security and Privacy.* Available at http://www.ipc.on.ca/images/Resources/up-bio_encryp_execsum.pdf.

Clarke, R. (2001). *Biometrics and Privacy.* Available at http://www.rogerclarke.com/DV/Biometrics.html.

Clarke, R. (2002). *Biometrics' Inadequacies and Threats and the Need for Regulation.* Available at http://www.rogerclarke.com/DV/BiomThreats.html.

— (2003). *Smart Gate: A Face Recognition Trial at Sydney Airport.* Available at http://www.rogerclarke.com/DV/SmartGate.html.

Creed, A. (2002). *Voice Recognition System Takes Pizza Orders Down.* Available at http://findarticles.com/p/articles/mi_m0NEW/is_2002_Jan_24/ai_82097010.

Daugman, J. and Malhas, I. (2004). Iris recognition border-crossing system in the UAE, *International Airport Review,* 2: 49–53.

Department of Foreign Affairs and Trade (2009). *The Australian e-Passport.* Available at http://www.dfat.gov.au/dept/passports/.

Department of Immigration and Citizenship (2008). *DIAC's Management of the Introduction of Biometric Technologies* (Audit report, 2007–2008), Australian National Audit Office.

Department of Homeland Security (2007). *U.S. Passenger Accelerated Service System* (*USPASS*). Available at http://www.globalsecurity.org/security/systems/inspass.htm.

EFF (2003). *Biometrics: Who's Watching you?* Available at http://www.eff.org/Privacy/Surveillance/biometrics/.

Electronic Privacy Information Centre (2007). *Radio Frequency Identification (RFID) Systems.* Available at http://www.epic.org/privacy/rfid/#introduction.

European Commission (2000). *Eurodac: A European Union-Wide Electronic System for the Identification of Asylum-Seekers.* Available at http://ec.europa.eu/justice_home/fsj/asylum/identification/fsj_asylum_identification_en.htm.

European Commission (2004). *First Annual Report of the Council and Parliament on the Activities of the EURODAC Central Unit* (SEC (2004) 557). Available at http://www.libertysecurity.org/IMG/pdf/first_annual_report_eurodac_2004.pdf.

European Commission's Joint Research Centre (JRC) (2005). *Biometrics at the Frontiers: Assessing the Impact on Society* (EUR21585 EN). Available at ftp://ftp.jrc.es/pub/EURdoc/eur21585en.pdf.

European Council (2000) *Council Regulation No 2725/2000 of 11 December 2000 Concerning the Establishment of 'Eurodac' for the Comparison of Fingerprints for the Effective Application of the Dublin Convention* ((EC) No 2725/2000). Available at http://eurlex.europa.eu/LexUriServ/LexUriServ.do?uri=OJ:L:2000:316:0001:0010:EN:PDF.

European Data Protection Supervisor (2007). *Summary Report on the EURODAC Audit.* Available at http://www.statewatch.org/news/2007/nov/edps-eurodac-short-summary.pdf.

Falavigna, D. (2000). *Speaker identification and verification.* Available at http://ditelo.itc.it/people/falavi/IdVe.html

FIDIS (2006). *Future of Identity in the Information Society.* Available at <http://www.fidis.net/fileadmin/fidis/deliverables/fidis-wp6-del6.1.forensic_implications_of_identity_management_systems.pdf>.

Fingerman, D. (2003). *Static Measurements and Moving Targets: Privacy, Biometrics and the Consumer-Bank Regulation Ship.* Available at http://danfingerman.com/papers/Biometrics_paper.pdf

Ford, W. and Baum, M. S. (1997) *Secure Electronic Commerce: Building the Infrastructure for Digital Signatures and Encryption,* New Jersey: Prentice-Hall, Inc.

Global Security (n.d.). *Hand Geometry and Handwriting.* Available at http://www.globalsecurity.org/security/systems/biometrics-hand.htm.

Greene, T. C. (2002). *Face Recognition Technology a Proven Farce.* Available at http://www.theregister.co.uk/2002/01/04/face_recognition_technology_a_proven/.

Grosche, S. S. (n.d). *An Overview of Biometrics.* Available at http://www.isg.rhul.ac.uk/msc/teaching/ic3/IC30405pt3.ppt.

Hill, C. J. (2001). *Risk of Masquerade Arising from the Storage of Biometric* (B.S. book, Australian National University). Available at http://chris.fornax.net/biometrics.html

Holloway, S. (2006). *RFID: An Introduction.* Available at http://msdn2.microsoft.com/en-us/library/aa479355.aspx.

Hong, L. and Jain, A. (1998). Integrating faces and fingerprints for personal identification, *IEEE Transactions on Pattern Analysis and Machine Intelligence,* 20: 1295–307.

Human Rights in Japan (2006). *Japan fingerprinting plans spark opposition.* Available at http://news.yahoo.com/s/nm/20060111/wl_nm/japan_fingerprinting_dcWedJan 11.

International Biometrics Group (2002). *How is Biometrics 'Defined?'*. Available at http://www.biometricgroup.com/reports/public/reports/biometric_definition.html.

International Civil Aviation Organisation (2004a). *Biometrics Deployment of Machine-Readable Travel Documents* (Technical Report Version 2.0.) Available at http://www.icao.int/mrtd/download/documents/Biometrics%20deployment%20of%20Machine%20Readable%20Travel%20Documents%202004.pdf.

Jain, A.K., Bolle, R.M. and Pankanti, S. (1999) *Biometrics: Personal Identification in Networked Society*, Boston, MA: Kluwer.

Jain, A.K., Ross, A. and Prabhakar, S. (2001) A hybrid fingerprint matching using minutiae and texture features, in IEEE (ed.), *Proceedings of the International Conference on Image Processing (ICIP),* Thessaloniki, Greece, IEEE, 282–85.

—— (2004) An introduction to biometric recognition, *IEEE Transactions on Circuits and Systems for Video Technology Special Issue on Image- and Video-Based Biometrics,* 14(1): 4–20.

Khan, I. (2006). *Vein pattern recognition – biometrics underneath the skin.* Available at http://www.findbiometrics.com/article/320.

LARAN (2004). *A basic introduction to RFID technology and its use in supply chain.* Available at http://www.printronix.com/library/assets/public/case-studies/rfid-laran-white-paper-english.pdf.

Lips, A.M.B., Hof, V.D.S., Prins, C., and Schudelaro, Ir.A.A.P (2004) *Issue of Online Personalisation in Commercial and Public Service Delivery*, Nijmegen, Netherlands, Wolf Legal Publisher.

Maltoni, A., Jain, A.K., and Maio, D. (2003) *Handbook of Fingerprint Recognition*, New York: Springer-Verlag.

National Science and Technology Council (NSTC) (2006a).Privacy Overview. Available at http://www.biometrics.gov/Documents/BioOverview.pdf.

—— (NSTC) (2006b), *Privacy and Biometrics Building a Conceptual Foundation.* Available at http://www.biometrics.gov/docs/privacy.pdf.

Newell, F. W. (1991). *Ophthalmology Principles and Practice,* 7th edn., Saint Louis, MI: Mosby, C.V.

Nuger, K.P. and Wayman, J.L. (2004b). An introduction to biometric authentication systems, In J. L. Wayman, J. Anil, M. Davide and M. Dario (eds.) *Biometric Systems: Technology, Design and Performance Evaluation,* Nottingham: Springer, 1–20.

Pankanti, S., Bolle, R. M., and Jain, A. (2000). Guest editors' introduction: Biometrics – the future of identification. *Computer,* 33(2), 46–9.

Ploeg, I. V. D. (2003). Biometrics and privacy (A note on the politics of theorizing technology), *Information, Communication, Society,* 6(1): 85–104.

Ratha, N. K. and Connell, J. H. (2001). Enhancing security and privacy in biometrics-based authentication systems, *IBM Systems Journal,* 40(3), 614–34.

Riisnæs, R. (2007) *Digitale sertifikater og sertifikattjenester - roller, oppgaver og ansvar: en tillitsorientert tilnærming til sertifikatutstederens villedningsansvar,* Bergen: Fagbokforlag.

Roethenbaugh, G. (1998) Biometrics explained, *International Computer Security Association,* Available at http://www.icsa.net/services/consortia/ cbdc/explained.shtml.

Sasse, A. (2004) *Usability and Trust in Information Systems* (Report for the Cyber Trust & Crime Prevention Project), UK Office of Science and Technology Foresight Program. Available at http://citeseerx.ist.psu.edu/viewdoc/download?doi=10.1.1.122.3472&rep=rep1&type=pdf.

Security, D.O.H. (2007) *Hand Geometry and Handwriting.* Available at http://www.globalsecurity.org/security/systems/hand.htm.

Simons, S. P. (2002). *Voice recognition market trends.* Available at http://ccrma.stanford.edu/~jhw/bioauth/andre/VoiceRecognitionmktApr02.pdf.

Smart Card Alliance (2002). *Smart Card and Biometrics in Privacy Sensitive Secure Personal Identification System* (white paper). Available at http://www-03.ibm.com/industries/government/doc/content/bin/bio1.pdf.

——(2007) *About Smart Cards: Introduction.* Available at http://www.smartcardalliance.org/pages/smart-cards-intro-primer.

Stakelbeck, F. W. (2004) *Biometrics: A Viable Solution for Financial Institutions.* Available at http://www.philadelphiafed.org/bank-resources/publications/src-insights/2004/third-quarter/q3si3_04.cfm.

Working Party on Information Security and Privacy (2004) *Biometric-based Technologies* (DSTI/ICCP/REG(2003)2/FINAL), Report for Directorate for Science, Technology and Industry Committee for Information, Computer and Communications Policy. Available at http://www.olis.oecd.org/olis/2003doc.nsf/LinkTo/NT000070D6/$FILE/JT00166988.PDF.

Wayman, J. L. (1999) Technical testing and evaluation of biometric identification devices, in A.K. Jain and J. Wayman, (eds.), *Biometrics: Personal Identification in Networked Society,* Boston, Kluwer Academic Press, 345–68.

——2001) Fundamentals of biometric authentication technologies, *International Journal of Imaging and Graphics,* 1(1): 93–133.

——2005) Iris recognition, in W. James, J. Anil, M. Davide and M. Dario (eds.), *Biometric Systems,* London: Springer, 63–95.

—— (n.d) *The Functions of Biometric Identification Devices.* Available at http://www.engr.sjsu.edu/biometrics/publications_tech.html.

Williams, M. (2006) *Scottish school is first to use palm-vein bio*metrics, *Digital'World Tokyo,* Available at http://www.netcharles.com/orwell/articles/scottish-school-to-use biometrics.html.

Wilson, S. (2005) *Lockstep Submission to Senate Privacy Inquiry.* Available at http://www.aph.gov.au/senate/committee/legcon_ctte/privacy/submissions/sub11.pdf.

—— (2006) *The use of biometrics to authenticate transactions.* Available at http://www.tedgp.asn.au/press/biometrics.pdf.

Wolfgang, R. and Wolfgang, E. (2003) *Smart Card Handbook,* 3rd edn., Hoboken: Wiley.

Woodward, J.D., Horn, C., Gatune, J. and Thomas, A. (2003) *Biometrics: A Look at Facial Recognition,* Santa Monica: RAND.

Zunkel, R.L. (n.d.) *Handgeometry Based Verification.* Available at http://www.cse.msu.edu/~cse891/Sect601/textbook/4.pdf.

Part II

3 Rational concerns about biometric technology

Security and privacy*

3.1 Introduction

Biometric technology (hereafter also referred to as 'biometrics') raises multiple rational concerns wherever it is applied. Based on the general understanding of biometrics established in Chapter 2, this chapter considers the complexities involved in implementing biometric technology by focusing on the security and privacy concerns this technology raises.

To what extent does biometric technology affect privacy and security? What is the special nature of biometric data compared with other forms of personal data? Is the increasing use of biometrics just a question of 'balance' or a 'trade-off' between privacy and security? Does biometrics actually raise any special legal problems in the context of privacy and security? What are the factors that influence security and privacy with respect to various biometric applications? This chapter focuses on these sorts of questions. In tackling such questions, the chapter also aims to clarify some of the misconceptions that inform aspects of the legal discourse surrounding biometrics.

Most people's knowledge of biometrics has been obtained from popular culture. In science-fiction movies such as Spielberg's *Minority Report*, people are regularly subjected to eye scans when they take public transport, enter office buildings, or simply walk in the street. In this film, eye scans are conducted for the purpose of identification, control and/or advertising. In reality, however, we are still currently far from the apparent efficiency and ubiquity of such scanning. Nonetheless, seductive claims have been made about the ability of biometrics to defeat terrorism and organised crime. Biometrics figures increasingly as the centrepiece technology in implementing counterterrorist policy. There seems to be an assumption among some people that technology can be applied as a sort of 'sticking plaster' to patch-up poorly conceived policy. This is especially the case with regard to biometrics, where a number of unrealistic claims are being made.

Most forms of technology inspire not only hope but also fear. The development of innovative technology has almost always raised new legal concerns. This is certainly true in the case of biometric technology. The increasing use of biometrics has led to fears that a surveillance society will be created, with scant room for personal privacy and autonomy. Doubts have also been raised about the level of security that an increased use of biometrics might actually deliver. It is further feared that the loss of privacy may lead in turn to a host of other problems, such as increasing social stigma, discrimination in employment, barriers to gaining health insurance and the like. With the growing use of biometrics, it is of paramount importance that discussions about the ethical, social and legal implications relating to the technology take place. In discussions so far, privacy and security concerns have often figured prominently (Woodward, 1997a; Cavoukian, 2003) – and for good reason, as this chapter reveals.

To understand the complex nature of biometric technology is not an easy task, especially if we extend such thinking to consider long-term implications. Concerns have been expressed relating to the utilisation of biometric technology, especially with respect to its usage by commercial entities and public organisations.[1] It has been pointed out that the power to identify and discriminate among individuals can be seriously misused across a variety of levels (Ashbourn, 2006). At the same time, organisations are enthusiastically rushing to incorporate biometric technology in their operations. Given this reality, it is important that we ensure that a complete and objective discussion occurs with respect to the technology and its implications.

To begin, this chapter outlines the special nature of biometric technology and biometric data. It then discusses the relationship between biometric technology and privacy. Following from this, the relationship between biometric technology and security is analysed in light of technology assessments and case studies. Following a general analysis of biometric technology, it goes on to evaluate the combination of factors which may affect privacy and security as it relates to certain biometric applications. The final section presents conclusions.

3.2 Special nature of biometric technology and biometric data

3.2.1 The biological nature of biometric data and technology

Compared with knowledge-based or token-based methods of authentication or identification, biometric technology is unique in the sense that it uses part of the human body or behaviour as the basis of the authentication or identification method. What is the significance of the fact that a body-related characteristic is used as an identifier or verifier? To answer this question, we need to first investigate what biometric data are.

3.2.1.1 Genetic and health-related data

The raw information at the heart of biometrics is personal by its very nature. It is intimately connected to the individual concerned (the 'information subject'). Consider the most popularly used and known form of biometric information – fingerprints. It has been claimed that even a fingerprint that is too smudged for ordinary identification could provide forensic scientists with sufficient DNA to construct a 'DNA fingerprint',[2] thus providing investigators with a powerful new tool in the search for evidence of a crime.

A single piece of biometric data often consists of two categories of information: information usable for authentication and information not usable for authentication. Information usable for authentication could include information such as genotypic information, randotypic information (sometimes called phenotypic, but without genetic parts), behavioural information and information about 'unchanging marks'.[3] Genotypic information is completely determined by genetics. With respect to data protection, genotypic information seems to be most controversial since it might reveal relationships to other persons, race or even potential disease. For instance, Dr. Marvin M. Schuster has discovered a 'mysterious relationship' between an uncommon fingerprint pattern, known as a digital arch, and a medical disorder called CIP. Based on a 7-year study, he found that 54 per cent of CIP patients have this rare digital arch fingerprint pattern (Elise and Hendricks, 1996). While still controversial within the scientific community, several researchers have reported a link between fingerprints and homosexuality (Cytowix, 1996; Hawkes, 1996). Certain chromosomal disorders, such as Down's syndrome, Turner's syndrome, and Klinefelter's syndrome are known to be associated with characteristic fingerprint patterns (Mordini and Petrini, 2007). Moreover, from examining a person's retina or iris, a medical expert can determine that the person may be suffering from common afflictions such as diabetes, arteriosclerosis and hypertension; furthermore, unique diseases of the iris and the retina can also be detected (Bickley, 1995).

However, the informational status of the biometric templates that are generated and applied in identification/authentication systems is somewhat unclear. As indicated above, a biometric template is digitised data of a person's physical or behavioural characteristics, and does not constitute the raw information or image itself. The template is generated through the application of a given algorithm. There is as yet no solid proof that the templates themselves actually contain medical information, though they are likely to do so. A template is as unique as the raw biometric data from which it is generated. It is possible to reconstruct from a template the part of the raw biometric data that is used for creating the template (Jain et al., 2005; Bromba, 2006). Generally, templates will only contain information necessary for comparison. However, what is necessary for comparison is neither fixed nor predetermined. As the biometric template should retain the special features of the raw biometric data as an identifier, it unavoidably becomes necessary to include some relatively unique and permanent features which are related to genetic information or health (Bromba, 2006). However, it is not certain if the information captured in the template would be sufficient for medical diagnostic purposes. Nonetheless, it is still reasonable to claim that there is a general link between biometric information and genetic and/or health information. The latter has been widely recognised as sensitive information pertaining to individuals and, quite often, their relatives.

Not all commentators on biometrics share the view that biometric data are intrinsically sensitive. According to one observer,

> [W]ith almost all biometric devices, there is virtually no personal information contained therein. From my fingerprint, you cannot tell my gender, you

> cannot tell my height, my age, or my weight. There is far less personal information exposed by giving you my fingerprint than by showing you my driver's license.
>
> (Wayman 1998: 265)

Another observer has claimed that '[b]iometrics is not a branch of medicine but rather a special form of mathematical and statistical science' (Ploeg, 1999: 43). The same observer goes on to state that 'we should perhaps not expect to be able to determine any intrinsic meaning of biometric data, or the biometric body in general, but investigate quite specifically what uses and practices biometrics will become part of' (Ploeg, 1999: 43).

At first glance, these statements seem to make sense, but they are based on the assumption that technology will stop developing. Presently, there is no verified report concerning the easy and rapid disclosure of health information directly from biometric data; moreover, possible linkages between biometric data and health information are only reported in relation to certain kinds of biometric data. Yet as the technology develops, it would be quite reasonable to predict that such disclosure and linkage may be possible in the future. The potential is clearly present. As a rule, if something can happen, it will happen. Hence, the long-term problem here is whether the data controller (i.e., the person or organisation in possession of the biometric data) *will* make such linkages.

It could be countered that even if biometric data have the potential to allow the disclosure of sensitive information, such data are not designed to be used in that way, so there is no need to worry. However, biometric features make it difficult to escape from situations in which biometrics might be misused if placed in the hands of individuals or governments (either with or without malicious intent). 'Function creep' can occur; indeed, many privacy advocates contend that function creep is inevitable. For example, Simon Davies has stated:

> The history of identification systems throughout the world provides evidence of 'function creep' – application to additional purposes not announced, or perhaps even intended, at the commencement of the scheme. [. . .] The existence of a relatively high-integrity scheme would create irresistible temptations to apply it widely, and inter-relate many hitherto separate collections of personal information.
>
> (Davies, 1994: 44)

The addition of a new function to an existing application can be valuable or detrimental to society, but the point here is that the potential of biometric data cannot be restricted by the purposes for which they are/were originally used. Knowing that certain medical disorders are associated with specific biometric patterns, researchers might actively investigate whether biometric patterns can be linked to behavioural characteristics, or predispositions to medical conditions. By comparing selected biometric data captured at enrolment with subsequent entries and current data, biometric technologies may detect several medical conditions

(Mordini and Petrini, 2007). There is no absolute guarantee that biometric data will not be used to reveal health information, though it would take a significant technological shift to go from current biometric systems to ones that reveal such information (Feldman, 2003).

3.2.1.2 Relative uniqueness, universality and stability

The commonly held idea that biometric technologies are capable of identifying individuals through one-to-many matching across large, shared databases is based on the belief that biometric identifiers are unique and universal. It has been established that each person is supposed to have unique fingerprints, facial features, irises and DNA. For instance, fingerprints have been used in forensic research for many years as purportedly unique identifiers of criminals. However, some recent cases have revealed that identification by use of fingerprints has been overturned on appeal at court (Cole, 2000). In fact, the 'uniqueness' of a fingerprint in forensic science remains an *assumption* without watertight proof (Cole, 2000). The belief that latent fingerprints can be matched to a single person is claimed to be 'the product of probabilistic intuitions widely shared among fingerprint examiners, not of scientific research. There is no justification based on conventional science, no theoretical model, statistics, or an empirical validation process' (Stoney, 1997: 340).

Nevertheless, it is worth pointing out that there is not yet any solid proof that this *assumption* is incorrect either. Hence, it may be true that fingerprints *per se* are unique. Yet it does not necessarily follow that the latent fingerprint is unique too, nor does it necessarily follow that the thumbprint template which simply extracted certain features of a raw thumbprint image can be as unique as its origin. The template may be based on a blurred, dirty and/or incomplete image of the thumbprint, which may affect the accuracy of the collected biometric information, making the biometric template's uniqueness more difficult to guarantee. Even DNA, which is widely recognised as the most accurate biometric identifier, has been to the object of criticism. While it is true that each individual (except identical twins) has a unique sequence of genes, (Holladay, 2002) only a subset of a particular gene is used in the forensic DNA-identification process for identification purposes. Hence, Professor Alec Jeffreys, a pioneer in developing modern DNA testing techniques, has pointed out that DNA testing is not an infallible proof of identity:

> [m]odern commercial DNA profiling compares a number of genetic markers – often 5 or 10 – to calculate a likelihood that the sample belongs to a given individual. Jeffreys estimates the probability of two individuals' DNA profiles matching in the most commonly used tests at between one in a billion or one in a trillion, 'which sounds very good indeed until you start thinking about large DNA databases.' In a database of 2.5 million people, a one-in-a-billion probability becomes a one-in-400 chance of at least one match.
>
> (Lawless, 2004)

It is also not guaranteed that the fuzzy biometric template, which actually uses just a part of the DNA sequence, will be 100 per cent unique. Thus, the 'uniqueness' of biometric data is not absolute, it is *relative*. The biometric templates generated from them are even less unique due to their 'fuzzy' nature. This also affects the stability of the biometric data.

The universality of biometrics is also relative. One problem with the widespread use of biometrics is that there are few biometrics (apart from DNA) that everyone possesses. Not everyone will have a particular biometric trait, or an individual's biometric trait may be significantly different from the 'normal' expected trait. For example, some people may be missing fingerprints due to skin disease, which may cause further problems when enrolling a large population in a fingerprint-based register. Discrimination concerns may also be raised in such a case. Therefore, a large-scale biometric scheme will usually need to utilise more than one biometric, for example, both fingerprint and face, to ensure that all people can be enrolled.

Unlike passwords or tokens, biometric identifiers are by their nature intended to be stable over time; without such stability, their utility will be quite limited (Schneier, 2000). Fingerprints, irises and DNA are widely recognised as stable biometrics, while faces, keystroke and voice patterns give rise to greater scepticism with respect to their stability, as they may change as people grow older. However, the stability of even the so-called stable types of biometric data is not absolute. For instance, the image of a fingerprint pattern is 'plastic' and does not remain as stable as is commonly imagined. Each time you place your fingerprint on a finger scanner, the pattern may appear to be the same from a short distance away, but there are actually small differences in the pattern due to dryness, moisture and elasticity conditions of the skin. Moreover, cuts and scratches can alter the pattern. It is thus likened somewhat to 'fuzzy' decryption (Dorizzi, 2005). The iris, another popular biometric measurement, though regarded as highly accurate, unfortunately also suffers from difficulties in consistently obtaining a valid image. The iris is often occluded by eyelids and eye lashes. In addition, data collection can also be hindered by specula reflections in uncontrolled lighting situations (Retica Systems, Inc., 2005). Similar problems also apply to other relatively stable biometric identifiers.

Consequently, it is relevant to discuss the nature of biometric data in a legal sense. Should biometric data be regarded as 'sensitive' personal data for the purposes of, say, EU data protection law? Is it necessary to differentiate a raw digital image from the template as such? What will be the legal implications if biometric data are to be treated as sensitive personal data?[4]

3.2.2 The automatic nature of biometric data and biometric technology

3.2.2.1 Fuzzy unicode of individual

Biometric data have been compared with various other more traditional bio-centric forms of identification such as a photograph and thumbprint that use ink print.[5] It would appear that biometric data are less easily (Schneier, 2000) (or at least not logically) distinguished from these images with respect to technical or moral values. But is this true?

Unlike a primitive image from which one can make distinction by various superficial means, biometric data are regarded as more reliable and accurate, though it has been recognised that there is presently no perfectly accurate biometric technology (Wayman et al., 2004). Relatively stable biometric data are associated with relatively unique biometric features. As a fuzzy match is deployed during verification or identification, the main characteristics of certain biometric features are digitalised regardless of superficial changes. For example, hand geometry technology uses a 32,000-pixel CCD digital camera to record the three-dimensional shape of the hand from silhouetted images projected within the scanner. The scanner does not register surface details such as fingerprints, lines, scars, dirt or fingernails (findBIOMETRICS.com, n.d). The scanner typically takes more than 90 measurements of the length, width, thickness and surface area of the hand and four fingers. Superficial changes that may affect correct identification are thus controlled for at the outset. In this sense, biometric data are akin to a fuzzy unicode of each individual, by which the body becomes an object, the identity of which is determined by mathematical means. Indeed, as has been commented, 'it is possible that the expanding technologies may eventually mean that the most important identity information may be that contained in a digital body' (Harte, 2004: 57). This unicode is deemed to be relatively accurate and reliable and is controlled by data controllers. Since it is a digital representation of a human being, like all other computerised data, it is easily reproduced, transmitted, analysed and re-used, while the data subjects have little if any de-facto control over it, and little if any knowledge of it or of how it will affect them in the real world. As Feldman warns: 'There is a danger that the more we focus on biological characteristics, the less we remember the intangible aspects of a person's character. As a result perhaps we should be wary of moving toward a society that constantly reduces us to our biologic characteristics' (Feldman, 2003: 666).

Furthermore, the fuzzy nature of the unicode differentiates it from other existing personal codes such as personal numbers or passport codes, because it is regarded as relatively stable and permanent. Certainly, while personal numbers and passport codes are unique for each person, when compromised they are technically very easy to change and they have no 'physical' linkage to a certain individual. It has been stated that 'increasingly, the way to keep information secure is to offer up a piece of yourself . . . to be recorded and used to verify your identity' (Theodore, M., cited in Alterman, 2003, 143). The concern here is this 'piece of yourself' that is offered up may be crucial in retaining control over one's identity, and may be hard to recapture once it is put in the form of a proprietary digital format (Theodore, M., cited in Alterman, 2003). Although biometrics is not designed to reduce people to a set of identifiers, the slippery slope leading from identification to identifying behaviour, and therefore personality, does appear to be a risk or even a natural inclination.

3.2.2.2 Possible linking and tracking

John D. Woodward has pointed out that 'if facial recognition or other biometric databases become interlinked, then the threat to information privacy has the potential to increase significantly' (Woodward, 2001a: 7). Biometric identifiers provide a possibility for disparate databases to be interlinked worldwide in an automatic way. This possibility depends to some extent of course on standardisation. Currently, the interoperability of biometric identifiers is still weak,[6] but there is a trend towards increased interoperability. For instance, the International Civil Aviation Organisation has recently adopted a global, harmonised standard for the integration of biometric identification information into passports and other machine-readable travel documents.[7] In addition, the US National Institute of Standards and Technology (NIST) has published a 'Common Biometric Exchange File Format' (CBEFF) aimed at promoting interoperability of biometric-based application programs and systems developed by different vendors (National Institute of Standards and Technology, 2001).

Thus, the balkanisation of biometric information is on its way to becoming a thing of the past and it can be reasonably expected that the linkage and tracking ability of biometrics will be developed and utilised to the fullest extent possible. It is not difficult to imagine a future situation in which an individual must use one particular standard biometric to pay tax, enter the workplace, go shopping, travel and obtain medical service. Such use of a biometric 'key' would open up the possibility that each of these records and transactions will be linked, thereby allowing governments or businesses to compile a comprehensive profile of the individual's actions.

Biometric ID systems have the potential to locate and track people physically. Of course, tracking can be accomplished without biometrics. For example, RFID, personal numbers, passwords and IP addresses can all be used as identifiers for tracking purposes. Initially then, tracking potential seems not to be a special characteristic of biometric technology. Nevertheless, such technology does create a heightened level of concern as it may facilitate surreptitious tracking.

Traditional authentication methods using, for example, passwords or tokens rely on either something you know or something you have, while RFID tags usually have to be on something you wear or carry. The collection of such information requires to some extent that the data subjects *do* something, but the biometric features of individuals are not secrets, and are something that a person inherently possesses. Certain biometric features like fingerprints and facial images can be collected without the cooperation of the data subjects. However, due to its inaccuracy, particularly in large-scale matching, most current biometric technology does not have the capacity to facilitate large-scale tracking. Such tracking may be more readily feasible by using RFID technology. Nevertheless, the potential still exists for improvements in biometric technology to allow large-scale tracking in the future or to realise large-scale tracking with the help of RFID. We see an example of this mutuality in the recent development of biometrically enhanced passports that are fitted with RFID tags.

3.3 Privacy and biometric technology

The issue of privacy is central to concerns about biometric technology. To evaluate the various privacy concerns requires in the first instance an understanding of what privacy is and what it entails. Among the most influential explications on the privacy concept are the following:

- The right to be let alone (Warren and Brandeis, 1890);
- A state of limited accessibility: secrecy, solitude and anonymity (Gavison, 1980);
- An interest in control of information about oneself (Westin, 1967).

As the variation in these explications demonstrates, privacy is a multi-faceted concept that is difficult to define using a single, simple formulation. However, this difficulty should not imply that privacy concerns lack importance. As has been pointed out, 'in one sense, all human rights are aspects of the right to privacy' (Volio, 1981: 184). Moreover, in the international law of human rights, 'privacy' is clearly and unambiguously established as a fundamental right to be protected. In Article 12 of the 1948 Universal Declaration of Human Rights (UDHR),[8] it is asserted that 'no one shall be subjected to arbitrary interference with his privacy, family, home, or correspondence, nor to attacks upon his honor and reputation'. The words of Article 17 of the 1966 International Covenant on Civil and Political Rights[9] are almost identical to those used in Article 12 Universal Declaration of Human Rights. On the other hand, Article 8 of the 1950 European Convention on Human Rights and Fundamental Freedoms (ECHR) states that,

- Everyone has the right to respect for his private and family life, his home and his correspondence.
- There shall be no interference by a public authority with the exercise of this right except such as is in accordance with the law and is necessary in a democratic society in the interests of national security, public safety or the economic well-being of the country, for the prevention of disorder or crime, for the protection of health or morals or for the protection of the rights and freedoms of others.

Engaging in extensive debate over the exact meaning of the privacy concept is unnecessary for the purpose of this chapter. It suffices to note that there are two main groups of privacy-related interests that are directly pertinent when labelling the issues that have arisen in contemporary discussion on the ethical and legal implications of biometrics. The first group of interests falls under the rubric of 'informational privacy' and concerns the control of personal information. These interests give rise to attempts to establish rules governing the collection and handling of personal data (Electronic Privacy Information Centre & Privacy International, 2005). The ability to control personal information about oneself is closely related to the dignity of the individual, self-respect and sense of personhood (Amnesty International Canada, 2001). Personal information may include autobiographical, health, legal and financial information. Information privacy lies at the very heart of the discussion over the risks and benefits of biometrics.

The second interest group falls under the rubric of 'physical privacy' with respect to the body itself. This sense of privacy transcends the physical and is aimed essentially at protecting the dignity of the human person. It concerns protection from intrusive searches and seizures, particularly the protection of persons' physical bodies against invasive procedures, such as drug testing and body-cavity searches (Electronic Privacy Information Center & Privacy International, 2005). It also relates to the right of the person to consent to or refuse treatment unless this right is abrogated through incapacity or for the greater good of the community. Privacy of DNA is an emerging area in this field. The widespread use of biometric technology may invade our physical privacy in several ways.

Furthermore, it is also relevant to introduce the issue of property rights with respect to privacy, which concerns the appropriation and ownership of interests in human personality (Rothstein, 1997). Individual property law has been a central feature of legal systems throughout history. Property notions are not necessarily inherent to privacy interests but in some jurisdictions, the notion of, say, informational privacy is partly derived from the assumption that all information about individuals is in some fundamental way their own, and is theirs to communicate or retain as they see fit (Department of Communications and Department of Justice, 1992). It might be useful for the law to use the doctrine of property to protect individuals' biometric information.[10]

3.3.1 Information privacy and biometric technology

Regardless of whether an individual voluntarily provides a biometric identifier or is forced to surrender it, they are nevertheless giving up information about themselves (Woodward, 1997a). Once collected, control over the biometric data is shifted from the data subject to the organisation that has access to the data. As biometric data are intimately linked with individuals in a relatively unique way, the data are usually considered 'personal' (Hert, 2005). Information privacy is therefore the most significant concern with respect to biometric technology. Losing control over personal data is the main challenge biometric technology poses to informational privacy. Biometric technology raises some of the same issues that arise when government agencies or private firms collect any personal information about citizens. However, such loss can occur in specific ways that distinguish biometric data from losses involving other personal data, making concerns about biometric technology of particular importance with regard to privacy protection. Therefore, it will be also relevant to explore the types of special legal problems that the technology poses, along with an analysis of various privacy implications resulting from the use of biometric technology.

3.3.1.1 Unnecessary collection

A central principle of rules grounded in informational privacy is that the collection of personal information should be limited to those data that are necessary and relevant to a legitimate purpose (Bygrave, 2002). One of the primary privacy concerns with respect to the potential massive use of biometric technology is that more personal information than is necessary and relevant will be collected, used and disclosed. As Lyon pointed out, there is a tendency to rely on technological enhancements and surveillance systems even when it is unclear that they work or address the problems that they were intended to resolve (Cockfield, 2003). The concern is that the excessive collection of biometric information will result in heightened monitoring of individuals. As mentioned in Section 3.2, it is difficult to predict exactly what biometric technology may bring, but it is clear that it has a broad potential to provide an extremely convenient and cost-effective method of gathering and analysing biometric data. The collection of biometric data therefore becomes the first threshold available for us to control the potential subsequent negative consequences of biometric technology.

From such data, it is possible to obtain health, racial and medical information about individuals which is not necessary for authentication or identification. This possibility also raises concerns about the possible disclosure and/or compromise of such information, which are usually considered sensitive. An interesting point of view concerning the health-related nature of biometric technology is that this technology benefits those who do not go to the doctor, and helps them to detect diseases earlier (Moo-Young, 2001). This claim may make sense to some extent, but the problem here is that unless the data subjects are clearly informed about this potential and consent to it, the technology effectively makes a particular sort of medical check-up compulsory. This would undermine individuals' privacy interests in relation to their own health information, including their interests with respect to being able to choose not to know certain details of their health status. Moreover, in many jurisdictions, particularly European, the data controllers in such a case ordinarily have no unqualified legal right to collect and keep such health information.[11] They do not ordinarily have an unqualified right to share the information with other interested organisations (such as insurance companies).

Another feature of biometric data is that they can identify people. However, when the purpose of collecting the biometric data is just for authentication and there is little or no benefit in having stronger user identification, it is difficult to justify the collection of strong unique identifiers. The potential fear is that when biometric technologies become pervasive, they will be deployed in situations where there are few or no benefits to be gained from strong user verification/identification. For example, when ascertaining whether I am old enough to go to a bar and drink alcohol, it is unnecessary to know who I am, when all I need to do is simply demonstrate that I am of legal age.

Unnecessary collection can also occur when collection of biometric information (such as facial recognition) becomes pervasive. Although companies and government agencies claim that their databases contain only data on wanted criminals, people actually have no idea whether all of the people in the database are criminals, as the quality control over these databases is far from perfect. As Agre has stated, 'we have to trust your word that the only people whose images are stored in the databases are wanted criminals, and we have to trust your word that you throw away all images that fail to match the database' (Agre, 2003: 5). Finally, even if the only people in the database today are criminals, the forces pushing the technology and its uses down a slippery slope to ever-expanding surveillance are nearly overwhelming (Agre, 2003).

This development places basic human rights at risk. In the words of Prins:

> Fundamental rights are likely to be violated in case biometrics is used for applications merely requiring a low level of security. In the end, organisations and government agencies must demonstrate that there is a compelling interest in using biometric technology and that an obligatory fingerprint requirement is reasonably related to the objective it required for.
>
> (Prins, 1998: 161)

As demonstrated above, unnecessary collection of biometric data occurs when excessive health data are collected, when identification is adopted unnecessarily, or when innocent individuals' biometric data are collected and kept in the criminal database. The significant fact to note here is that these situations are deemed to be inevitable as long as biometric data are collected, because they are rooted in the inherent nature of biometric data. The more widely and frequently the technology is accepted, the more difficult it will be to limit it later on.

One way of avoiding unnecessary collection of biometric data is to set strict legal standards to ensure that the collection of biometric data is worth the sacrifice in privacy. The challenge is to identify data that are truly relevant and avoid collection of data for its own sake. To adopt the terminology of Prins, there should be a 'compelling interest', and the collection of data should be 'reasonably related to the objective'. Yet what exactly constitutes a 'compelling interest'? And in what way can data be regarded as 'reasonably related' to the objective? These are quite abstract terms which Prins leaves largely unexplained but which nevertheless need elaboration. Therefore, the first legal questions posed by biometric technology are: How does one decide on the necessity of collection? If the collection of biometric data is to be restricted due to its invasiveness of privacy, where should one draw the line?[12]

3.3.1.2 Unauthorised collection

As with other personal data, biometric information can be gathered or stored easily and surreptitiously without the subject having control or knowledge of the data-sharing process. Moreover, biometric technology together with the use of RFID augments the possibility that covert collection of biometric information on data subjects will occur directly. Once the biometric information is stored in an RFID tag carried by the data subject (for example, in a biometric passport without sufficient security protection), collection of biometric data from the RFID tag could be realised without the knowledge of the data subject. Only certain biometric patterns (e.g., facial, voice and/or gait) can theoretically be collected without the data subject's knowledge, though with the help of RFID-enhanced cards (which are now being widely used for storing biometric data (Cavoukian, 2004)), all kinds of personal data (including biometric data) might be collected, tracked and profiled without the data subject's knowledge or consent.

Here, our primary focus will be on facial-recognition technology which is being increasingly used for covert collection of biometric information. The deployment of facial-recognition technology by the Tampa police at the 2001 Super Bowl in Florida and the possibility of scanning passports from a distance are good examples to consider (Alterman, 2003). Given the indiscriminate nature of a number of surveillance systems, covert surveillance of the general public may become the norm. The use of biometric technology has the great potential to enhance law enforcement's ability to gather data without the participation or even awareness of the individual. This necessarily involves the collection and analysis of various body characteristics – resulting in a type of 'biological' surveillance.

Some commentators suggest that 'the benefits of implementing facial-recognition technology are far more important than the benefits of right to privacy in public places' (McCoy, 2002: 471). With terrorist attacks having deeply affected the world, there is an intense willingness to embrace any technology that might offer even a slight hope of guaranteeing public safety. This appears to include a blind faith in untested biometric technology and an eagerness to sacrifice privacy to public authorities and their agents without any proof that such a sacrifice is even justified. In fact, as I demonstrate, there is no evidence that the technology is completely effective or that it would be successful in thwarting terrorists or other criminals.

In the US legal context, the legal doctrine of 'reasonable expectation of privacy' developed pursuant to the Fourth Amendment to the Bill of Rights in the Constitution has frequently been invoked in commentary on the legality of the use of facial-recognition systems and other biometric technologies in public spaces. Some critics, including judges, have argued that 'a subjective expectation of privacy does not, by itself, give rise to Fourth Amendment protection'.[13] The legal problem that is raised here is this: Can a person legitimately have an expectation of privacy in a public space?

Representations which uniquely identify a person by means of physical characteristics have inherent moral value as objects connected with one's body, and hence with one's autonomy and self-esteem (Alterman, 2003). Although people knowingly expose their faces and voices in public, or leave fingerprint samples on cups in coffee shops, they do not really anticipate that their body will be used to create identifying representations that they have not endorsed.[14] The unauthorised collection of biometric information means that individuals are not in a position to know what types of personal information are collected and used. Even if they have been informed about the collection, they are usually not in a position to challenge conclusions made about them. This suggests that mandating the use of biometric information without the free consent of the data subject, even when the information is gleaned from public spaces, is a serious derogation of an individual's right to privacy. The mandatory collection of biometric information accordingly becomes quite problematic when there is insufficient legal justification for collection. It therefore behoves us to be cautious about the legal basis on which mandatory or even opt-out biometric systems are justified, and to consider from a legal perspective the conditions under which individuals might accept such systems.

3.3.1.3 Unauthorised use and function creep

According to Cavoukian, 'the threat to privacy arises not from the positive identification that biometrics provide, but the ability of third parties to access this in an identifiable form and link it to other information, resulting in secondary use of that information without the consent of the data subject. This erodes the personal control of an individual over the uses of his or her information' (Cavoukian, 1998). The unauthorised secondary use of biometric data is perhaps the greatest risk that biometric technology poses to informational privacy.

Moreover, biometric data offer strong evidence of one's identity since they represent relatively unique biological characteristics which distinguish one person from another. As biometric data can usually be linked to only one individual, they therefore act as a powerful, unique identifier that brings together disparate pieces of personal information about an individual (Cavoukian, 1998). As a relatively unique identifier, biometric data not only allow individuals to be tracked, but create the potential for the collection of an individual's information and its incorporation into a comprehensive profile by linking the various databases together. The automatic nature of biometric identifiers makes them easy to copy and otherwise be shared among countless public and private sector databases. For example, an article in a hotel trade publication points out that

> with the use of this (biometric) technology a front desk clerk could know instantly at check-in that Mr. John Smith during his last stay purchased: three Cokes from the mini-bar, two martini's in the lounge, ate dinner at the hotel restaurant where he ordered the special and since his last visit has moved from Chicago to Atlanta
>
> (Rinehart, 2001)

The record of Mr. Smith's alcohol consumption may be used by his insurance company who may be curious about Mr. Smith's alcohol consumption and want to rank his risk of contracting heart or liver disease. Information in this profile may be used out of context to the detriment of the data subject, and unjust decisions about them would be made simply by automatically analysing a profile, which may contain incomplete or inaccurate data. All of this could be done without the consent or knowledge of the data subjects.

As indicated above, 'function creep' is unavoidable (Davies, 1994). A widely cited example is the US Social Security Number, which is used for a broad range of purposes (Woodward 1997b). It has been claimed that 'any high-integrity identifier (like biometrics) represents a threat to civil liberties, because it represents the basis of a ubiquitous identification scheme, and such a scheme provides enormous power over the populace. All human behaviour would become transparent to the state, and the scope for non-conformism and dissent would be muted to the point envisaged by the anti-Utopian novelists' (Clarke, 1994: 34).

From a legal perspective, the issue of 'function creep' is not new; however, the development of biometric technology raises further challenges for legislation to tackle. Central databases, data matching/linking and profiling are technical factors that enable 'function creep'. Lack of auditing or censuring may also be relevant human factors related to unrestricted 'function creep'. Privacy advocates believe that any identification scheme can be carried out with a hidden agenda and that the slippery slope effect can be relevant to several factors such as the motivations of governments and business, and whether safeguards exist. However, it is noteworthy that the special nature of biometric data makes 'function creep' more attractive, while the legal measures possible to control function creep are still quite limited. Many relevant issues do not appear to have been dealt with by legislation related to the use of biometric technology. What technical measures should be allowed with respect to the collection and use of biometric data? How can individuals detect events involving unauthorised use of data? The unauthorised use of personal data has usually been justified as necessary in order to improve sales and marketing for the benefit of the customer. There is currently limited legislative or other regulatory guidance to specify what types of information can be collected or whether private or public organisations are prevented from sharing or selling biometric information to others. At the same time, many commentators seem to agree that attempts by the biometrics industry to create guidelines in order to self-regulate do not constitute an entirely satisfactory solution; there are too many dangers associated with these self-imposed guidelines (McCoy, 2002). This is particularly pertinent with regard to the establishment of national and international biometric information databases.

3.3.1.4 Loss of anonymity

Anonymity has been frequently linked with autonomy and a person's sense of freedom.[15] The ever-increasing quantity of personal data available online makes it more convenient to track and profile individuals by government or private organisations. Consequently, anonymity may turn out to be the only tool available for ordinary people to defend themselves against being profiled. People desire anonymity for a variety of reasons. Certainly there are some people who wish to have anonymity in order to avoid detection from law enforcement and undertake illegal activities. But most people want to have anonymity simply because it is fundamental to their sense of freedom and autonomy. Professor Clarke has stated: 'The need to identify oneself may be intrinsically distasteful to some people. For example, they may regard it as demeaning, or implicit recognition that the organisation with whom they are dealing exercises power over them' (Clarke, 1994).

Will widespread use of biometric technology substantially undermine a person's ability to remain anonymous? It has been argued that it is possible to use biometric technology for anonymous authentication (Impagliazzo and More, 2002; Grijpink, 2005). In spite of this, it has also been pointed out that if one really wants to remain anonymous, biometric technology should not be the appropriate technology of choice, as biometrics is by its nature generally inconsistent with anonymity (T. Dunstone, personal communication, 2006).[16] Biometric systems are created to identify or authenticate people, and it will generally not be a difficult task to link, directly or indirectly, a biometric identifier to other personal data. Moreover, a biometric identifier is usually more difficult to change or make than, say, an address or phone number. While facial recognition technology is not sufficiently advanced to accurately identify members of the public on the street,[17] the combination of biometric data in public and private databases will result in people being identified in ways not previously required. This is seen as an affront to personal dignity and an invasion of privacy. Loss of anonymity is one of the principal reasons that privacy advocates object to the use of biometric technology, as there is a significant possibility that anonymity will be restricted. If personal information could be linked and identified using the biometric data, one's ability to remain anonymous would be severely diminished (Clarke, 1994).

Woodward has argued that 'to the extent there is less individual anonymity today than in decades or centuries past, biometrics is not to blame' (Woodward, 1997a). He goes on to claim that while a biometric identifier is very accurate, 'it is not the first nor is it the only identifier used to match or locate information about a person' (Woodward, 1997a). He therefore concludes, 'it is not obvious that more anonymity will be lost when biometrics is used' (Woodward, 1997a). These arguments seem to make sense at first glance (as they use a fact as their premise): yet such a conclusion is misleading for the following two reasons:

First, Woodward underestimates the reach of biometric data. He uses 'name', 'social security numbers' and 'account numbers' as examples of 'other numerical identifiers' to compare with biometric data, and infers that since there were many other identifiers before biometrics, the latter should not be blamed for lack of anonymity. There is no existing identifier, such as name or social security number, that can be readily equated with biometrics. A social security number is not universal at all, as its use is often restricted to a particular jurisdiction. As for account numbers, it is not usual to see people use these as authentication methods other than for obtaining financial service, and such numbers can also be changed and/or restricted to a certain location and time period. As a matter of fact, no existing identifier can expose so much about us as biometric data can, nor is there any other identifier that is *supposed to be* so universal, long-lasting[18] and intimately linked to us as biometrics. To say that the use of biometrics will not cause further loss of anonymity is overly optimistic.

Second, Woodward concludes that because biometric technology is not the only identifier that may erode anonymity, it should not be blamed. This is akin to saying that because A is not the only person who commits a crime, he/she should therefore not be punished. Despite the fact that there exist in the world many methods of eroding anonymity, it cannot be denied that biometric systems are detrimental to anonymity.

However, there are various degrees of anonymity, which vary from context to context. Absolutely anonymous and semi-anonymous are two categories that have different legal implications in the biometric context. Due to the identification abilities of biometric technology, it is actually almost impossible to have absolute anonymity when biometric information is collected, as matching and linking will be practically unavoidable and easy to accomplish. Semi-anonymity may be possible, provided that the biometric system is carefully designed from the start. So, some of the issues that need to be considered are: If the achievement of absolute anonymity is a reasonable goal, should biometric technology be adopted at all? Or should the data subject be given a choice not to use the technology? What are the legal mechanisms/justifications that can be applied to restrict the excessive use of biometric technology? If semi-anonymity is adequate for the protection of privacy, how should biometric technology be used? What are the legal requirements necessary to ensure that anonymity remains a possibility? And assuming that privacy-enhancing technologies are usually quite helpful in ensuring anonymity, should they be recommended by law?

3.3.2 Physical privacy and biometric technology

Physical privacy is the right to be free from unwanted, unreasonable intrusions or searches of one's body. Physical privacy is concerned with bodily integrity and is indirectly concerned with emotional integrity together with human dignity. Issues revolving around physical privacy include schemes for compulsory immunisation, blood transfusion without consent, compulsory provision of samples of bodily fluids/tissues and compulsory sterilisation (Clarke, 2000). Physical privacy is also defined as freedom from contact or monitoring by others, which might include searches, video surveillance and other forms of monitoring.[19] Physical privacy as it relates to biometrics is usually not a focus of discussion, but the relationship between an individual's bodily integrity and information (on which the demarcation between the rights to physical privacy and informational privacy is based) cannot be clearly divided. Consider the example of DNA: there is no clear point that can be drawn to distinguish the point at which bodily matter becomes information. To say that the use of data derived from the body merely involves the data and not the person denies the constitutive and enduring relationship between the data and one's identity as an embodied person (Ploeg, 2001).

Breaches of bodily integrity in the context of a 'search' may not be identical to a physical invasion of the body's boundaries. Compared with other medical examinations, the collection of biometric information with current technology is significantly different with respect to its level of 'intrusiveness', even when such collection is compulsory. However, if the information gathered with respect to a person's identity is also regarded as an intrusion of one's 'physical privacy', then biometric technology will have a tremendous impact. The new intensive forms of linking, tracking, controlling and manipulating of persons through their biometric information become possible indications that some form of physical privacy may be at stake. Such forms of surveillance may not violate 'bodily' integrity in the traditional sense associated with physical intrusion, but physical privacy might still be jeopardised, especially once the collection of body-related biometric information such as DNA becomes easy and inconspicuous.

In the biometric context, the question remains: does the collection of biometric data constitute an intrusion into a person's body? The collection of most forms of biometric data requires some infringement of the data subject's personal space. Iris and fingerprint scanning requires the close proximity of biometric sensors to body parts such as eyes, hands and fingertips. The adoption of other types of biometric technology may involve relatively invasive processes, such as substance-abuse testing, body and genetic screening, and may therefore be regarded as an intrusion into a person's physical privacy, even if the collection of biometric data is unsuccessful for various reasons. In the context of law enforcement and forensic identification, the use of fingerprinting, DNA and various other forms of bodily searches jeopardise physical privacy (Ploeg, 2001). Most countries have laws and regulations which are intended to regulate the use of such searches, in order to protect the individual's rights against infringement by state powers and law enforcement. It is noteworthy that in the US legal context (see particularly the Fourth Amendment to the Bill of Rights in the Constitution), 'searches' may include the gathering of physiological information from individuals.[20]

In most West European countries, a distinction is usually made with respect to searches *on* the body (such as a search of clothes, skin, fingertips and face) and searches *in* the body (Ploeg, 2001). However, when no actual touching is involved, as is the case with facial recognition or voice recording, it becomes quite difficult to devise categories which define the types of searches falling under the categories of 'inside' or 'outside'. Even methods of DNA collection can be quite inconspicuous. Nevertheless, it can still be argued that inconspicuous methods of biometric collection are a serious breach of physical privacy and body integrity (Schotsmans, 1998). The reason for this is that such searches do not have a specific goal of finding traces related to a crime, but are instead conducted for the purpose of generating identity information specific to an individual (Schotsmans, 1998). This difference in purpose actually renders the collection of physical biometrics a more serious breach of integrity and privacy. It indicates that there may be a presumption that someone is guilty until proven innocent. This would be contradictory to the generally accepted legal doctrine that a person is innocent until proven guilty. This kind of 'search' will bring a large number of innocent people into surveillance schemes.

In discussions on the relationship between biometrics and privacy, moral and legal concerns about physical privacy usually take a backseat to concerns about informational privacy. The reason for this is clear: 'physical intrusion' resulting from the use of biometric technology usually results from the collection of physical information. The fact that an individual is subject to 'intrusion' by biometric technology is not the focus of most legal and social commentators, because the harm resulting from 'physical intrusion' is not regarded as important as the loss of control over one's biometric information.[21] The purpose and subsequent use(s) of biometric data would therefore appear to be more relevant to this discussion. Yet for some people of specific cultural or religious backgrounds, the mental harm resulting from physical intrusion may be quite serious (Woodward 1997b).

Existing rules on the collection of physical data by government agencies usually specify under what conditions a person can be required to provide fingerprints and/or bodily tissues. If consent is required, rules are in place to regulate the scope of consent. If forced searches are allowed, specifications are usually provided as to how and by whom the search will be performed. Therefore, the legal questions surrounding the issue should be: Under what conditions should the collection of biometric information be permitted? If required, what exactly should be the extent of coverage of the consent? When compulsory collection of biometric information is required, who is eligible to conduct it? What sorts of procedures are required? If 'notice' should be a requirement for the collection of biometric information, what kind of information should be in the 'notice'? What exactly should be filed and stored? And what should be made accessible to whom and for what use?

Some authors consider as an invasion of physical privacy the possible harm to hygiene which could be caused by biometric technologies and the use of biometric sensors (Woodward, 2001b). However, this view stretches the notion of privacy too far. Biometric technology may infringe an individual's right to full health, but hygienic problems do not directly violate the right to privacy as such.

3.4 Security and biometric technology

Biometric technology has frequently been linked with security objectives. For example, biometrics are extolled as the most secure and convenient form of authentication because biometrics 'cannot be borrowed, misplaced, forged, stolen, duplicated, or forgotten' (QT Worldtel, Inc., 2003: 4). There has also been discussion concerning the balance between security and privacy, including the effect on civil liberties which results from the use of biometrics justified on the premise of enhancing public safety. Yet, as pointed out by Clement, the so-called 'trading off' between privacy and security is an inappropriate way of looking at the issue, as it is 'a distraction that prematurely concedes and obscures a dangerous presumption'(Clement et al., 2002: 195). Strong convictions held on the basis of faith in the efficacy of biometric technology might actually be a romanticised illusion. Human beings have an almost blind faith in all things scientific (Wayman, 2000), and biometrics is certainly cloaked in a 'scientific' mantle.

3.4.1 Limitations of technology

No biometric technique is completely accurate. Facial recognition, the primary biometric selected by International Civil Aviation Organisation in 2002, has a very low accuracy percentage in uncontrolled lighting situations, and the false positive rate is unknown in large-scale applications (Ezovski, 2005). According to the already noted large-scale test results from the UK Passport Service in 2005 (conducted in preparation for the introduction of their ID card), it was found that the successful rate for verification in iris-based biometrics rose to 96 per cent to quota participants, and was significantly higher than that for disabled participants 91 per cent. The iris verification success rate was higher for younger participants than it was for older participants (Ezovski, 2005). Fingerprint verification achieved a success rate of 81 per cent for quota participants and 80 per cent for disabled participants. One of the factors influencing failure rates was that the single fingerprint device used for verification occasionally did not record sufficient details from a finger (Ezovski, 2005).

In the real world, the accurate identification of a suspect from a larger group in uncontrolled conditions utilising biometric system will arguably be very difficult, and the system will also be affected by such things as age and whether any of the participants wear glasses. Even iris recognition, which has been widely accepted as based on a very accurate biometric, is still not sufficiently accurate for common deployment (Gomm, 2005). Most biometric technology has not yet been demonstrated to be successful in large-scale applications. A report released by the European Commission in March 2005 warned that, technologically speaking, there is a current lack of independent empirical data (European Commission's Joint Research Centre (JRC), 2005). The possibility that even a small percentage of false positives would result in a large number of law-abiding individuals being mistakenly labelled as criminals is also a concern. Bruce Schneier, a specialist on security issues, observes that even with a 99.9 per cent accuracy rate, frequent false positives would occur (numbering perhaps hundreds or thousands), at sites such as airports where large numbers of individuals are processed. Accuracy is also an issue for those anti-terrorism systems utilising data mining, matching or profiling. Some critics are concerned that if inaccurate information is used in a profile (Cavoukian, 2003), data that were accurate for a moment in time may be preserved by a computer and then that moment extended through a data match or profile, often to the detriment of the data subject (Reichman, 1987).

It has been argued that biometric technology may be more accurate than human beings in checking identity (T. Dunstone, personal communication, 2006).[22] Unfortunately, no solid evidence has yet been provided to support this claim. Even if such a claim were found to be true, it does not necessarily follow that biometric technology can perform its task satisfactorily in the light of the considerable costs in money and effort that would be required. Nor can it be an adequate substitute for traditional authentication methods. As Dr. Ted Dunstone has emphasised, it is just a convenient alternative.[23]

Professor Andela Sasse, a biometrics expert, recently advised UK parliamentarians that biometric technologies were 'a lot less mature' than manufacturers made them out to be (Australian Democrats Press Releases, 2006). Biometric technology is based on the assumption that human pattern, fingerprints, irises and faces will stay the same over time, which is not true. Moreover, 'even if the underlying biological traits of interest are truly unique, it does not follow that our machinery will be able to measure them faithfully' (Wilson, 2005: 4). A number of technical problems arise from the fact that biometric data are only *relatively* unique and stable.

There exist a number of other practical problems associated with biometric technology. In Germany, where the e-passport scheme was initiated, complaints have arisen concerning price, privacy and technical difficulties. Teeth and smiles can confuse the facial recognition system, and the distance between the chin and forehead on the photo must be between 32 and 36 mm. It is not easy for people to abide by all these specifications (Laitner and Williamson, 2005). Furthermore, a perception of discrimination might develop among certain groups of people who are unable to use the biometric system for reasons of ethnicity, physical disability, medical condition etc.

Biometric technology has long been recognised as a useful weapon to combat fraud. However, the computer systems that are used to process biometrics are vulnerable to the same sorts of manipulation as other computers. People can access, erase, or alter what is stored there. 'In the end, security depends upon people . . . but the weak link is the systems, procedures, and people who implement them' (Norman, 2003). In addition, there are reports of cases in which a biometric system has been fooled without difficulty. For instance, studies have shown that thin fingerprint-pads adhered to fingers have managed to fool scanners (Higgins, 2003). More recently, a German computer security consultant has shown that he can clone the new electronic passports that the United States and other countries have begun to distribute (Zetter 2006). More sophisticated methods of biometric fraud may also appear with further development of technologies. It goes without saying that to steal or reproduce a fingerprint is still more difficult than stealing a key or a smart card, but it must be noted that it is not always necessary to steal the real finger or iris to compromise the system.

Data accuracy is one of the privacy concerns recognised in the privacy and data protection laws of Europe, Australia and the United States. The issue of accuracy is of particular importance with respect to biometric information. As demonstrated above, the accuracy of current biometric technology is questionable for a variety of reasons. Thus, the legal standard for determining the 'accuracy' threshold for biometric data becomes quite problematic. Should there be different standards for the different contexts in which biometric technology is utilised? How should such a standard be set? Or should one be set at all? For other personal data, it is much easier to check the accuracy of the personal data by the data subject. However, it becomes quite difficult for data subjects to ensure the accuracy of biometric data even when they are given the right to access such information and are given the right to revise it. Extra technical help would be required for such capacities to be instituted. In all likelihood, verification of the accuracy of biometric information will be the responsibility of the data controller. In practice, this becomes quite problematic. Just as it was pointed out by Feldman, the organisation using the biometric technology usually has little interest in its accuracy and reliability (Feldman, 2003). The potential damage and resulting consequences to the individual resulting from a criminal investigation using inaccurate biometric data would be far greater than the damage resulting from the use of inaccurate biometric data in the commercial sector.

3.4.2 Misconceptions relating to biometric technology

The limitations of biometric technology should not result in a complete indictment of the technology. A more serious problem arises from misconceptions concerning the levels of security that biometric technology can guarantee. The difficulty of challenging a false biometric match is particularly troubling in situations that involve government agencies or criminal investigations. For example, over-reliance on digital images of fingerprints led the FBI to wrongly suspect an Oregon lawyer of involvement in the 2004 Madrid train bombings (Leyden, 2004). In that case, the suspect was lucky enough to be released when the Spanish investigators matched the fingerprints to an Algerian, forcing the FBI to admit it was wrong. However, in cases where no other match is found, and there exists a false match and an overconfidence in the technology, innocent people might remain in jail.

Another possibility is that 'the reliance on such flawed security measures might ultimately compromise security further by reducing vigilance and paying less heed to other warning signs' (Roy, 2005). The Ressam case indicates that it was human skill alone that prevented a terrorist attack (Schneier, 2000). When Ressam attempted to enter the United States, he had an authentic Canadian passport issued under a false identity. The computer system cleared him by his ID, but custom agents felt he was suspicious because he was sweating, fidgety and avoided eye contact. Hence, a pertinent point we need to raise is that the ability to accurately identify an individual does not mean that we really know what an individual might do, unless he is already on a suspect list and we identify him from the database. As for numerous potential terrorists with clean backgrounds and authentic ID, biometric identification can do nothing. The ability to ascertain 'who you are' does not mean that we necessarily know what an individual has already done and what they might later do.

As described above, biometric technology raises greater concerns than other forms of identification or authentication methods with regard to our blind misconception of the technology's reliability. Individuals are the weakest part of the whole process and should be protected. From a legal perspective, the greatest challenge for biometric technology will be to delineate procedures for individuals to challenge a biometric determination made in the course of a denial of rights or privileges. It is noteworthy that both individuals and organisations are involved in the biometric technology application process; there should be a back-up system for potential mishandling of information. How should such a system be designed? Should third parties be involved? Who should bear the burden of proof? And who should be liable for any resulting losses?

3.4.3 Security problems posed by biometric technology

As noted previously, and contrary to popular belief, biometric technology is far from mature. In practice, a variety of errors will inevitably occur. These errors are likely to be compounded by the frequent absence of 'fall-backs', as is the case in the event of identity theft (Wilson, 2005). No security system is perfect, and a truly secure system always contains an effective fall-back measure in the event of a critical security breach. Generally, once a biometric is compromised, it is compromised forever. In the event of biometric identity theft, there would appear to be no alternative but to withdraw the user from the system.

It has been reported that some research has shown it possible to transform a biometric iris template so that it assumes a new format unique to a particular application. Thus, a template generated in a format corresponding to a particular application A could not be misappropriated and used to authenticate a user for application B (Braithwaite et al., 2002). In addition, there are also reports of research on cancellable biometrics (Cowley, 2005). Instead of enrolling with your true finger (or other biometric), the fingerprint is intentionally distorted in a repeatable manner and this new print is used. If for some reason your original fingerprint is stolen, an essentially 'new' fingerprint can be issued by simply changing the parameters of the distortion process. This technology may enhance the security level of biometric technology, but several problems still remain:

- It might not protect against replay attack, if the attacker has copied the user's actual biometric character (e.g., by photographing the iris) (S. Wilson, personal communication, 2006).[24]
- Some information may be discarded by using different formats in extracting the biometric template. If each template from a single biometric character is different, then each template has fewer bits of entropy than it would have if it were only one. That is, each template is 'fuzzy' and this will erode accuracy, resulting in higher false match rates.[25] Generally, an accurate and whole image is required for biometric identification (Clarke, 2006).[26]
- These methods are still very much at an experimental level, and are not ready for commercial deployment (S. Wilson, personal communication, 2006; T. Dunstone, personal communication, 2006).[27]
- The degree of correlation that might exist between one template and another is not known with certainty. If an attacker can obtain a template (and/or the original biometric character), they may be able to predict what the next generated template will look like (S. Wilson, personal communication, 2006).

Another kind of measure that may be used against misappropriation is biometric encryption: when an individual enrols his/her biometric sample, the system randomly generates a digital key (which is unknown to the user and the system operator). The encryption algorithm then binds this key securely to the biometric sample to create an encryption template (Cavoukian, 2007). The key and the biometric sample are then discarded and cannot be regenerated from the stored encryption template. When the individual enrols in the system again, if the template he provides matches the part of biometric template in the encryption template, the encryption algorithm retrieves the key, this key can be used in a conventional cryptosystem[28] (Cavoukian, 2007). In the end, the biometric is again discarded; the system only stores the biometrically encrypted key not the biometric itself, which helps to alleviate privacy concerns to some extent. Although this kind of technology can not deal with replay attacks either, it helps when the system is compromised and a new key can be generated as the key is independent of the biometric. However, such systems are still quite immature, it has been reported that there is currently only one commercially available biometric encryption system (Cavoukian, 2007).

In addition to these problems, it is clear, based on many existing biometric applications and statements made by advocates, that building up a centralised personal database with links to identification and verification systems is supposed to be a fundamental part of the entire biometric system. This also creates a great 'honey net' for crackers. The implementation of a centralised system would require widespread access from various remote locations. This may generate significant numbers of failures and render the system vulnerable to compromise, which could be accomplished by hackers 'physically accessing one of the sites, by finding some communication-based vulnerability, or by bribing or corrupting someone with access to the system' (Kent, 2006). Through such access, identity theft or alteration of data could be achieved with little difficulty. Moreover, with such a complex and centralised 'security' system, a failure at one location is likely to cause cascading effects throughout the whole system. Such failures might occur either through physical attack on the infrastructure or a cyber attack (Kent, 2006). It has been noted that especially in the absence of costly dedicated networks, an Internet-based system would 'inevitably be the target of malicious attacks as well as subject to unintentional or incidental damage' (Kent, 2006). In other words, the so-called 'security' system would actually generate less security and more vulnerability.

Would greater security be achieved by storing biometric templates in a portable device? It has been argued that the best method to avoid central storage and to achieve security and preserve privacy would involve the storage of biometric information on a portable device such as a mouse, mobile, laptop computer or smart card. However, this solution has been criticised as 'a worrying gimmick, closely equivalent to writing the PIN on the back of your credit card' (Wilson and Prints, 2004). A majority of commercial fingerprint detectors can be fooled by replica prints. So if you lose your phone or smart card, a clever thief will find your biometric security information very conveniently deposited across the keypad (Wilson and Prints, 2004). A robust liveness detection system is needed to combat such fraud, yet in commercial practice, it remains uncommon in fingerprinting systems (Wilson, 2005).

Another major security concern is that biometric technology introduces a new dimension to identity theft (Clarke, 2001). For instance, the risk of identifying theft may be increased when a national ID card with a biometric identifier is used. Criminals could masquerade as someone else when the card is issued – this could become a very effective form of identity theft (O'Neill, 2005). A widely used biometric identifier used on various occasions may actually facilitate easier identity theft in one location. Once this occurs, it will be extremely difficult to issue another biometric identifier or prove that the theft actually occurred. Biometric systems currently in use are successful because the number of people enrolled is limited. When the system fails, human administrators are available to assist in the authentication process. Creating an automated system on a national scale is beyond the capability of any existing technology. Merging the existing systems into a single central database would cause the reliability of those systems to be lost. Furthermore, biometric databases would be subject to new forms of abuse which may be more difficult to correct and would entail significant consequences for individuals whose biometric identifier is compromised (Rotenberg, 2002).

The issue of what happens to collected biometric information, where and how it is stored, and who has access to it, extends beyond the state and public sector; it is pertinent to private sector actors too. Is there sufficient trust between the entity (public or private) deploying the biometric technology and the individuals who provide biometric data with respect to issues of access, storage and use? Is the biometric data used for a specific goal and can individuals know that databases are not being linked together? What kind of security measures are in place to protect the biometric information? Is the organisation (public or private) that holds the biometric information legally compelled to employ affirmative security measures to protect the biometric information? Who should be liable if biometric information is compromised? And what types of legal remedies are available to protect the individual's legal rights? Such questions over data management represent significant challenges which will have to be addressed in order to properly integrate biometrics into security processes.

3.5 Summary

The challenge in confronting any radically developing technology is that it must be approached with a broad understanding of its associated practical and technological limitations. In this chapter, I have attempted to explore the characteristics, threats and limitations of biometric technology in order to evaluate it on a rational basis.

Biometric information is a special category of personal information which is intimately linked to the physical body. Biometric technology therefore has the potential to create relatively unique and stable digital representations of an individual's identity. As we embrace biometric technology, we must pause to consider some of the implications of its widespread implementation, especially with respect to potential risks to privacy. Such risks include unnecessary and unauthorised collection of personal data, function creep and the loss of anonymity. These risks are inevitable and have the tendency to become more serious as application of the technology spreads. Further, limitations of the technology and misconceptions concerning its capabilities affect the effectiveness and quality of biometric identification systems and introduce security risks that have not received sufficient attention. Moreover, biometric technology could be discriminatory and introduce concerns about permanent identity theft. Therefore, the sacrifice of privacy incurred as a result of using biometric technology does not necessarily provide us with the 'security' we hope for.

Biometric technology has risen to prominence despite much criticism. Halting the deployment of this technology would be unrealistic; a more practical solution would be to adopt appropriate legal and policy measures to regulate the use of biometrics. The subject of the next parts of the book is the formulation and implementation of such measures.

Notes

* This chapter is based on the author's article, Liu 2008.

1 This discussion of biometric technology will therefore be discussed in those two contexts. Purely private/domestic use of biometric technology such as on home PCs or private PDAs is outside the scope of this discussion. This delimitation is based on the consideration that in the private/domestic context, the use of biometric technology does not involve the same degree of privacy concerns as in the other two contexts.

2 DNA fingerprinting is a technique by which an individual can be identified at a molecular level from certain repeating sequences in the DNA present on different chromosomes. These genetic markers are known to vary from individual to individual (except in the case of identical twins) Brinton & Lieberman (n.d).

3 This is the classification used by Bromba (2006).

4 For further discussion, see Chapter 4.

5 See, for example, *Messing v. Bank of America, N.A.* 373 Md. 672, 821 A.2d 22 (2003).

6 According to the author's interview with Prof. Roger Clarke, Xamax Consultancy Pty. Ltd., Canberra, 5 August 2006.

7 The 188 Contracting States that adhere to ICAO are obligated to issue only ICAO-standard Machine-Readable Passports (MRPs) by 01.04.2010. See ICAO (2004b).

8 United Nation (UN) General Assembly resolution 217 A (III) of 10 December 1948.

9 UN General Assembly resolution 2200A (XXI) of 16 December 1966; in force 23 March 1976.

10 See Chapter 7 for more discussion about property right and biometrics.

11 See further the discussion of European law in Chapter 4.

12 These legal issues are discussed in Part III.

13 *United States v. Smith*, 978 F.2d 171 (5th Cir. 1992) at 177.

14 For more discussion about the individual's privacy right over information that is neither intimate nor sensitive and is drawn from public spheres, see Nissenbaum (1998).

15 For more discussion about the concept of anonymity, see Nicoll and Prins (2003).

16 Dr. Ted Dunstone is the chair of the Biometrics Institute in Australia, Sydney. See also Crompton (2002).

17 For example, large-scale test results from the UK Passport Service in 2005 (preparing for their ID card) found that facial recognition had an overall success rate of only 69% accuracy in verifications. This success rate dropped for people over 60 years of age, as well as for people whose appearance had changed in the period between enrollment and subsequent verification. For people who had been registered at mobile centres, the success rate dropped to 48% (UK Passport Service, Biometrics Enrolment Trial, Management Summary, May 2005, p.8). If facial recognition is used for identification, which is more challenging; the success rate will be even lower.

18 Due to the limitations of biometric technology, no biometric today provides lasting signatures on electronic transactions, though it is supposed to be capable of such. I return to this in Section 3.4 (about security and biometrics).

19 See, for example, *Griswold v. Connecticut*, 381 U.S. 479 (1965).

20 *United States v. Smith*, 978 F.2d 171 (5th Cir. 1992).

21 To my knowledge, there are no documented cases of biometrics actually causing direct physical harm to a person.

22 Interviewed by author in Australia, Sydney, 14 August 2006.

23 Interview with Dr. Ted Dunstone, Chair of the Biometric Institute in Australia, Sydney, 14 August 2006.

24 E-mail correspondence with Mr. Stephen Wilson, Chairman of Lockstep Consulting.

25 Id.

26 E-mail correspondence with Prof. Roger Clarke, Xamax Consultancy Pty. Ltd., Canberra.

27 This is still true according to the recent report on biometrics.

28 In a conventional cryptosystem, the user inputs a simple PIN or password to unlock an encryption key. Here biometrics is used to generate this key.

References

Agre, P.E. (2003) *Your Face Is Not A Bar Code: Arguments Against Automatic Face Recognition in Public Places.* Online. Available at http://polaris.gseis.ucla.edu/pagre/bar-code.html.

Alterman, A. (2003) A piece of yourself: ethical issues in biometric identification, *Ethics and Information Technology*, 5: 139–50.

Amnesty International Canada (2001) *Protecting Human Rights And Providing Security*, AI comments with respect to Bill C-36, Ottawa, Canada.

Ashbourn, J. (2006) *The Social Implications of the Wide Scale Implementation Of Biometric and Related Technologies*, Background paper for the Institute of Prospective Technological Studies, DG JRC – Sevilla: European Commission, January 2005.

Australian Democrats Press Releases (2006) *Smartcard for Surveillance?* Online. Available at http://www.democrats.org.au/news/index.htm?press_id=5202&display=1.

Bickley, L.S. (1995) *A Guide to Physical Examination and History Taking.* Philadelphia: Lippincott.

Braithwaite, M., von Seelen, C., Cambier, J., Daugman, J., Glass, R., Moore, R. & Scott, I. (2002) *Application-Specific Biometrics Templates.* Available at http://www.cis.upenn.edu/~cahn/publications/autoid02.pdf.

Bromba, M. (2006) *On the Reconstruction of Biometric Raw Data From Template Data.* Available at http://www.bromba.com/knowhow/temppriv.htm.

Bygrave, L.A. (2002) *Data Protection Law: Approaching its Rationale, Logic and Limits,* The Hague: Kluwer Law International.

Cavoukian, A. (1998). *Privacy and biometrics: An oxymoron or time to take a 2nd look.* Paper presented at Computers, Freedom and Privacy 98 Conference. Available at http://www.ipc.on.ca/scripts/index_.asp?action=31&N_ID=1&P_ID=11551&U_ID=0.

Cavoukian, A. (2003) *National Security in a Post 9/11 World: The Rise of Surveillance . . . The Demise Of Privacy?* Available at http://www.ipc.on.ca/images/Resources/up-nat_sec.pdf.

—— (2004) *Tag, You're It: Privacy Implications of Radio Frequency Identification (RFID) Technology.* Available at http://www.ipc.on.ca/images/Resources/up-rfid.pdf.

—— (2007) *Biometric Encryption: A Postive-Sum Technology That Achieves Strong Authentication, Security AND Privacy.* Available at http://www.ipc.on.ca/images/Resources/up-bio_encryp_execsum.pdf.

Clarke, R. (1994) 'Human identification in information systems: management challenges and public policy issues'. *Information Technology & People*, 7(4): 6–37.

—— (2000). *Beyond the OECD Guidelines: Privacy Protection for the 21st Century.* Available at http://www.rogerclarke.com/DV/PP21C.html#Priv.

—— (2001) *Biometrics and Privacy.* Available at http://www.rogerclarke.com/DV/Biometrics.html.

Clement, A., Guerra R., Johnson J., and Stalder, F. (2002) National identification schemes (NIDS): A remedy against terrorist attack? In K. Brunnstein and J. Berleus (eds.), 195–205, *Proceedings of the Sixth Conference on Human Choice and Computers: Issues of Choice and Quality of Life in the Information Society*, Boston: Kluwer.

Cockfield, A. J. (2003) 'Who watches the watchers? A law and technology perspective on government and private sector surveillance', *Queen's Law Journal*, 29(364): 373–406.

Cole, S. (2000) Myth of fingerprints: A forensic science stands trial, *Lingua Franca*, 10(8): 54–62.

Cowley, S. (2005) *IBM works toward replaceable biometrics*. Available at http://www.csoonline.com.au/index.php/id;260154133;fp;8;fpid;8.

Crompton, M. (2002) Biometrics and privacy, *Privacy Law and Policy Reporter*, 9(3), 53–8.

Cytowix, R.E. (1 Sept 1996) All in the genes, *Washington Post*. p.N.

Davies, S.G. (1994) Touching big brother: How biometric technology will fuse flesh and machine, *Information Technology & People*, 7(4): 38–47.

Department of Communications and Department of Justice (1992) *Privacy & Computers*, Ottawa: Information Canada.

Dorizzi, B. (2005) Biometrics at the Frontiers: Assessing the Impact on Society. Technical Impact of Biometrics, Background Paper for the Institute of Prospective Technological Studies, DG JRC – Sevilla. Available at http://cybersecurity.jrc.es/docs/LIBE%20Biometrics%20March%2005/TechnologicalImplications_Dorizzi.pdf.

Elise, H. and Hendricks, M. (1996) In short – health and medicine, *John Hopkins Magazine*, 48(2). Available at http://www.jhu.edu/jhumag/696web/hlthmedc.html.

Electronic Privacy Information Center & Privacy International (2005) *Privacy and Human Rights: An International Survey of Privacy Laws and Developments*, Washington, DC: Electronic Privacy Information Center.

European Commission's Joint Research Centre (JRC) (2005) *Biometrics at The Frontiers: Assessing the Impact on Society* (EUR21585 EN). Available at ftp://ftp.jrc.es/pub/EURdoc/eur21585en.pdf.

Ezovski, G.M. (2005) Biometric Passports: Policy for International and Domestic Deployment *Journal of Engineering and Public Policy*, 9. Available at http://www.wise-intern.org/journal/2005/Ezovski.pdf

Feldman, R. (2003) Considerations on the emerging implementation of biometric technology, *Hastings Communications & Entertainment Law Journal*, 25: 653–82.

Gavison, R. (1980) Privacy and the limits of law, *Yale Law Journal*, 89(3): 421–28.

Grijpink, J. (2005). Two barriers to realising the benefits of biometrics, *Computer Law & Security Report*, 21(2), 138–45.

Gomm, K. (2005) *U.K. Passport Agency: 'Iris Recognition Needs Work'*. ZDNet Asia. Available at http://www.zdnetasia.com/news/security/0,39044215,39283306,00.htm.

Harte, A. (2004) Privacy and identity in a changing world, *Australasian Psychiatry*, 12(1): 55–7.

Hawkes, N. (26 Feb 1996) Fingerprint clues to health, *London Times*.

Hert, P.J.A.de (2005) *Biometrics: Legal Issues and Implications* (Background paper for the Institute of Prospective Technological Studies, DG JRC – Sevilla). Available at http://www.statewatch.org/news/2005/apr/jrc-biometrics-paul-de-hert.pdf.

Higgins, P.T. (2003) Fingerprint and hand geometry, in J. K. Brownlow (ed.), *Biometrics*, New York: McGraw-Hill, 24–41.

Holladay, A. (2002) *Identical Twins Have the Same DNA (Almost)*, Available at http://www.wonderquest.com/twins-dna.htm.

Impagliazzo, R., and More, S. M. (2002). *Anonymous credentials with biometrically-enforced non-transferability*. Available at http://portal.acm.org/ft_gateway.cfm?id=1005150&type=pdf&coll=&dl=ACM&CFID=15151515&CFTOKEN=6184618.

Jain, A.K., Ross, A. and Uludag, U. (2005). Biometric Template Security: Challenges and Solutions. Available at http://www.ee.bilkent.edu.tr/~signal/defevent/papers/cr1805.pdf.

Kent, S.T. (2006) Ids – Not That Easy: Questions about National Wide Identity System.

Available at http://www7.nationalacademies.org/ocga/testimony/IDs_Not_That_Easy.asp.

Laitner, S. and Williamson, H. (2005) E-passport is no laughing matter. *Financial Times*. Available at http://www.ft.com/cms/s/d51ee8d0-50c6-11da-bbd7-0000779e2340.html.

Lawless, J. (2004) Fingerprint Privacy Concerns, CBS News. Available at http://www.cbsnews.com/stories/2004/09/08/tech/main641998.shtml.

Leyden, J. (2004, May 26) FBI apology for Madrid bomb fingerprint fiasco, *The Register*. Available at http://www.theregister.co.uk/2004/05/26/fbi_madrid_blunder/.

Liu, Y. (2008) Rational concerns about biometric technology: security and privacy, in R. Subramanian (ed.), *Computer Security, Privacy and Politics: Current Issues, Challenges and Solutions*, Hershey: IGI Global, 85–120.

McCoy, S. (2002) Comment: O big brother where art thou? The constitutional use of facial-recognition technology, *John Marshall Journal of Computer & Information Law*, XX(3): 471–85.

Moo-Young, R. (2001) Eyeing the future: Surviving the criticisms of biometric authentication', *North Carolina Banking Institute*, 5 (Spring): 421–35.

Mordini, E. and Petrini, C. (2007) Ethical and social implications of biometric identification technology, *Ann Ist Super Sanità*, 43(1): 5–11.

National Institute of Standards and Technology (2001) *Common Biometric Exchange File Format* (CBEFF) (NISTIR 6529). Available at http://www.itl.nist.gov/div895/isis/bc/cbeff/CBEFF010301web.PDF.

Nicoll, C. and Prins, C. (2003). Anonymity: challenges for politics and law, in C. Nicoll, J. E. J. Prins, and M. J. M. van Dellen (eds.), *Digital Anonymity and the Law*, The Hague: T.M.C. Asser Press, 287–95.

Nissenbaum, H. (1998). Protecting privacy in an information age: the problem of privacy in public, *Law and Philosophy*, 17, 559–96.

Norman, D. (2003) Don Norman's jnd.org: recommended readings (reviewing Bruce Schneier, *Secrets & lies: Digital Security in a Networked World)*. Available at www.jnd.org/recommended_readings.html.

Schotsmans, P. (1998) Ownership of the body: a personalist perspective, in H. ten Have, J. V. M. Welie and S. F. Spicker (eds.), *Ownership of the Human Body: Philosophical Considerations on the Use of the Human Body and its Parts in Healthcare*, 159–72, Dordrecht: Kluwer Academic Publishers.

O'Neill, R. (2 Aug 2005) IDologists, *Sydney Morning Herald*. Available at http://www.smh.com.au/news/next/idologists/2005/08/01/1122748570079.html?oneclick=true.

Ploeg, I.V.D. (1999) Written on the Body: Biometrics and Identity, *Computer and Society*, 29(1): 37–44.

—— (2001) *Biometrics and the Body as Information: Normative Issues of the Socio-Technical Coding of The Body*. Available at https://www.bmg.eur.nl/smw/publications/vdp_02.pdf.

Prins, C. (1998) Biometric technology law, making our body identify for us: legal implications of biometric technologies, *Computer Law & Security Report*, 14(3): 159–65.

QT Worldtel, Inc. (2003). Centralised out-of-band authentication system. Paper presented at Southeast Europe Cybersecurity Conference. Available at http://www.cybersecurity-cooperation.org/documents/QT_COBAS_White_Paper.pdf.

Reichman, N. (1987) Computer matching: toward computerised systems of regulations. *Law and Policy*, Oct edn, 391–410.

Retica Systems Inc. (2005) *Eye Biometrics*. Available at http://www.retica.com/site/biometrics/index.html.

Rinehart, G. (2001) Biometric payment: the new age of currency. *Hospitality Upgrade Magazine*. Available at http://www.hotel-oline.com/News/PressReleases2000_1st/Mar00_BiometricCurrency.html.

Rothstein, M.A. (1997) *Genetic Secrets: Protection Of Privacy and Confidentiality in the Genetic Era*, New Haven: Yale University Press.

Rotenberg, M.A. (2002) Joint Hearing on Identity Theft Involving Elderly Victims before the Special Committee on Aging. Available at http://www.epic.org/privacy/biometrics/testimony_071802.html.

Roy, B. (2005) A case against biometric national identification systems (NIDS): trading-off privacy without getting security, *Windsor Review of Legal and Social Issues*, 45(3), 59–61.

Schneier, B. (2000) *Secrets And Lies: Digital Security in a Networked World*, New York: John Wiley & Sons, 141–145.

Stoney, D.A. (1997) Fingerprint identification: scientific status, in D.L.Faigman, D.H. Kaye, Saks and J. Sanders (eds.), *Modern Scientific Evidence: The Law and Science of Expert Testimony*, St Paul, MN: West Publishing, 368–99.

Volio, F. (1981) Legal personality, privacy, and the family, in L. Henkin (ed.), *The International Bill of Rights: The Covenant on Civil and Political Rights*, New York: Columbia University Press, 186.

Warren, S.D. and Brandeis, L.D. (1890) The right to privacy, *Harvard Law Review*, 4: 193–211.

Wayman, J.L. (1998) Biometric identification and the financial services industry (Congressional testimony of May 20), in J. Wayman (ed.) *National Biometric Test Center: Collected Works*. Available at www.engr.sjsu.edu/biometrics/nbtccw.pdf.

—— (2000). *When Bad Science Leads to Good Law: The Disturbing Irony of the Daubert Hearing in the Case of U.S. v. Byron C. Mitchell*. Available at http://www.engr.sjsu.edu/biometrics/publications_daubert.html.

Wayman, J.L., Jain, A.K., Maltoni, D. and Maio, D. (2004) *Biometric Systems: Technology, Design And Performance Evaluation*, London: Springer.

Westin, A.F. (1967) *Privacy and Freedom*, New York: Atheneum.

Wilson, S. (2005) Lockstep submission to senate privacy inquiry. Available at http://www.aph.gov.au/senate/committee/legcon_ctte/privacy/submissions/sub11.pdf.

Wilson, S. and Prints, T. (14 Aug 2004). Telltale prints, *New Scientist*. Available at www.newscientist.com/aricle.ns?id=mg18324604.200.

Woodward, J.D. (1997a) *Biometrics: Identifying Law & Policy Concerns*. Available at http://www.cse.msu.edu/~cse891/Sect601/textbook/19.pdf.

—— (1997b) Biometrics: privacy's foe or privacy's friend? *Proceedings of the Institute of Electrical and Electronics Engineers*, 85(9): 1480–1492.

—— (2001a) Super bowl surveillance: facing up to biometrics, (Rand Issue Paper 209), Available at http://www.rand.org/pubs/issue_papers/2005/IP209.pdf.

—— (2001b). *Army Biometric Applications: Identifying and Addressing Socio-cultural Concerns*. Santa Monica, California, RAND.

Zetter, K. (2006, August 3) Hackers Clone E-Passports, *Wired*. Available at http://www.wired.com/news/technology/0,71521-0.html?tw=rss.technology.

Part III

4 Regulation of biometrics by privacy and data protection law in Europe

4.1 Introduction

The European legal framework for addressing threats by biometrics to privacy and related interests has strong roots in doctrines and rules for the protection of human rights. These rules are manifest primarily in the European Convention on Human Rights and Fundamental Freedoms (ECHR), particularly the provisions of Article 8 dealing with the right to respect for private life, family, home and correspondence. As shown further in this chapter, respect for people's private lives implies legal protection for physical privacy and informational privacy. At the same time, the collection and further processing of biometric information, particularly by automated means, raises specific concerns about data protection which are covered by a range of legal instruments dealing specifically with data protection. Central among these instruments is the EU Data Protection Directive and the Council of Europe's Convention on Data Protection. Thus, this chapter devotes considerable attention to the provisions of the Directive, Data Protection Convention and the ECHR.

As noted in Chapter 1, several reports on biometrics and privacy/data protection issues in the European context have been published in recent years.[1] These reports show that uncertainties remain about the application of privacy and data protection laws to the use of biometric technology, including the specific criteria that make the collection and processing of biometric data legitimate and lawful. The uncertainties sometimes result in inconsistencies or even contradictory legal practices of national data protection authorities with respect to implementation of biometrics in similar or equivalent situations.[2]

This chapter aims to address these uncertainties. In doing so, it does not give a comprehensive study of all the relevant legal issues pertaining to biometrics in the European legal framework, but rather aims to address the most controversial legal problems raised by the use of biometric technology, within the context of privacy and data protection law. Based on the study of relevant reports on biometrics, as well as legal rules on privacy/data protection, this chapter continues the discussion of the physical privacy and informational privacy risks of biometrics mentioned in Chapter 3. It also touches on the permissibility of recognizing a property interest in biometric data. Section 2 briefly sketches the European legal framework of human rights law and data protection law relevant to biometrics. This framework forms a basis for the further assessment of the legal implications of biometrics discussed in the same section. In Section 3, the specific legal issues raised by biometrics that have been inadequately addressed or left untouched by the current laws are evaluated. This is followed by an in-depth discussion of legal cases concerned directly with biometrics in Norway and Sweden. A summary of findings is presented in the final conclusions (Section 4), and some proposals for basic rules are formulated.

4.2 Sources of privacy protection

Europe has been a leader in terms of attempting to provide comprehensive privacy protection for individuals in the face of new forms of ICT (Levin and Nicholson, 2005). The first national data protection legislation in the world was adopted in Sweden in 1973.[3] Today, virtually all European countries have broad privacy or data protection statutes.[4] Regulation of personal data in Europe stems from the rights to privacy and protection of personality which are deeply embedded in European legal tradition. Section 4.2.1 gives a brief overview of the main legal codes on privacy and data protection which play a role in addressing privacy-related concerns raised by biometrics.

4.2.1 Article 8 of the ECHR

Privacy is understood as a basic human right, and is expressed as such in the Convention for the Protection of Human Rights and Fundamental Freedoms. The Convention was adopted under the auspices of the Council of Europe (CoE)[5] in 1950. All CoE member states, including all member states of the EU, have ratified the Convention and are bound by its guarantees. Article 8 of the Convention is most relevant for privacy and data protection. It states:

1. Everyone has the right to respect for his private and family life, his home and his correspondence.
2. There shall be no interference by a public authority with the exercise of this right except such as is in accordance with the law and is necessary in a democratic society in the interests of national security, public safety or the economic well-being of the country, for the prevention of disorder or crime, for the protection of health or morals or for the protection of the rights and freedoms of others.

Although the instrument does not expressly refer to the protection of personal information, data protection issues have been addressed in case law. The European Court of Human Rights (ECtHRs) in Strasbourg has repeatedly expressed the view that the protection of personal data is of fundamental importance to a person's enjoyment of his or her right to respect for private and family life as guaranteed (Hustinx, 2005; Hert and Gutwirth, 2009). The ECtHR held, for example, in *Leander v. Sweden* (1987)[6] that the storage and release of personal information and the refusal to allow Mr. Leander an opportunity to refute it amounted to an interference with his right for private life guaranteed by Article 8(1).[7] Gathering personal information may also be found to constitute an interference with the right(s) in Article 8(1), thus requiring justification under Article 8(2). The Court held in *Klass and others v. Germany* (1978)[8] that interception of telephone communication may be an interference with private life and correspondence and thus violate Article 8(1).[9] In *Copland v. United Kingdom* (2007), the Court also affirmed that the monitoring of e-mails sent from work should be protected under Article 8, as should information derived from the monitoring of personal internet usage.[10] The Court has held in the case of *Marper v. United Kingdom* (2008)[11] that indefinite storage of DNA may violate Article 8. The latter case is dealt with in detail in Section 4.3.1.3.

The Court usually employs a three-step analysis to determine whether a state party to the Convention has violated Article 8 (Harris et al., 2009). First, the Court decides whether the case involves private life, family life, home or correspondence. Second, it discerns whether a public authority has either interfered with the right *or* failed to take steps to protect that right. Governments bear both negative and positive obligations for privacy protection. They are not only required to not interfere with privacy rights directly but are also required to take positive steps to 'respect' or protect a person's rights under Article 8 from interference. In practice, this obligation has led to a requirement that states provide protection from privacy-invasive actions by private sector bodies. Consequently, Article 8 may require public sector bodies to make legal regulations for certain biometric applications within the private sector.[12] Third, the Court decides whether the state's acts or omissions in a particular case are justified by Article 8(2). A margin of appreciation is given to the state to reach its own assessment as to justification for its acts or omissions, particularly consideration of whether the case involves intimate matters, positive obligations, urgent measures or national security (Yourow, 1996). Based on the study of the Court's case law dealing with data protection, Bygrave has noted that interference under Article 8(1) is likely to be found when the relevant data reveal details about the data subject's personality, are processed without the subject's consent or knowledge, or the processing potentially casts the data subject in a negative light or could result in a restriction of their freedom of choice (Bygrave, 1998). These factors are to a large extent incorporated into the data protection principles in both the Council of Europe's Convention on Data Protection and the EU's Data Protection Directive.

4.2.2 Convention No. 108 on Data Protection

In 1981, the CoE specifically addressed the issue of data protection in its adoption of the Convention for the protection of individuals with regard to the automatic processing of personal data. This instrument, also referred to as Convention 108, has been signed by 44 of the 47 CoE member countries and ratified by 41 of them.[13] It was the first legally binding international instrument in the data protection field and remains the only convention dealing specifically with data protection. The Convention enumerates a series of basic principles for use of personal data, including transborder flows of such data. It is potentially open for ratification by states that are not members of the CoE; concomitantly, it is also envisaged to be potentially more than an agreement between European states (Bygrave, 2002). The Convention is not self-executing; instead, it contains an obligation for the contracting parties to adopt national legislation in line with its principles (Hustinx, 2005). An additional protocol that entered into force on 1 January 2004 requires parties to set up supervisory authorities that are to exercise their functions in complete independence, along the lines stipulated in equivalent provisions of the EU Data Protection Directive.[14] It also extends the Convention's rules on transborder data flow, again in line with equivalent provisions of the Directive.

The Convention is intended primarily to cover computerised ('automated') processing of data on physical persons in both the private and public sectors. It subjects the processing of certain types of 'sensitive' data (data on a person's race, political beliefs, health, religion and sexual life) to more stringent conditions than other personal data (Article 6). The Convention also enshrines the individual's right to know that information is stored on him or her and, if necessary, to have it corrected (Article 8).

The Convention consists of three main parts: substantive law provisions in the form of basic principles, special rules on transborder data flow and mechanisms for mutual assistance and consultation between parties. The major part of the Convention is Chapter 2 that sets out basic principles for processing personal data. These principles embody a common minimum of the standards that were previously promulgated in a large number of national laws passed by CoE member states and in early CoE resolutions on data protection.[15] The principles may be summed up as follows (Bygrave, 2009):

- Fair and lawful processing: 'Personal data shall be obtained and processed fairly and lawfully' (Article 5(a));
- Purpose specification: Personal data shall be stored for 'specified and legitimate purposes and not used in a way incompatible with those purposes' (Article 5(b));
- Minimality: The amount of personal data collected and stored shall be limited to what is necessary for achieving the purpose for collection/storage (Article 5(c) and (e));
- Adequate information quality: Personal data shall be 'adequate, accurate and relevant' in relation to the purposes for which it is stored (Article 5 (c) and (d));
- Sensitivity: 'Personal data revealing racial origin, political opinions or religious or other beliefs, as well as personal data concerning health or sexual life', are subject to more stringent protection (Article 6);
- Security: 'Appropriate security measures shall be taken for the protection of personal data stored in automated data files' (Article 7);
- Transparency: Any person shall be able to 'establish the existence of an automated personal data file, its main purposes, as well as the identity and habitual

residence or principal place of business of the controller of the file' (Article 8(a)) and to 'obtain at reasonable intervals and without excessive delay or expense confirmation of whether personal data relating to him are stored in the automated data file as well as communication to him of such data in an intelligible form' (Article 8(b));

- Rectification: 'Any person shall be enabled to obtain . . . rectification or erasure of such data if these have been processed contrary to the provisions of domestic law giving effect to the basic principles set out in Articles 5 and 6 of this convention' (Article 8(c)).

4.2.3 Data Protection Directive

Directive 95/46/EC on the protection of individuals with regard to the processing of personal data and the free movement of such data was enacted primarily in order to remove potential obstacles to the free flow of information in the EU and difficulties in having to comply with different data protection standards in the various EU member states.[16] The Directive has constituted the most important point of departure for national privacy and data protection initiatives within the EU since its adoption in 1995. As noted in Chapter 1, the Directive has been incorporated into the 1992 Agreement on the European Economic Area (EEA) such that states which are not EU members but party to the EEA Agreement (i.e. Norway, Iceland and Lichtenstein) are legally bound to abide by the Directive. It has also greatly influenced the shaping of privacy and data protection regulatory regimes outside Europe (Bygrave, 2004, 2009). The Directive became effective in October 1998, by which time EU member states were required to have transposed most of its provisions into national legislation (Article 32(1)). For some commentators, the Directive marks the high-water mark of legal protection for information privacy. It is distinguished by its breadth in the data, activities and geographic area to which it applies (Cate, 1999).

As indicated above, a primary objective of the Directive is to harmonise national data protection laws of member states. However, the Court of Justice (CJ) has confirmed that the Directive concerns not only promotion of the internal market but the protection of privacy and other basic human rights. The Court has also held that the interpretation of the Directive must turn partly on relevant case law from the ECtHRs pursuant to the ECHR.[17] Also, noteworthy is the recognition in the EU of data protection as a basic human right distinct from the right to privacy. This is evident from the EU's Charter of Fundamental Human Rights which contains separate provisions on data protection (Article 8) and privacy protection (Article 7).[18]

The Directive covers the processing of personal data in both the public and private sector. However, exceptions are specified for public security, state security, criminal law and processing in a purely private domestic context (Article 3(2); see also Article 13(1)). Processing of personal data is defined as 'any operation or set of operations which is performed upon personal data, whether or not by automatic means, such as collection, recording, organisation, storage, adaptation or alteration, retrieval, consultation, use, disclosure by transmission, dissemination or otherwise making available, alignment or combination, blocking, erasure or destruction' (Article 2(b)). Personal data are defined as 'any information relating to an identified or identifiable natural person ('data subject'); an identifiable person is one who can be identified, directly or indirectly, in particular by reference to an identification number or to one or more factors specific to his physical, physiological, mental, economic, cultural or social identity' (Article 2(a)). More stringent rules apply to the processing of certain categories of sensitive personal data, i.e. 'data relating to information regarding racial or ethnic origin, political opinions, religious or philosophical beliefs, trade union membership, physical or mental health or condition and sexual life' (Article 8(1)).

The Directive extends the eight principles outlined in the Council of Europe's Data Protection Convention and similar principles set out in the Organisation of Economic Co-operation and Development (OECD) Guidelines for the Protection of Privacy and Transborder Flow of Personal Data.[19] Specifically, the Directive requires that individuals be informed before personal data are disclosed for the first time to third parties for the purposes of direct marketing and to be expressly offered the right to object to such disclosures or uses (Article 14(b)). In Article 6, the Directive provides rules based on the fair and lawful processing principle and purpose specification principle. It then creates a 'legitimacy' principle, found in Article 7, prohibiting any data processing operation that does not have a legitimate purpose (or unambiguous consent of the individual) within defined criteria. The Directive further grants individuals a qualified right to challenge any decision significantly affecting them that is based completely on the automated processing of data relating to them, including decisions involving creditworthiness or work performance (Article 15(1)). Individuals also have the right to monitor and challenge the use of personal information after it is processed. Individuals are given a right of access to the data about them, to have it corrected, to receive confirmation of the purposes of the processing and to obtain the identities of third party recipients (Article 12). The Directive further promotes the free flow of information between EU member states, prohibiting privacy-based restrictions on such flow (Article 1(1)). However, transfer of personal data to other countries is prohibited unless the latter are found to provide an 'adequate' level of protection, subject to certain exceptions (Articles 25–26).[20]

Enforcement is ensured by requiring member states to provide a judicial remedy for infringements of rights (Article 22). Member states are also required to designate an independent public authority to monitor the application of the Directive within their territory and take investigative and interventionist action if necessary (Article 28). Article 29 creates a 'working party on the protection of individuals with regard to the processing of personal data'. The chief task of the working party is to advise the Commission on key portions of the Directive and its implementation (Article 30).

4.3 Privacy challenges of biometric technology on privacy law

As highlighted in Chapter 3, the use of biometric technology poses privacy risks that raise much the same basic concerns about informational and physical privacy that prompted the creation of specific regulatory regimes for privacy protection in Europe, particularly from the 1970s and onwards. Various critical legal issues must be clarified in order to find out to what extent those regulatory regimes can provide effective legal remedies to the privacy risks posed by biometrics.

In the following sections, I investigate how the present privacy laws can help solve these risks and determine to what extent the laws should be improved. On their face, the legal instruments in Europe provide a comprehensive framework for the legal protection of privacy and related interests in the face of the various risks posed by biometrics, as described in Chapter 3. Instead of tediously illustrating how all of the existing provisions may be useful for dealing with the privacy risks posed by biometrics, I focus on the most problematic and significant provisions in the data protection instruments when biometrics is concerned. By 'significant', I mean the provisions that are crucial in determining the applicability of the regulatory regime to the development and application of biometrics; by 'problematic', I mean the provisions (or those aspects of them) that may suffer from shortcomings in regulating biometrics adequately. The legal point of departure for my analysis is the basic European data protection framework (as supplemented by the traditional human rights framework), in particular the basic practices or principles of data protection as spelled out in Convention 108, the Data Protection Directive and case law pursuant to Article 8 of the ECHR. To begin, I study the scope of the privacy instruments.

4.3.1 Scope of the privacy instruments

4.3.1.1 Coverage

Biometric technologies involve physical or behavioural characteristics of a person that are created to identify or authenticate people. The legal significance of biometric data, including raw biometric images and biometric templates, has been discussed in several policy documents concerning biometrics. In the progress report of Convention 108 on biometrics, it is clearly pointed out that Article 8 ECHR has particular relevance for biometrics, because the collation and subsequent processing of biometric data might interfere with private life. The report states that the right to respect for private life implies respect for a person's body, and that during the process of collection and use of bodily features, human dignity should be fully respected. It is also mentioned that the need for fair fallback alternatives for handicapped people or people whose physical characteristics do not fit technical standards, in order to avoid possible discrimination concerns (The Consultative Committee of Convention 108, 2005). The progress report further states that if biometric data are automatically processed, the principles of Convention 108 apply, whether or not private life is at stake (The Consultative Committee of Convention 108, 2005). The Article 29 Data Protection Working Party's (WP29) report on biometrics clearly confirms that both raw biometric images and the templates derived from raw biometrical images are personal data. Hert's report for the European Commission on biometrics also notes that 'it is sufficient for the

Directive that data make it possible to identify a person, it is not necessary to know the name of the person to speak of "personal data" in the sense of the Directive' (Hert, 2005: 13). Hence, it is not generally controversial to suggest that the use of biometric technology and the processing of biometric data fall within the scope of the ECHR and the more specific data protection instruments. The general difference is that the ECHR applies directly to the public sector and indirectly to private sector entities (individuals and/or organisations),[21] whereas the specific data protection framework applies to both the public and private sectors and deals with conflicts between citizens directly. Another basic difference is that the ECHR may apply also to cases that do not involve the processing of *personal* data, whereas such data must be involved to attract the application of the specific data protection framework. However, whereas application of Article 8 ECHR to biometric data is largely dependent on a finding that private life is at stake, Convention 108 and the Data Protection Directive will apply to the automatic processing of biometric data regardless of such a finding (though the data must be 'personal' in the sense that they can be linked to an identifiable individual person).

Compared with the provisions in Convention 108 and the Data Protection Directive, the wording in Article 8 ECHR is quite nebulous with respect to its requirements concerning the processing of personal data. Although Convention 108 gives effect to Article 8 ECHR with regard to automatic processing of personal data, which includes biometric data, the general wording in Article 8 about 'interference with private life' does not give a straightforward answer as to what concrete processing of personal data is an interference and what is not. Concrete case applications are used in general as translations of this provision. As noted above, an analysis by Bygrave (1998) shows that a variety of factors help determine whether interference is occasioned by processing of personal data. These factors include the presence or otherwise of consent to the processing, knowledge of the processing, the data subjects' reasonable expectation of how the data would be used and what kind of consequences the processing has for data subjects' privacy and related interests. The first list of these factors has so far been the most salient in the sense that all instances of data processing that have been found by the ECtHR as constituting an interference have been non-consensual.

As Bygrave notes, one cannot conclude from the case law pursuant to Article 8 ECHR that non-consensual processing necessarily amounts to an interference or that processing based on data subject consent will always be held as not amounting to an interference (Bygrave and Schartum, 2009). The same considerations must surely apply with respect to the processing of biometric data.

As mentioned above, the recent case of *Marper v. United Kingdom* (2008) has demonstrated that biometric data are covered by the ECHR. This case aimed to determine whether the United Kingdom could justify its system of indefinitely retaining the DNA of *unconvicted* people. It is the only ECtHR decision to deal directly with a form of biometric data. In an old case, *Lundvall v. Sweden*, (1985),[22] there is discussion by the European *Commission* of Human Rights[23] on the use of personal identification numbers, which could be compared with biometric data insofar as these are also used as a unique identifier. In that case, the Commission observed the following: (1) the use of personal identity numbers is not prohibited by the ECHR, and (2) the manner of using the personal identity number is relevant to Article 8. The Commission added that:

> it is therefore conceivable that the use of personal identity numbers as a way of storing data in different registers and the matching of such registers could raise an issue under Article 8 of the Convention.[24]

This comment could be regarded as especially relevant to the use of biometric data since one aspect of the invasive potential of biometric data is that they could easily facilitate data matching and linking. This comment shows that the Commission was aware of the potential threat of such data. However, there is no further explanation on this matter from the Commission. The Court has yet to deal specifically with the use of personal identity numbers or other unique identifiers in the context of matching and linking of data registers.

4.3.1.2 Derogations

Article 8(2) of ECHR sets out three main and cumulative criteria for justifying the interference with the right(s) under Article 8(1): the interference must be (1) 'in accordance with the law', (2) 'necessary in a democratic society' and (3) in furtherance of at least one of the aims listed in paragraph 2 (i.e. 'national security, public safety, economic well-being of the country, prevention of disorder or crime, protection of health or morals or protection of the rights and freedoms of others'). Hence, it is not enough to only find that a certain action amounts to an interference with Article 8(1) right(s); it is equally important to find out if this action could be justified under Article 8(2). Article 9(2) of Convention 108 and Articles 3(2) and 13 of the Data Protection Directive (see Article 9 of the latter with respect to freedom of expression) have provisions with a similar effect as Article 8(2) ECHR and allow for derogations from their respective sets of data protection principles.

The issue of balancing interests is of great importance for the application of biometric technology. Historically, a large percentage of biometric applications was developed for governmental purposes. Most biometrics were created and developed for enhancing security at borders and for law enforcement purposes. Consequently, the use of biometric technology by governmental agencies is very likely to be justified by claiming a need for 'national security', 'public safety' and 'prevention of disorder of crime' because biometric technology in general has a reputation as 'useful and reliable' for enhancing national security (Hert, 2005). However, as demonstrated in Chapter 3, the use of biometric technology does not necessarily provide us with enhanced security all the time. Nevertheless, as biometric technology is still frequently considered very reliable, this justification might be considered reasonable. There are, though, some other specific criteria from the ECHR case law that still need to be fulfilled even when the security or public interest flags are legitimately raised. The study of ECHR case law is helpful for interpreting similar provisions in Convention 108 and the Data Protection Directive. Supposing that the 'legitimate aim' is easily fulfilled when using biometric technology for surveillance under the claim of 'national security' or 'public interest', I will explore how the other two criteria can also be fulfilled.

4.3.1.2.1 'IN ACCORDANCE WITH THE LAW': SUFFICIENTLY PRECISE LAW FOR SERIOUS VIOLATIONS

This criterion demands consideration of whether a law exists to authorise the interference. As Mountfield has stated:

> The convention requires a positive basis for an interference with rights. The concept of legality requires first that there must be an identified and established legal basis in domestic law for restrictions on Convention rights. In the absence of such a basis, no derogation from rights can be justified, no matter how worthy its objects.
>
> (Mountfield, 2001: 7)

The requirement that the derogation should be 'in accordance with law' is in accordance with the 'transparency principle'. The famous *Sunday Times v. United Kingdom* (1980) case[25] has demonstrated that 'certainty of law' is quite important. In addition, the law needs to be 'accessible', which implies that the norm must be published. Internal guidelines from government departments or public authorities may not fulfil the accessibility requirement unless they are published or at least unless their content is sufficiently knowable (Mountfield, 2001). The Court in *Sunday Times* stated:

> There are two requirements that flow from the expression 'prescribed by law'. Firstly, the law must be adequately accessible: the citizen must be able to have an indication that is adequate in the circumstances of the legal rules applicable to a given case. Secondly, a norm cannot be regarded as a 'law' unless it is formulated with sufficient precision. To enable the citizen to regulate his conduct: he must be able, if need be with appropriate advice, to foresee, to a degree that is reasonable in the circumstances, the consequences which a given action may entail.[26]

This view on the requirement of certainty, accessibility and precision was developed further in, *inter alia*, the *Rotaru v. Romania* (2000) case.[27] In this case, the ECtHR decided that the secret collection of personal data by the agents of the state was a violation of Article 8. The Court noted that no provision of domestic law laid down any limits on the exercise of those powers; it did not define the kind of information that could be recorded, the categories of people against whom surveillance measures such as gathering and keeping information could be taken, in which circumstances such measures could be taken or the procedure that should be followed. There was no explicit, detailed provision in Romanian law concerning the persons authorised to consult the files, the nature of the files, the procedure to be followed or the use that could be made of the information obtained. The

Court also noted that, although the relevant law empowered the relevant authorities to permit interferences necessary to prevent and counteract threats to national security, the grounds for allowing such interferences were not laid down with sufficient precision. The Romanian system for gathering and archiving information did not provide any safeguards, nor were supervision procedures provided by the law. That being so, the Court considered that the domestic law did not indicate with reasonable clarity the scope and manner of exercise of the relevant discretion conferred on the public authorities. Hence, the Court concluded that the holding and use by the Romanian Intelligence Service of information on the applicant's private life had not been 'in accordance with the law'.

There is a similar kind of reasoning in the decision in *Kruslin v. France* (1990), which concerns the covert interception of a telephone conversation.[28] The Court emphasised that this act is 'a serious interference with private life and correspondence' and must be based upon a law that is 'particular[ly] precise'. The Court pointed out that 'it is essential to have clear, detailed rules on the subject', and the French law failed to specify safeguards against possible abuses of the telephone-tapping system. The judgment in this case highlights that when interference with rights under Article 8(1) is serious, there is a need for comprehensive regulation of the activities that incur the interference.[29]

Regarding the collection and use of biometric data, the above judgements may be quite relevant for the interpretation of the general derogations of the data protection legislation. In the public context, 'consent' often plays a minor role; compulsory collection is often the norm. Nevertheless, as the above decisions make clear, in cases involving surreptitious surveillance of persons, the ECtHR has viewed the violation of privacy as serious.[30] The progress report of Convention 108 about biometrics states that 'surreptitious processing of . . . [biometric data] would be contrary to the principle of fair processing and would therefore only be allowed if the criteria of its Article 9 are fulfilled'.[31] It went on to state that 'a law would be needed describing the exact ambit of the exceptions to the general rule of fair processing'.[32]

According to this argument, the governmental agencies should have sufficiently detailed and precise rules for the collection and processing of the biometric data, even when they are collecting or using them for the purpose of public interest or national security. Thus, in cases where no specific laws regulate the covert collection of biometric data (e.g., in the context of a screening (negative identification) process, using a video surveillance system incorporated with facial recognition technology), there will not be justification of the collection under Article 8(2).

Do we need to claim, therefore, that only surreptitious surveillance is a 'serious' violation of privacy rights? Are there any other situations that might also be considered as 'serious' violations of privacy rights and therefore need more precise and comprehensive laws for their justification? In his analysis of the case decisions in *Klass*[33] and *Malone*,[34] Bygrave notes that although the surveillance in both cases was secret, 'the court did not specifically address the issue of the extent to which it viewed the secrecy of the information and its processing as a necessary element' (Bygrave, 1998: 256) of the privacy interference of Article 8(1) ECHR.

Accordingly, it may be argued that 'the court would have found interference even if the information and its processing were known to the data subject all along' (Bygrave, 1998: 256). The main theme emphasised in these cases is that there should be adequate legal safeguards to protect against arbitrary interference in the right(s) guaranteed by Article 8(1). Therefore, should we regard the 'function creep' of biometric information also as a 'serious violation' of privacy? In the progress report of Convention 108 on biometrics, the Consultative Committee of the Convention held that if biometric data are used for purposes incompatible with the original purpose of collecting these data, 'Article 9 of the convention [i.e. Convention 108] would demand that such a measure be proved to be necessary in a democratic society for the purpose of public security', and that 'the derogations to the criterion of compatibility should be laid down in a law specifying the way such data can be preserved and used for this new purpose' (The Consultative Committee of Convention 108, 2005, para 104). This opinion is in accordance with the court judgment mentioned in *Kruslin v. France* (1990) which emphasises that the derogations be necessary and prescribed by law. It may be argued that the use of biometric technology for surreptitious surveillance, or other use of such technology involving 'function creep', is only justified when there is comprehensive and precise law or regulations for collection and processing that prohibits abuse of biometric information.

4.3.1.2.2 NECESSITY IN A DEMOCRATIC SOCIETY

One commentator states that the development of the national DNA database in the United Kingdom.

> [d]emonstrates the fundamental shortcoming of the law in protecting liberties and privacy. Providing data are 'lawfully' collected there can be no objection whatsoever – but what if the laws themselves are contrary to the standards of a democratic society? What if the cumulative collection of personal data is such that democracy slides into authoritarianism and authoritarianism into totalitarianism?
>
> (Bunyan, 2004)

Partly because of this possible limitation of law, the criterion 'in accordance with the law' is not the only criterion for justifying interference under Article 8(2); the 'necessity' requirement also plays an important role. However, the Strasbourg judges are hesitant and reluctant to apply the check of 'necessary in a democratic society' and they clearly prefer the much 'safer' testing of the legality requirement mentioned above (Hert, 2005).

Compared to the requirement of 'in accordance with the law', the criterion of 'necessity' is more abstract. In *Leander v. Sweden* (1987), the criterion of 'necessity' has been interpreted as that which 'corresponds to a pressing social need' and is 'proportionate to the legitimate aim pursued'.[35] In evaluating the proportionality aspect of 'necessity' in the context of an interference that is occasioned by the processing of personal information, the Court generally asks whether the public interest in processing the information outweighs the individual's right to privacy. There is thus a kind of balancing test applied.

At the same time, the Court accords the nation state a 'margin of appreciation' for making the judgement of what is necessary in the particular circumstances of the case. This doctrine allows the Court to take into account the fact that the Convention will be interpreted differently in different signatory states. It defers to a national legislator, an administrative or judicial body or another authority the possibility of having varying views on matters and to appreciate these in different ways for cultural, historical and/or philosophical reasons.[36] The extent of the margin of appreciation varies from case to case and depends on different factors. In the *Leander* case, the Court referred to the doctrine in the following way:

> The Court recognises that the national authorities enjoy a margin of appreciation, the scope of which will depend not only on the nature of the legitimate aim pursued but also on the particular nature of the interference involved. In the instant case, the interest of the respondent state in protecting its national security must be balanced against the seriousness of the interference.[37]

The Court also held that when the measures taken are to safeguard national security, the margin of appreciation available to state parties in assessing what is necessary for fulfilling that aim is wide.[38] This may be considered an obstacle for protecting privacy rights in the biometric context whenever the purpose of 'national security' is claimed. Nevertheless, in the *Klass* case, the Court stressed that this does not mean that the contracting states enjoy unlimited discretion to subject persons within their jurisdiction to secret surveillance.[39] The Court 'being aware of the danger such a law poses of undermining or even destroying democracy on the ground of defending it, affirms that the Contracting States may not, in the name of the struggle against espionage and terrorism, adopt whatever measures they deem appropriate'.[40] This argument is of particular importance to the application of biometric technology, which is frequently claimed to combat terrorism.

The court takes into account various factors to identify whether the measure is 'necessary' enough to derogate from the privacy right in Article 8(1). In *Z. v. Finland* (1997),[41] the ECtHR held that disclosure of an individual's HIV status was only justified where there was an overriding public interest. The Court emphasised that communication or disclosure of such information without consent of the patient called for most careful scrutiny on the part of the ECtHR.[42] To evaluate the necessity of the disclosure, the Court stated that the government enjoyed a margin of appreciation in striking a fair balance between the interest of publicity of court proceedings and interests in confidentiality of personal data. Besides weighing the various interests concerned, the ECtHR examined each measure in turn in order to find out whether the measure chosen is appropriate, and found at the outset that the decision-making process did not give rise to misgivings and that remedies were apparently available for challenging the seizure and for having the limitation on the confidentiality order quashed.

In the *Klass* case, the court claimed that 'whatever system of surveillance is adopted there must be adequate and effective guarantees against abuse'.[43] This is also reaffirmed in *Malone v. United Kingdom* (1984)[44] which states that there should be adequate legal safeguards to protect against arbitrary interference with Article 8(1) rights. However, to what extent such safeguards are adequate and effective is relative: '[i]t depends on all the circumstances of the case, such as the nature, scope and duration of the possible measures, the grounds required for ordering such measures, the authorities competent to permit, carry out and supervise such measures and the kind of remedy provided by the national law'.[45]

The existence of 'confidential measures' and 'an independent control body' has been highlighted as relevant for fulfilling the requirement of 'adequate safeguards'.[46] In *Klass*, based on the consideration of the German G-10 law[47] and detailed examination of other German legislation, the Court concluded that the German legislation was justified as necessary in a democratic society, in the interests of national security and for the prevention of disorder or crime. The G-10 law, which concerns surveillance, was described by the Court in the following terms:

> A series of limitative conditions have to be satisfied before a surveillance measure can be imposed. Thus, the permissible restrictive measures are confined to cases in which there are factual indications for suspecting a person of planning, committing or having committed certain serious criminal acts; measures may only be ordered if the establishment of the facts by another method is without prospects of success or considerably more difficult; even then, the surveillance may cover only the specific suspect or his presumed 'contact-persons' . . . under the law there exists an administrative procedure designed to ensure that measures are not ordered haphazardly, irregularly or without due and proper consideration.[48]

These conditions are in accordance with the other case decisions of the ECtHR. Although the G-10 law is only domestic legislation in Germany, these requirements may be a good example of when the derogation of 'national security' can be justified more generally. With regard to biometric surveillance, it implies the following:

- General surreptitious collection and use of biometric data is prohibited; it should be only targeted at one or several suspected persons where there is credible evidence for the suspicion. This outlaws the possibility of using a biometric system for general covert collection ofbiometric data, or large-scale tracking and linking of people's biometric data for the purpose of surveillance.
- There is no other alternative measure that can fulfil the same aim without considerable difficulties. This indicates that when another authentication or identification method can achieve the goal, surveillance using biometrics should be avoided.
- Administrative procedures should be designed to ensure that any invasive surveillance measures used are not ordered haphazardly, irregularly or without

due and proper consideration. This is a requirement in accordance with the previous criterion of 'in accordance with the law'.

- Technical and administrative measures have been adopted to ensure that the biometric data collected are not disclosed to unauthorised people.
- An independent control body has been established to supervise the implementation of a privacy policy for biometric systems.

4.3.1.3 Case study: S and Marper v. United Kingdom (2008)[49]

The case concerns two applicants whose DNA and fingerprint samples were taken by the police when they were arrested. The first applicant, Mr. S, was arrested and charged for attempted robbery at the age of 11 in 2001; he was acquitted later in the same year. The second applicant, Marper, was arrested on 13 March 2001 and charged with harassment of his partner. The case was formally discontinued as he and his partner had become reconciled. However, both the applicants' DNA and fingerprint samples were still stored in two databases of DNA and fingerprint information established in the United Kingdom as crime fighting tools, in which criminals and suspected criminals were to be registered. Both applicants asked for their fingerprints and DNA samples to be destroyed, but in both cases the police refused. Their information was retained in the databases as a potential resource for future criminal investigations under the Police and Criminal Evidence Act, 1984 and the Criminal Justice and Police Act, 2001. In England and Wales,[50] the police were then allowed to hold fingerprints and DNA samples of persons investigated for a crime for an indefinite period. The English Court of first instance upheld the police decision, and subsequent appeals to the Court of Appeal and the House of Lords were dismissed. The case then was brought before the ECtHRs (Grand Chamber). The applicants claimed that their registration was contrary to their right to privacy, as they had never been found guilty of any misdoings. The Court agreed and unanimously found that there had been a violation of Article 8 of the Convention.

This is a landmark case deciding that the blanket and indiscriminate nature of the powers of retention of the fingerprints, cellular samples and DNA profiles of persons suspected but not convicted of offences, as applied in the case of the applicants, breaches the right to privacy according to Article 8 of ECHR. The case focused on whether it was proportional to store cellular samples and biometric information such as fingerprints and DNA of innocent people which was collected legally by the police. The case discussion is particularly relevant to biometrics with respect to the discussion on:

1. The relevance of Article 8(1) in ECHR concerning the collection of fingerprints and DNA information: is there an interference with privacy? What of the biometric information of minors?
2. Possible justification of the interference under Article 8(2): were the measures taken proportional and what margin of appreciation shall be given to the states?

Below I look into the detailed arguments in this case judgement, and find out what it indicates for general collection of biometric information. Due regard will be given to the evaluation of proportionality of the measure.

4.3.1.3.1 IS THERE AN INTERFERENCE WITH PRIVACY?

To begin with, the Grand Chamber considered whether the retention of DNA samples and fingerprints amounts to an interference with Article 8(1). The Court found, in line with its previous case law, that the concept of 'private life' is a broad term *not susceptible to exhaustive definition*. It covers the physical and psychological integrity of a person. It can therefore embrace multiple aspects of the person's physical and social identity. The mere storing of data relating to the private life of an individual amounts to an interference within the meaning of Article 8. However, the Court also pointed out that in determining whether the personal information retained by the authorities involves any of the private life aspects mentioned above, the Court will have due regard to the *specific context* in which the information at issue has been recorded and retained, the nature of the records, the way in which these records are used and processed and the results that may be obtained. Concerning the retention of cellular samples and DNA profiles, the Court maintained that an individual's *mental distress* over the possible *future use* of private information retained by the authorities is legitimate and relevant to a determination of the issue of whether there has been interference.[51]

The Court observed that the DNA profiles contain substantial amounts of unique personal data, and that their processing by automated means allows the authorities to go well beyond neutral identification. The data could be used for *familial searching* with a view to identifying a possible genetic relationship between individuals. The DNA profiles' capacity to provide a means of identifying genetic relationships between individuals is in itself sufficient to conclude that their retention interferes with the right to the private life of the individuals concerned. The Court considered the use of the DNA database in searching for relatives as particularly sensitive. Concerning the fingerprint records, the Court admitted that they constituted personal data, which was decided to be comparable to personal photographs or voice samples; hence, the general approach taken by the convention organs in respect of photos and voice samples should also be followed in respect of fingerprints. The Court accordingly considered that the retention of fingerprints on the authorities' records in connection with an identified or identifiable individual may raise important private life concerns.[52]

4.3.1.3.2 CAN THE INTERFERENCE BE JUSTIFIED?

Since an interference with the right to privacy was established, the Court had to decide if the interference was justified under Article 8(2) of the ECHR. The arguments from both sides were as follows:[53]

The applicants argued that retention of that information was not justified, because the purpose for retention was vague and open to abuse; there were insufficient safeguards against misuse or abuse; the government failed to demonstrate the necessity of the provisions for crime protection; and the blanket nature of the retention irrespective of the offences involved. The retention resulted in stigmas which were particularly detrimental to children as in the case of S and to members of certain ethnic groups that are over-represented on the database.

The government argued the interference is proportional. They tried to demonstrate the efficacy of the measures with regard to crime, and that the data were stored securely and used for only limited purpose. Storage and retention were in accordance with the law and for a legitimate purpose. The retention resulted in no stigma and produced no practical consequence for the applicants unless the records matched a crime scene profile. So a fair balance was struck between privacy and the general interest of the community and fell within the state's margin of appreciation.

The Court adopted three steps to evaluate whether the retention of DNA and fingerprints was justifiable under Article 8(2). In reaching its decision, the Court considered CoE recommendations on privacy, EU directives on privacy, the practices in other EU countries and a report by the Nuffield Council on Bioethics (2007). In particular, it noted the marked difference in the law between England/Wales and Scotland.

The first step was assessment of whether the interference is in accordance with the law. The Court recalled its well-established case law that the wording 'in accordance with the law' requires the measures taken to have some basis in domestic law and to be compatible with the rule of law. This requires that the law must be adequately accessible and foreseeable.[54] The Court agreed with the applicants that at least five of the purposes for the retention were rather general and may give rise to extensive interpretation. The Court emphasised that:

> It is essential in this context . . . to have clear, detailed rules governing the scope and application of measures, as well as minimum safeguards concerning, inter alia, duration, storage, usage, access of third parties, procedures for preserving the integrity and confidentiality of data and procedures for its destruction, thus providing sufficient guarantees against the risk of abuse and arbitrariness.[55]

The Court noted that the UK law (except Scotland) was wide-ranging in nature and that there was a lack of safeguards set down against potential misuse. Accordingly, the requirement of 'in accordance with the law' was not met by the UK law.

The Government argued that the power of retention applies to all fingerprints and samples taken from a person in connection with the investigation of an offence and does not depend on innocence or guilt. The court found this difficult to reconcile with the obligation imposed by section 64(3) of the Police and Criminal Evidence Act (PACE) 1984 to destroy the fingerprints and samples of volunteers at their request, despite the similar value of the material in increasing the size and utility of the database.

The court then went on to the second step to assess whether there was a legitimate aim and agreed with the government that the retention of fingerprint and DNA information pursues the legitimate purpose of the detection and prevention of crime.

The third step, which is the most difficult, was to evaluate the necessity of such a measure. In response to the government's argument about the margin of appreciation, the Court emphasised that when considering how much discretion to give a state, it would firstly examine the importance of the right in question. In this case since there was a particularly important matter of identity at stake less discretion would be given. Secondly, the Court noted that a lack of consensus in other European states may also result in limiting the discretion. Hence, the Court considered that the margin of appreciation in this case should be limited. It also countered the government's arguments that as a pioneer in new techniques it should be given a greater margin of discretion, by noting that pioneers of new technology have an even greater responsibility to ensure rights are protected.

To evaluate the necessity of the measure, the Court mentioned:

> The interests of the data subjects and the community as a whole in protecting the personal data, including fingerprint and DNA information, may be outweighed by the legitimate interest in the prevention of crime.[56]

It noted that the United Kingdom (except for Scotland) was the only one of the contracting states to have an indefinite period of retention for all recordable offences. Further, it found that the retention was not the least intrusive measure that the state could have taken to achieve its aim. The Court also agreed with the applicants about the stigma attached to the data retention where the applicants are not guilty of a crime. The presumption of innocence and the effect on minors make indefinite retention especially harmful. Additionally, the Court found a difference of treatment between those suspected of a crime and those who volunteered samples, which undermined the government's argument that retention was not related to criminal suspicion. Finally, the Court concluded that the blanket and indiscriminate nature of retention meant that the legislation was disproportionate. The government should not be simply citing crime prevention as a generality but looking to other states and considering how to maintain a database that is proportionate and balanced.

4.3.1.3.3 RELEVANCE TO BIOMETRIC INFORMATION

The above argument shows that the Court holds, in general, a similar view on the application of Article 8 ECHR to the DNA profiles and fingerprint information as this book has proposed. This includes recognizing that the mere retention of DNA profiles and fingerprints could amount to privacy interference; the interpretation of 'in accordance with law' when biometric information is concerned; using a balancing test and least drastic means test to evaluate the proportionality issue. The concern about the social stigma attached to the undermining of the presumption of innocence has also been mentioned in this book. To avoid repetition, I will only point out some of the points that have not been adequately addressed in my previous analysis.

Margin of appreciation In this case, there is a very interesting discussion about the discretion that should be given to member states with regards to the application of Article 8, which is particularly relevant to the processing of biometric information. The crucial question was whether the interference with privacy fell within the margin of appreciation – a margin that varies depending upon 'the nature of the Convention right in issue, its importance for the individual, the nature of the interference and the object pursued by the interference'.[57] To the extent that the Court is influenced by the rough consensus among other member states, its decision might also be regarded as a way of disciplining a 'club member' that has stepped too far out of line.

Generally speaking, according to the results of the application of the present legal framework upon biometric applications, member states have a considerably large 'margin of appreciation' in determining if a specific biometric system is in conformity with the data protection directive and the domestic legislation (FIDIS, 2007). In the context of biometrics, such margin of appreciation is enhanced, according to the view of FIDIS, because the national data protection laws contain, in most countries, no specific provision or criteria on the processing of biometric data (FIDIS, 2007). As a result, the DPAs tend to use different criteria which sometimes result in conflicting decisions on similar cases. Although the WP29 has tried to provide specific guidelines on the use of biometrics in its working document on biometrics in 2003, the problems which were left unclearly addressed give the national DPAs great discretion to determine the specific context.

However, in *Marper*, the ECtHR emphasised that:

> The breadth of the margin varies and depends on a number of factors including the nature of the Convention right in issue, its importance for the individual, the nature of the interference and the object pursued by the interference. The margin will tend to be narrower where the right at stake is crucial to the individual's effective enjoyment of intimate or key rights. Where a particularly important facet of an individual's existence or identity is at stake, the margin allowed to the State will be restricted. Where, however, there is no consensus within the Member States of the Council of Europe, either as to the relative importance of the interest at stake or as to how best to protect it, the margin will be wider.[58]

Accordingly, the importance of the right concerned and the relative consensus among most member states concerning retention of DNA indicate that the United Kingdom shall be given less discretion. In contrast with the fact that the DPAs have a great margin of appreciation in deciding whether the biometric data should be processed and how it should be processed, it may be that when more biometric-related privacy cases are brought before the ECtHR the member state governments are likely to have less discretion than before. But for now the national

DPAs may, in practice, still have a large margin of appreciation because specific guidelines are lacking from either the case law of the ECtHR or, say, the policy and practice of the WP29.

Another important factor in this case is that biometric data fingerprints and DNA profiles are used in forensic science. Nowadays, there exist large automated fingerprint identification systems and many DNA databases for criminal cases. But as the data controllers are government departments and the police, the collection of biometric data is usually mandatory and not subject to authorisation of DPAs. The Court's decision on the restriction of collecting and storing biometric data from innocent people and minors may have special value for future forensic use of biometric information in Europe.

The effect on minors The collection of biometric information from children and teenagers has arisen in many different countries for various reasons.[59] Generally, the privacy concern relating to the collection of data belonging to this specific group of people is focusing on the quality of the biometric data.[60] But in this case, the Court emphasised more the social stigma and psychological implications provoked by the blanket indiscriminate retention of the relevant biometric data in the case of children. Unlike other ordinary collection of biometric data for identification or authentication for access control or membership checks, in this case the retention is by the police for possible future use of criminal detection. This may be an important factor which raises heightened concerns about the social stigma and psychological implications to children.

Will the Court's concern about the retention of children's biometric information in this specific case also extend to collection of children's biometric information in other circumstances? The answer is uncertain; nevertheless, similar objections about collecting biometric data have also been raised. For instance, some people's response to fingerprints taken in schools is that the collection may stigmatise those who have their fingerprints taken. They may feel it to be indicative of mistrust and suspicion and is identified with being 'treated like criminals' (Information Commissioner's Office, 2007). There are also concerns that fingerprinting in schools may teach children to give up important personal information, particularly biometrics, to those in authority, and that this is perfectly routine and mundane. It has been suggested that fingerprinting in schools is part of a concerted attempt to 'soften up' the younger generation for increased state privacy intrusion, including initiatives such as ID cards and DNA testing (Information Commissioner's Office, 2007). The response from the Information Commissioner's Office to these arguments is quite negative. It argued that those are actually not data protection concerns, and tried to avoid responding directly to the specific concern to children by focusing on a discussion on the general risks and data protection concerns of biometrics (Information Commissioner's Office, 2007). The general issues considered from a data protection perspective are the proportionality of introducing a biometric system and the requirement to obtain the signed consent of the student users (and their parents or guardians in the case of minors) giving them a clear and unambiguous right to opt out of the system without penalty.

The UK government response As a signatory of the ECHR, the UK Government was required to abide by the decision. That being so, in May 2009, the UK Government announced a consultation on proposals to delete innocent people's computerised DNA records and fingerprints after 12 years if they have been accused of a serious violent or sexual offence, or six years for a lesser offence (GeneWatchUK, 2009). This consultation concluded in November 2009 with a new policy response from the government which lays out the new rules on point. Data coming from adults arrested but not charged or convicted of an offence shall be deleted after 6 years, as will be the case of juveniles arrested but not charged or convicted of 'serious offences'. Juveniles arrested but not charged or convicted of other offences shall be kept for 3 years. Finally, any juvenile, without previous convictions, convicted of any offence that is not 'serious' will have their data deleted after 5 years.[61]

4.3.2 Biometrics and informational privacy

4.3.2.1 *Legal status of biometric data*

4.3.2.1.1 ANONYMISATION OF BIOMETRIC DATA

The European Data Protection Directive does not apply to anonymous data; data are only personal in the sense that they concern an identified or identifiable individual person (Article 2(a)). The notion of 'identifiable' in the Directive is very extensive. Recital 26 in the preamble indicates that data can be considered personal if the controller himself or any other person is able to identify the person(s) behind the data using reasonably likely means. Convention 108 and the OECD Guidelines have some similar provisions concerning the 'identifiability' requirement for personal data. When it comes to biometric data, the Consultative Committee of Convention 108 states that 'as soon as biometric data are collected with a view to automatic processing there is the possibility that these data can be related to an identified or identifiable individual' (The Consultative Committee of Convention 108, 2005:16). WP29 also states:

> [i]t appears that biometric data can always be considered as 'information relating to a natural person' as it concerns data, which provides, by its very nature, information about a given person. In the context of biometrical identification, the person is generally identifiable, since the biometric data are used for identification or authentication/verification at least in the sense that the data subject is distinguished from any other.
>
> (Article 29 Data Protection Working Party, 2003: 5)

However, in a footnote, the Working Party states, 'in cases where biometric data, like a template, are stored in a way that *no reasonable means* can be used by the controller or by any other person to identify the data subject, those data should not be qualified as personal data'.[62] Accordingly, Professor Grijpink states that when biometric information is used for verification, stored in an offline medium, such as a smartcard or local database, and does not contain additional identifiable information such as names, or addresses, it can be regarded as truly anonymous. Therefore, claims Grijpink, it is safe to exclude it from the constraints set by the data protection legislation (Grijpink, 2001).

Is it possible to use biometric data anonymously in certain applications, and therefore exclude them from the data protection constraints? When are data 'anonymous' in the sense of the Directive? According to the latest interpretation from WP29:

> 'Anonymous data' in the sense of the Directive can be defined as any information relating to a natural person where the person cannot be identified, whether by the data controller or by any other person, *taking account of all the means likely reasonably to be used either by the controller or by any other person* to identify that individual. 'Anonymised data' would therefore be anonymous data that previously referred to an identifiable person, but where that identification is no longer possible.
>
> (Article 29 Data Protection Working Party, 2007: 21).

Anonymisation is seen as particularly beneficial when processing sensitive personal data, and the Directive 95/46/EC makes a specific reference to the logical exemption of anonymised data from the scope of the provisions, claiming that 'the principle of protection shall not apply to data rendered anonymous in such a way that [the] data subject is no longer identifiable'. This is the legal basis of Grijpink's arguments. Nevertheless, it is critical to point out that these arguments are also based on another argument that the 'biometric template without anything in common with the source, cannot be regarded as a personal detail because it cannot be traced back to the person from whom the measured value originated, or it can only be done with disproportionate effort' (Grijpink, 2001:154–60; Grijpink, 2005: 139). Prins also makes similar arguments that the Directive does not apply to templates on smartcards, because this kind of data is not about identified or identifiable persons (Prins, 1998).

This then seems to be contradictory to the view from Hert's report (Hert, 2005). I agree that a biometric template is also personal data under the Directive. According to Article 2(a) of the latter, 'an identifiable person is one who can be identified directly or indirectly, in particular to his physical, physiological, mental, economic, cultural or social identity'. Furthermore, it is not necessary to know the name of the person to speak of 'personal data' in the sense of the Directive. In the preparatory works for the Directive (Bygrave, 2002), the European Commission stated that the term 'personal data' should be 'as general as possible, so as to include all information concerning an identifiable individual',[63] and they added that 'the definition of personal data would also cover data such as appearance, voice, fingerprints, or genetic characteristics'.[64] Biometric data are often *in se* personal data, allowing the identification of individuals directly or indirectly. Therefore, only if special techniques are used, may it be possible to render biometric data anonymous (Kindt and Dumortier, 2005). Whether or not they are stored with a name will not affect their identifiability. On the other hand, perhaps what matters here is the supposed 'offline' situation, which hinders the linkage and comparison to some extent. However, it may not follow that linkage and comparison cannot be undertaken or realised in a local database. Biometric templates as such may be less invasive than biometric templates linked directly with traditional personal data, yet it is doubtful that 'great effort' is needed to transfer or copy one biometric template from the offline system to an online system, and making it identified through comparison.

In addition, just as Wilson has pointed out, when the papers refer to 'anonymous' biometrics, it seems to mean a template that is used without needing to *know* (or record) the owner's identity. The museum entry example given by Grijpink is illustrative.[65] The problem with this notion, of course, is that if the technique is used by the same person at multiple places and multiple applications, it is always the *same* template that is captured.[66] It is not like picking up a different temporary personal code each time. The same 'code' gets used each time, and that code (template) really maps to the biological person. So even if it is 'anonymous' for the purposes of the application, all domains using the same technique are very linkable by the data controller who holds this code, because there is no choice of code (S. Wilson, personal communication, 26 June 2007). On the other hand, it is also important to think about the question in two dimensions. The first dimension is likelihood and the second is consequence (or degree of harm done). Usually we think of 'risk' as the product of likelihood and consequence (S. Wilson, personal communication, 26 June 2007). Even if the likelihood of 're-linkage' or 'comparison' is low, the consequence of this is serious. In circumstances when such data are left unprotected, the risk is actually high.

The Norwegian Data Protection Authority has pointed out that not storing the biometric information with other kinds of traditional personal information such as names, addresses, identification numbers, etc., does not mean biometric information can be regarded as 'anonymous' according to the national data protection law (Datatilsynet, 2006). This kind of application can only be regarded as less invasive than applications that save the biometric information directly together with other traditional kinds of personal information. According to Datatilsynet, '[t]he use of biometric identifier for commonplace purposes implies therefore a real restriction in individuals' possibility to be anonymous in society' (Datatilsynet, 2006: 6).

Moreover, the choice of such biometric applications actually indicates that users usually do not have a high-security requirement for the situation, for example, access control of a school restaurant or a gym. There are obvious, less-invasive alternatives for authentication. Accordingly, these applications may not fulfil the proportionality test of the data protection requirements.

Grijpink gave an interesting example in which he draws a parallel between an anonymous fingerprint on a glass in a restaurant and a biometric template. However, they are actually different because a fingerprint saved in digitised form in a computer is much easier to preserve and reuse. It will not take much effort to identify the data, since unlike normal medical information, biometric data are created to identify people. Indeed, just as was claimed in the progress report, 'the option of making the [biometric] data anonymous is not available as biometric data, by their very nature, form an instrument to identify individuals, particularly when they are automatically processed'.[67]

Additionally, it will be relevant to mention the discussion in the UK case *R. v. Department of Health, ex parte Source Informatics Ltd* (2000).[68] The High Court, Queen's Bench Division, held that the definition of 'processing' under the Directive encompasses the process of anonymisation, concerning transforming identifiable data into non-identifiable data. Although the Court of Appeal overturned the decision of the High Court, stating 'common sense and justice alike'[69] appear to favour the contention that anonymisation of data is unobjectionable under English law. The Office of the Information Commissioner in the United Kingdom issued a guidance note on the concept of 'personal data', which took the opposite view to the Court of Appeal.[70] It states that '[i]n anonymizing personal data the data controller will be processing such data and, in respect of such processing, will still need to comply with the provisions of the Act'.[71] Therefore, the provisions of the Directive apply to this processing of personal data even if they are not applicable to the resultant, anonymised data. This view could be useful here. If we apply this rule here, then as long as we recognise that raw biometric image is personal data, the process of extracting the template and storing it separately from other sources can be regarded as 'the process of anonymisation' (if this can be called 'anonymous' as Grijpink claims), and this process is still covered by the Data Protection Directive. Nevertheless, we have to agree that the kind of application mentioned by Grijpink is somewhat privacy enhancing, and it manages to make the biometric information less directly identifiable. However, it is doubtful that it amounts to the anonymous level that can be excluded from data protection law.

4.3.2.1.2 ARE BIOMETRIC DATA SENSITIVE PERSONAL DATA?

The value of being sensitive personal data has been highlighted in various legal documents. In *Z. v. Finland* (1997), the 'highly intimate and sensitive nature of information'[72] was considered an important factor that 'calls for the most careful scrutiny on the part of the Court, as do the safeguards designed to secure an effective protection'.[73] Bygrave pointed out the nature of the data in question is relevant in assessing the necessity of a data-processing operation that interferes with an Article 8(1) right. Generally speaking, the more intimate or sensitive the data are judged to be, the more stringent will be the application of the necessity criterion (Bygrave, 1998). Article 8 of the Data Protection Directive stipulates that the processing of 'special categories of personal data' shall be prohibited, unless it belongs to the listed exceptions in that provision. Article 6 of Convention 108 also recognises that 'special categories of data' demand appropriate safeguards. In the public sector, when trying to justify the mandatory collection and unauthorised processing of such data, careful scrutiny will be needed to fulfil the criteria of necessity. In the private domain, this will indicate that in most commercial applications *explicit consent* should be gained.[74]

Article 8 of the Directive states personal data that reveal 'racial or ethnic origin, political opinions, religious or philosophical beliefs, trade-union membership, and the processing of data concerning health or sex life' are sensitive personal data. A similar provision can be found in Article 6 of Convention 108 about special categories of data, in which data related to health and race are also included. The list in the Data Protection Directive is usually considered exhaustive (Bygrave, 2002). At the same time, personal data that can reveal the above listed information should be regarded as sensitive personal data.

Prins has stated that '[i]n certain situations the use of biometrical data could imply use of sensitive personal data' (Prins, 1998: 162). The WP29 has held that some biometric data could be considered sensitive in the meaning of Article 8 of Directive 95/46/EC when they reveal racial or ethnic origin or data concerning health. It gives the example of facial images as obvious sensitive data, but fails to give further clarification. Whether processing contains sensitive data or not is a question of discretion that can only be determined by specific biometric characteristics used in the biometric application itself (Article 29 Data Protection Working Party, 2003). The Working Party has not given any definite answer about what specific criteria should be fulfilled if biometric data is to be regarded as sensitive personal data. What are the specific criteria for determining the sensitivity of biometric data? When exactly should they be regarded as revealing racial or ethnic origin or data concerning health? The guidance here is nebulous. It may be tempting, as the Working Party has suggested, leaving it to the judges to decide whether the biometric data being used are sensitive or not. However, as the technology develops, this suggestion does not seem to be as appropriate, especially when taking into account that the understanding of biometric technology is still quite limited among people other than scientific experts.

As mentioned before, biometric information consists of information, some of which is usable for authentication.[75] Information usable for authentication may include information such as genetic information – unchanging marks that are generally considered sensitive personal information. As highlighted in Chapter 3, it is reasonable to infer that with the development of science, more sensitive information will be disclosed by the application of biometric information and this information may or may not be relevant for the purpose the data was collected.

How about the biometric template, then? There may be two methods by which one may trace sensitive information from the biometric template: first, to extract the raw biometric data from the biometric template; second, to get sensitive information directly from the template.[76] So are these methods technically possible? The possibility of reconstructing raw data from a biometric template has been discussed by many scientific experts. Supposing some relevant information is available, it has been demonstrated that it is indeed possible to reconstruct those parts of the raw data used for authentication.[77] These parts, as we analysed above, may include sensitive information, so we could reasonably infer the biometric template also includes relevant sensitive information.

The second possibility of using the biometric template directly to trace sensitive information has not been discussed so far as the author has been able to discern,[78] though it has been discussed as a security issue by some researchers (Bromba, 2006). Hence, it would be presumptuous to make an assumption so early without consulting scientific expertise. The progress report of Convention 108 by the Consulting Committee[79] states, 'the choice of data . . . extracted in generating a template should avoid revealing sensitive data as, in general, these data will not be able to verify the data subject's identity or identify him or her'. This requirement seems to be contradictory to our above analysis because it can be read as indicating that it is possible to separate the 'sensitive part' of biometric data from the part used for the 'extracting template', which is aimed for verification or identification. However, according to our above analysis, the so-called 'sensitive information' is included in the 'information useful for authentication'; therefore, it would be difficult to exclude them. It has been pointed out that in almost all biometric raw data, considerable additional information is included, in particular about health. Although one would expect that in a biometric template the additional information is considerably reduced, such an assertion still needs additional verification.[80]

As I have mentioned, there is an inevitable link between health information and some biometric information, either in raw data or in the biometric template. Should it be safe to conclude that they are sensitive? The idea of treating the biometric information in general as sensitive information has not received much support. The concern here is that biometric information *per se* does not directly disclose health information, and it would take a significant technological shift to go from current biometric systems to systems that easily reveal disease or other health information. Furthermore, at present, only some of the biometric information such as fingerprint, iris and DNA has been proved to be able to disclose health-related information.

It might be relevant to mention the interesting case decided by the Norwegian Data Protection Tribunal concerning CCTV-footage (fjernsynsovervåking) inside buses. It concluded that the mere holding of the footage (lagrete kameraopptak) should not be considered as personal information (personopplysninger). Only if somebody is about to process footage to identify some individual will this processing amount to personal information.[81] The decision is based on the idea that the footage itself (like a blood sample),[82] before analysis or using certain equipment, cannot be regarded as 'identifiable', as it is the 'carrier' of the personal information, not personal information itself directly. Unlike human tissue, the footage becomes 'data' when it is linked with an individual, while human tissues can only be the carrier of personal data. By the same token, it might be reasonable to conclude that biometric information *per se* may not be regarded as sensitive information unless some measures are intended to be adopted or have been adopted for using the biometric information to disclose health or genetic information. Hence, it might be argued generally that biometric information is only *personal information*. However, when someone is processing it with the intention or consequence of generating sensitive information such as health, genetic or racial information, it becomes *sensitive personal information*.

This argument may make sense in theory; however, its practical value remains to be seen. To some extent it may restrict the private data controllers from using biometric information to generate health or racial information without explicit consent from the data subjects, thereby restricting 'function creep'.[83] Nevertheless, problems may still exist as it may be more practical and likely that the commercial entities or public sectors do not really need to generate such information *from* the biometric information, but simply use it as unique, popular and convenient key data to get such information directly. This is what has been highlighted as the potential *linking and tracking* ability of biometric data. However, this nature may arguably make biometric data *sensitive data* in general. As Bing pointed out in his famous paper on the classification of personal information:

> Certain key data (especially the personal number) are not sensitive *per se*, but derive sensitivity from the information to which one gains access through the key. This means that, in order to determine the sensitivity of key data, it is not sufficient to consider the grading this data element has been given isolated; one must also take into account what information one thereby may connect to the nexus-person. This may provide a basis for data security deliberation the submission of the key represents in itself a threat to the protection of highly sensitive information, an increased risk of undesired access to personal information.
>
> (Bing, 1972: 107–108)

Even if the health and genetic-related nature of biometric data is disregarded, biometric data still differentiate themselves from other common personal data by their potential ability to be the 'key data'. Moreover, these key data are unlike 'personal numbers' or other identification numbers, as they are not restricted by national jurisdictions or certain groups; they are numbers that are used around the world, especially when considering the possibility of realizing general interoperability of biometric data in the future. This opinion has been adopted in the California Supreme Court decision in *Perkey v. Department of Motor Vehicles* (1986) in which fingerprints were considered a kind of 'sensitive personal information'.[84] Although in the US legal context 'sensitive personal information' may have a slightly different legal meaning from its European counterparts, the Court's recognition of the sensitivity of fingerprints compared with names or addresses may still be of some value here.

According to a survey about the concept of sensitive data conducted in 2006 in the United Kingdom, up to 63 percent of respondents considered biometric data sensitive personal data, and more than one-third of respondents rated biometric, genetic and contact details as extremely sensitive, whereas only one-fifth of respondents rated data concerning racial or ethnic origin, political opinions or data concerning religious or philosophical beliefs as extremely sensitive.[85] It is suggested that the time is ripe to review the provisions of the Directive. In the twenty-first century, new concerns are arising due to technological developments (McCullagh, 2007). Considering the special characteristics of biometric data – that they are health and genetics related and that they can be used as relatively unique and universal 'key data' for getting all kinds of personal information – they should be considered 'sensitive personal data' in general, not just when regarded as health or racial data.

In concurrence with Simitis, the sensitivity of data varies from context to context, and a decision to simply include a new category of sensitive data should not be taken lightly (Simitis, 1999, cited in Bygrave, 2002). He asserts that 'sensitivity is no more perceived as an *a priori* given attribute. On the contrary, any personal datum can, depending on the purpose or the circumstances of the processing be sensitive' (Simitis, 1999). Bygrave also pointed out that as determination of sensitivity tends to be coloured by context, there must be a softening of any fixed, relatively *a priori* division of data into sensitive and non-sensitive categories (Bygrave, 2002). This view is also shared by Bing who states that '[attributing] sensitivity . . . cannot be done in the abstract, its determination will to a large degree depend on the actual content and the purpose or context to which the data is [*sic*] communicated or used' (Bing, 2008). The argument on the significance of context when determining sensitivity is more comprehensive than a definition-based approach and more likely to reflect the real concerns of different data subjects (McCullagh, 2007). However, in practice, the emphasis on context may be too abstract to implement without specific guidance. Hence, we probably still need to operate with a reference list containing specific categories of sensitive data, but this list ought to be understood as an indicative, flexible reference list. In practice, by putting the data on such a list, it may be easier to restrict their processing than rely simply on the more open-ended and vague criterion of 'context'. But some consideration must also be taken of the specifics of each case due to differences in the context and purpose of the data being used.

*4.3.2.2 Principle of proportionality**

4.3.2.2.1 UNDERSTANDING THE NOTION

The principle of proportionality is a basic principle in European Union law which is manifested in a wide range of legal instruments and judicial decisions (Craig and Burca, 2008). It is also a decisive factor in the legal review of biometric systems by the DPAs (Article 29 Data Protection Working Party, 2003). However, the WP29 gives little guidance in its working document on biometrics on how the proportionality principle must be applied to biometrics. The principle is manifest in, *inter alia*, Article 6(1) of the Data Protection Directive, the relevant parts of which are as follows:

- Personal data must be processed *fairly* and lawfully, and
- Personal data must be collected for specified, explicit and legitimate purposes and not further processed in a way incompatible with those purposes.
- Personal data must be *adequate, relevant* and *not excessive* in relation to the purposes for which they are collected and/or further processed.

A similar provision also appears in Article 5(c) of Convention 108, which states that personal data undergoing automatic processing shall be 'adequate, relevant and *not excessive* in relation to the purposes for which they are stored'.

The proportionality principle has its origins in public law, where it protects individuals against state interference, in the context of the safeguarding of fundamental human rights.[86] The principle has now also been used to decide upon interests of private parties that are in conflict. However, as has been pointed out previously, 'the precise content of the principle of proportionality is far from settled; not only does it appear to apply in different contexts, but it seems to mean different things to different courts and it also develops over time' (Ellies, 1999, introduction).

There is little doubt that the notion of proportionality is involved in the criterion of fairness that appears, for example, in Article 6(1)(a) of the Data Protection Directive (Bygrave, 2002). However, fairness is also an undefined and broad term. It could indicate that the data processing shall not intrude unreasonably upon the individual's privacy, autonomy and integrity and shall be transparent (Bygrave, 2002).

According to Bygrave, in EU law, the proportionality principle is generally recognised as having three components: *suitability, necessity and non-excessiveness* (Bygrave and Schartum, 2008). Suitability encompasses the notion of relevance – i.e. that the measure concerned is relevant to the purposes for which it is employed; this implies too that it is logically capable of achieving those purposes. As for necessity, this may be illuminated by referring to Article 7 of the Data Protection Directive, which describes, in a limitative way, the legitimate grounds on which processing of personal data is permitted. The last-mentioned ground for processing personal data (Article 7(f)) is as follows:

> Processing is necessary for the purpose of the legitimate interests pursued by the controller . . . except where such interests are overridden by the interests for fundamental rights and freedoms of the data subject which require protection under Article 1(1).

This may indicate that the most relevant explanation of the word 'proportional' may be reserved for actions whose benefits exceed their costs by a considerable margin, or put differently, 'too far' outweigh them. In contrast to the other provisions of Article 7, the provision in Article 7(f) is distinguished by its lack of any objective content; it could refer to any specific meaning arising out of a particular situation under consideration (Tranberg, 2007a). If this *balancing test* is taken to mean the weighing against one another of competing considerations, then it would seem to address the problem of proportionality in data protection law more directly than other abstract terms. In other words, when the measure chosen passes the balancing test, then it may be regarded as proportional. This explanation is consistent with the standard formula used by the CJ to decide issues of proportionality; this balancing test is often referred to as a 'test of necessity' (Tridimas, 1999).

However, an interest test can require a good deal more than balancing. A particular restraint may be justified in cost-benefit terms and still not be necessary under this test if a lesser restraint would have served the purpose just as well. This brings us to another important standard of proportionality. The proportionality principle aims to enjoin the data controllers from compromising important interests any further than is strictly necessary for the attainment of the (legitimate) purpose(s) concerned. Treating the data controllers' action as excessive or unnecessary when it goes beyond such a minimum would give the proportionality principle precisely that effect. As Tridimas has described:

> It must be established whether the measure is necessary to achieve that aim, namely, whether there are other less restrictive means capable of producing the same result.
>
> (Tridimas, 1999)

In connection with this *least drastic means* test, the proportionality principle requires that the measure chosen be the one that is most privacy-friendly to fulfil the legitimate purpose. In other words, if there is a choice between several appropriate measures, the least invasive means must be adopted. This principle should apply irrespective of the nature of the relevant measure, whether legislative or administrative, general or concrete (Tranberg, 2007a) However, a means may of course be the least drastic imaginable and still fail a balancing test (Struve, 1967). In practice, the judiciary does not always distinguish in its analysis between the balancing test and the least drastic means test; and more attention is given to the former test than the latter (Tridimas, 1999).

In the recent *Huber* case, decided by the CJ, the concept of 'necessity' in the Data Protection Directive has received a new interpretation. In this case, the Court decided that a centralised database containing personal data relating to EU citizens who are not nationals of the member state concerned is incompatible with the requirement of necessity under Article 7(e) of Directive 95/46/EC unless:

1. It contains only the data which are necessary for the application by the authorities of the legislation relating to the right of residence;
2. Its centralised nature enables that legislation to be more *effectively applied*[87] as regards Union citizens who are not nationals of that Member State.[88]

The first condition concerning the requirement of data necessary for the application could be understood as explicitly consistent with the usual requirement of proportionality. The second cumulative condition about 'effective' could be regarded as a development of the usual requirement of proportionality. It may indicate that the CJ took a relatively prudent view towards such central storage of personal data. It may also indicate that 'necessary' for at least the purposes of Article 7(e) is not to be treated as meaning 'indispensable'. Whether this understanding of the term applies to the other processing grounds laid down in Article 7 which hinge on 'necessity' is not entirely clear. In my view, it probably can be so applied, thus adding, it would seem a new element in proportionality evaluations under that Article.

Therefore, to evaluate the proportionality of a certain chosen measure we need to know such things as the nature of the purpose it aims to achieve, the availability and effectiveness of other less drastic measures, and the loss that the data subjects will suffer from implementation of such measures.

4.3.2.2.2 EUROPEAN ORGANISATIONS' OPINIONS ABOUT PROPORTIONALITY AND BIOMETRICS

The WP29, the Consultative Committee of the 108 Convention and the EDPS have commented on the proportionality principle in the biometric context.[89] The Consultative Committee seems to focus on proposing using biometric templates instead of raw biometric images, which contain less 'sensitive information' than the raw biometric data. This argument is still reasonable and useful today, although some of the facts being used are not accurate. For example, the requirement that the template 'should avoid any possible link with sensitive data' (The Consultative Committee of Convention 108, 2005) may not be realistic according to my previous analysis. Further, the statement that 'the original picture of the biometric feature cannot be reconstructed as no text can be reconstructed from a list of key words' (The Consultative Committee of Convention 108, 2005) is also incorrect. As I mentioned before, scientists have demonstrated that it is possible to reconstruct the original image from a biometric template, provided certain statistics are available. Hence, although the idea that a raw biometric image is more privacy-invasive than a biometric template is still true, biometric templates are not as safe as imagined either. This concentration on biometric template and raw data may be explained by the fact that in 2005, biometric technology's development was not yet well known and there still existed quite a lot of misunderstanding about the technology.

The WP29 emphasised that, when collecting personal biometric data, it is necessary not only to define the purpose, but also to consider questions of proportionality and legal basis from the point of view of the protection of the individual person's fundamental rights and freedoms, and in particular whether the purpose can be achieved in a less-invasive manner. The opinion from the Working Party focuses primarily on avoiding unnecessary storage or central storage of biometric data for authentication, not identification. Based on Article 20 of the Data Protection Directive concerning prior checking, the use of 'other types of application' of biometrics that require storage of biometric data for cases such as high security installations 'may need to be submitted to prior checking by the data protection authorities in accordance with national law' (Article 29 Data Protection Working Party 2003: 6). However, it fails to clearly demonstrate what is considered 'proportional' when biometrics is used for identification. Should 'identification' be included in the 'other types of applications' that need storage of biometric data, and then require prior checking? The WP29 is not clear in this respect. However, if the WP29 argues that authentication using central storage of biometric data should be checked, it follows that the identification application should also be checked as it is generally considered more privacy-invasive.[90] Additionally, if the prior checking decided the central storage of personal data is proportional, then it is acceptable.

In addition, the WP29 also asserts that biometric data may only be used if they are 'adequate, relevant and not excessive' (Article 29 Data Protection Working Party, 2003: 7). This implies a strict assessment of the necessity and proportionality of the processed data (Article 29 Data Protection Working Party, 2003). The working party states that the principle of proportionality is a decisive factor in the legal review of biometric systems by the DPAs. However there is no clear guidance on how this principle should be applied to the use of biometric data, and what should be considered as 'adequate, relevant and not excessive' when biometric technology is chosen.

Other criteria mentioned by the Working Party and the Consultative Committee are concerned with the storage length of the necessary biometric data after collection. It is highlighted that biometric data should not be stored longer than necessary (Article 29 Data Protection Working Party, 2003; The Consultative Committee of Convention 108, 2005) Factors such as the kind of biometric data, whether to choose finger print or hand geometry and whether the biometric data can be left unintentionally[91] are also touched upon. However, the criteria for deciding in a coherent manner upon the proportionality of the use of biometrics remain unclear.

The European Data Protection Supervisor highlights the sensitive nature of biometric data and requires that the processing of biometric data should only take place after a thorough assessment of its risks and should follow a procedure allowing full democratic control. They recall the requirements of the ECHR and its case law, particularly the requirements of being 'in accordance with the law' and 'necessary in a democratic society' for the protection of important interests. They note to the additional requirements as to the quality of the legal basis for interference (it must be provided for in an accessible legislation and be foreseeable), the proportionality of any measure and the need for appropriate safeguards against abuse (European Data Protection Supervisor, 2006a).

4.3.2.2.3 PROPORTIONALITY TEST OF BIOMETRICS

The DPAs confirm in their decisions the importance of the principle of proportionality in the biometric context.[92] Generally, most DPA reviews of biometrics remain focused on whether the use of biometric identification and authentication techniques is proportional according to Article 6 of the Data Protection Directive. However, the interpretation used by the DPAs varies. For instance, in 2005, the Belgian DPA rendered a negative decision on a biometric access system, while in 2001 the Dutch DPA was willing to accept the use of a similar system provided the controller implemented nine recommendations for the processing of the biometric data (Article 29 Data Protection Working Party, 2003). When considering whether to allow the use of fingerprints to admit children to a school restaurant, the French Commission nationale de l'informatique et des libertés (CNIL) refused such use because they held that digital fingerprints would pose too many dangers for misuse; consequently the system was considered excessive. However, for the same purpose, they accepted the use of hand geometry patterns in a school refectory because they would not leave 'traces' and could not be misused for any other than the original purpose.[93] However, it seems that the UK DPA has accepted the use of fingerprints in similar circumstances, emphasizing though that it is important that the information is only used for the purposes specified when it is collected, and that it should be appreciated by those operating biometric systems that high standards of security will be needed to safeguard them. Biometric data must also be destroyed when they are no longer needed (Information Commissioner's Office, 2007).

It is difficult to sort out a common view of what constitutes necessary or proportional use of a controversial technology in particular circumstances for identification or verification. This seems to be a subjective question that is to a large extent dependent on the judiciary or the data protection authority's discretion. However, despite the subjectivity, reasonable decisions can only be made after evaluation of various objective factors. It will be meaningful to find out the various factors the judiciary or authorities use to reach different conclusions on similar questions. Below we use cases from Norway and Sweden to illustrate possible relevant legal factors.

Cases from Norway[94] According to Article 33 of Norwegian Personal Data Act, 'The Data Inspectorate may decide that the processing of data other than sensitive personal data shall also be subject to licensing, if such processing otherwise will clearly violate weighty interests relating to protection of privacy'. Biometric information is considered as a special category of personal data that should apply to this provision. Biometrics cases do not amount to a large percentage of the annual civil cases in Norway. Nevertheless, with the widespread use of the technology, there are a handful of biometric privacy cases in which the data protection law has been applied and discussed. It is clear from these decisions that the interpretation and application of the proportionality principle is at the centre of each discussion.

The Norwegian Data Inspectorate (Datatilsynet)[95] has made several similar decisions to prohibit the use of biometric technology in various biometric applications in 2006, on the basis that they are not allowed according to Article 12 of the Personal Data Act of Norway. Article 12 states:

> National identity numbers and other clear means of identification may only be used in the processing when there is an objective need for certain identification and the method is necessary to achieve such identification.[96]

The decisions of the Inspectorate were appealed to the Data Protection Tribunal (Personvernnemnda) (PVN). While the PVN upheld some of the decisions of the Data Protection Inspectorate, some were reversed.

Reversed cases:

Case 1: Tysvær Municipality[97]

Tysvær municipality uses biometric access control for all the new laptops and some of the desktop computers in its administration. The Data Inspectorate held that using smartcards combined with passwords could offer the same kind of security level. However, the PVN held that the smartcard may have some risks that the fingerprint biometric system does not have. Hence, the PVN concluded that the use of fingerprints for computer access control is necessary, and should be allowed.

Case 2: Esso Norge[98]

Esso Norway planned to use fingerprint authentication technology for access control to its oil depots in five different cities; this was to be on a voluntary basis. The Data Inspectorate found that the fingerprint was not used for identification but authentication, which is not covered by Article 12; thus it should not be allowed. In addition, they did not think the biometric application fulfilled the necessity requirement of the law. The Data Inspectorate suggested manual control with entrance certificates could be a good alternative for the application. The PVN held that Article 12 applied to both 'identification' and 'authentication', and that using the fingerprint can both authenticate and identify people. The PVN stated that the security level guaranteed by the biometric technology cannot be duplicated by other alternatives. Hence, there was, in its view, an objective requirement for accurate identification, and it was also necessary to use accurate identification methods such as a biometric application. Moreover, held the PVN, other alternatives are provided if the employees do not want to use the biometric application. The criteria of Article 12 shall not exclude consent as a legal basis of processing.

Upheld case examples:

Case 3: REMA 1000[99]

Rema 1000, a chain store in Norway, was using fingerprints for working time registration of employees. Employees needed to input their employee number or ID number when using the system for authentication. The Data Inspectorate agreed that Rema 1000 has an objective need for accurate working time registration. However, they pointed out that fingerprints were used for authentication, not for identification, as Article 12 of the Norwegian Data Protection Law requires. The Data Inspectorate also argued that the legal requirement of necessity was not fulfilled. The PVN pointed out that Article 12 can apply both to identification and authentication, as seen in case 2. On the other hand, the PVN agreed with Rema 1000 that there was an objective need for accurate identification, but questioned whether the existence of this *objective need* fulfilled the 'necessity' requirement of the law for adopting such an invasive authentication method. According to the PVN, there are other alternatives REMA 1000 could use to satisfy its need.

Case 4: Oxigeno Fitness[100]

Oxigeno Fitness wanted to use fingerprints for member access control on a voluntary basis. The Data Inspectorate stated that they could see there was an objective need for accurate identification, but the biometric measure chosen did not fulfil the 'necessity' requirement. The PVN agreed with the Data Inspectorate and-stated that the training centre could choose less-invasive and less-accurate identification measures that are adequate for achieving the security needs of the training centre.

Shortly after making its decisions in these cases, the Data Inspectorate issued a statement in which it elaborated upon the reasons for the line it took on the use of biometric technology (Flesland, 2006):

- According to Article 12 of the Norwegian Personal Data Act, the *unique and precise* identification measure is only useful when there is an *actual objective* need for ensuring identification *and* the method is *necessary* for such identification. Fingerprints, the iris and other kinds of biometric identification fall within the scope of 'unique identification measure', therefore the adoption of such measures must fulfil these strict conditions. The DPA did not find that any of the above cases fulfil these requirements of law.
- The processing of biometric data should also fulfil the requirements of Articles 8, 9 and 11 of the Personal Data Act.[101]
- It is not meaningful to distinguish the raw biometric image and the template stemming from it, provided that this code can identify certain individuals.
- It is not meaningful to state that encryption is adopted in connection with the use of biometrics and therefore makes it difficult to be misused. These are just measures that need to be adopted for security reasons when the basic objective of using such a measure fulfils the law's requirements.

The Data Inspectorate went on to state that there may be some revisions of Article 12 in the Personal Data Act in the near future so that biometrics can be used as a *verification* method when necessary. The Inspectorate's attitude towards biometric technology is in accordance with the policy statements emanating from within the EU by the WP29 and the 108 Consultative Committee:

- As a unique identification measure, biometric technology needs special attention.
- The use of biometrics is not necessary when other less intrusive alternatives are available for achieving the reasonable security purpose.
- Both biometric images and biometric templates are identifiable and relatively unique, which means they fall into the category of personal data.
- Encryption can play a valuable role in protecting privacy.
- There is a difference between biometric 'identification' and 'authentication'.

(a) Identification and authentication

Some of the above opinions of the Data Inspectorate are confirmed by the Data Protection Tribunal (PVN) in its case analysis. Nevertheless, the PVN disagrees with the Inspectorate about the scope of Article 12. In case 2, it states that Article 12 can be both applied to identification and authentication because 'identification method' is a general reference to both kinds of technologies. It notes that 'identification method' could have different interpretations. First, it could be used as a key to find information about a certain person from a collection of information like a database. Personal numbers are a good example of such identification keys. On the other hand, the PVN notes, 'identification measures' could be used for authentication after the relevant information is found. According to the PVN, Article 12 should be interpreted as covering biometric technology in general as a kind of 'clear means of identification'. Hence, claims the Tribunal, even when such an 'identification measure' is used for authentication and not identification, it should still fall within the scope of Article 12.

In my opinion, the PVN's view is quite reasonable, and here I would like to discuss it further. The preparatory work of Article 12 states:[102]

> The provision generally regulates the use of personal numbers and other clear means of identification such as fingerprint and other biometric data. Thus the identification methods shall not be used when they are unnecessary. The requirement for necessity in the first paragraph will only be fulfilled when other and less accurate identification methods, such as name, address, and customer number, are not sufficient. It is also important to find out how important the accurate identification is for the collector, i.e., what is the consequence of mistakes. In addition the needs of society shall also be considered.[103]

Based on the above explanation, it is clear that the focus of Article 12 is whether 'accurate identification means' such as biometric technology are being used when they are unnecessary, and what factors may be relevant to determine 'necessity'. The term 'identification method' mentioned here may not actually refer to the technical notion of 'identification', which is in contrast to authentication. From the context, the term 'identification' is more likely to be used in a more general sense, as the legislator might not have the exact biometric technology in mind when the legislation was drafted. Moreover, although there is no explicit wording whether it is permitted to use an 'accurate identification method' for authentication instead of identification, it also needs to be pointed out that:

- There is no absolute prohibition against using accurate identification methods when it is necessary to do so; the main thrust of this provision is to restrict the use of 'accurate identification methods', not to prohibit the use of the technology.
- Although biometric technology could also be used for authentication in addition to identification, the use of the biometric technology *per se* as a kind of 'accurate identification means' is the focus of concern in this provision, not how it is being used. In other words, the provision is concerned only with the first step (whether to adopt this technology at all), not the second step (how it should be adopted).
- As a general rule, something not mentioned in a law does not necessarily mean it is prohibited. Hence, it should not be deduced from this provision that since only the word 'identification' is mentioned, it indicates 'authentication' is prohibited, especially when the purpose of the provision is to talk about the 'accurate identification technology', not its use.
- Last but not least, the purpose of the provision and the Personal Data Act more generally is to safeguard the individual's right to privacy (see also Article 1). Biometric technology being used for authentication is generally regarded as more privacy friendly than identification, because identification usually dis-

> closes more personal information than authentication. For biometric authentication, an individual will typically consent to their biometric data being collected and they retain control of these data, such as when biometric matching takes places on a smartcard in their possession. However, in biometric identification, a centralised database is used and there is the possibility of converting identification systems. Hence, to interpret the provision as allowing biometric technology for identification when necessary, but prohibiting biometric technology for authentication generally, will be contradictory to the purpose of this legislation.

Consequently, it would be problematic for the Data Inspectorate to prohibit the use of biometric technology based on the fact that it is used for authentication. Besides, in the discussion about the interpretation of Article 12, the PVN tends to have different opinions on the evaluation of other factors, which will be mentioned in the following discussions.

In a 2008 report on the possible revision of Article 12 in the Personal Data Act, Schartum has suggested treating the processing of biometric data differently according to the purpose for which the biometric data are being used (Schartum and Bygrave, 2008). Basically, Schartum proposes that use of biometrics for the purpose of identification be more stringently regulated than use of biometrics for the purpose of authentication. The emerging question raised by this report and, indeed, the above debate over the case decisions is whether it is necessary to differentiate between identification and authentication when regulating biometrics. To answer this question, the important sub-questions to address first are: What are the differences between identification and authentication when privacy is concerned? What will be the legal value of regulatory differentiation based on such differences?

The privacy risk of identification is to a large extent linked with the tracking and linking ability of a central database. Central storage of biometric information is one important reason for identification raising more privacy risks than authentication. In contrast, biometric authentication does not require central storage of biometric data, although in some applications central databases are still used for authentication.[104] In authentication contexts, the biometric data could be stored in an ID chip and carried by the individual, hence avoiding many privacy concerns raised by central storage.

The possibility of anonymity is also relevant to privacy concerns of identification and authentication processes. Identification is aimed at finding out who someone is. In other words, in the identification process, we need to establish an identity. By contrast, authentication is concerned with finding out if a person is who they claim to be. In this process, identity is claimed by the individual and we only need to verify the claim. Accordingly, it is impossible to remain anonymous when one person is identified; too much of their personal information will be disclosed. However, in authentication, anonymity is still possible, especially when the purpose of authentication is only to prove a kind of qualification, such as membership of a club. We do not need to know who the person actually is as long as they prove that they are one of the club members.

From the above two points, it is clear that the identification process is generally more privacy invasive than authentication. There is no question that biometric identification and authentication also have the above two characteristics. But this does not necessarily apply to biometric systems *in practice*. If the biometric information is stored on a smart card held by the individual, and the individual could remain anonymous to some extent when authentication is conducted, the biometric system concerned could be regarded as more privacy friendly than other biometric identification systems. Nonetheless, this does not necessarily mean that if the biometric information is collected for authentication, it is not stored in a central database, or the individual remains anonymous. On the contrary, in practice, many biometric authentication systems still keep a central database, and the individual does not remain anonymous. For instance, in some systems, a database of biometric templates are kept; when performing an authentication task, the system will first look up the correct identity and find the related template in the database and then compare the template to the newly provided biometric sample. This is also called 1-to-1 verification (P. Bours, personal communication, 9 September 2009).[105]

Sometimes the line between authentication and identification is very thin. Generally, a biometric system can be used for both authentication and identification. While in principle it is possible to technically restrict a system to authentication, this is heavily dependent on the implementation (P. Bours, personal communication, 9 September 2009). Many biometric systems are implemented as authentication systems instead of identification systems. For instance, when one person uses a fingerprint to get access to their computer, only one template is stored on that computer; although no identity is claimed, the person is still authenticating themselves. Moreover, even if a biometric system is originally restricted to authentication, it is never totally impossible to circumvent a system. How difficult it is will to a large extent depend on the implemented security measures. Hence, the privacy risks of a certain system cannot be determined by the purpose of the system; the security measures taken and the privacy-enhancing designs are both relevant.

As a starting point, the possible consideration for providing different regulation for biometric identification and authentication is that they raise different privacy risks. However, as it is pointed out above, although such privacy risks exist, they may be influenced by a lot of different factors. In fact, the consideration for the whole system, and the way it is implemented is much more important to privacy protection than just considering the intended purpose of the application. Hence, the legal value of providing differentiated regulation on the basis of whether a system is used for biometric identification or authentication is quite limited.

(b) Necessity

The notion of 'necessary' generally denotes something that is 'indispensable, vital, essential [and] requisite'.[106] The necessity issue is the main point for finding out if a biometric system fulfils the legal requirement of proportionality. Traditionally, the balancing test and least drastic means test are applied to evaluate the proportionality. These notions are also implied in the decisions in the above Norwegian cases, though they are not explicitly noted.

The Data Inspectorate stated the following in its case decision concerning the Oslo training centre:

> The requirement of necessity in the first paragraph [of Article 12] will only be fulfilled when other or less accurate identification measure such as name, address or customer number are not sufficient. It is also important to consider the importance of such accurate identification for the user and what kind of consequences a mistake can cause. In addition, social need can also be considered.

Obviously, the first sentence above concerns the 'least drastic means' test, whereas the latter entails the idea of a 'balancing test'. The PVN holds a similar view to the Data Inspectorate about how to decide the 'necessity' issue; however, different opinions about the details have lead to different results.[107]

In case 1, because of the simple recognition of the objective requirement for security, the discussion of necessity is focused on the existence of alternatives, whereas the balancing test is not the core of debate. The Data Inspectorate and the PVN have different opinions about the respective security level of the smartcard and biometrics. The PVN agreed that the smartcard, combined with a password, can provide a high security level; however, it cannot reach the same security level that biometrics can guarantee. This actually raises an interesting discussion about the security level of biometrics and smartcards.

As demonstrated in Chapters 2 and 3, biometric technology is still in its infancy. It has been pointed out that 'despite all the hype, no biometric system can preclude identity theft. Typical false positive error rates of 0.1% are rather higher than the public might imagine, and indicate that biometrics in practice are not at all immune to impostors' (Wilson, 2005: 19). Furthermore, smartcards are a well-developed technique and 'offer the only truly effective solution to the scourges of website ghosting, phishing and, in the longer term, spam' (Wilson, 2005: 11).

Smartcards and biometrics, by themselves, provide a considerable boost to the identification and authentication mechanism of any system. The advantages of using a biometric for identification are obvious when convenience is taken into account. As the PVN pointed out, biometric identification is preferred over traditional methods because the person to be identified is required to be physically present at the point of identification, and identification based on biometric techniques obviates the need to remember a password. However, the possibility of fraud with biometric data should not be underestimated. Many people seem to ignore this, believing that biometrics is the final and perfect solution to each and every identification problem.[108] The combined use of a smartcard and password can be more accurate, more difficult to spoof and does not use any health or genetic related information. Hence, while biometrics can be useful to add features to security protocols, they must be used carefully. There is no absolute better or worse option when biometrics is compared with smartcards and passwords. Because biometrics has specific properties and weaknesses, it must be treated as a specific part of the protocol and not as a replacement for another part such as keys, smartcards or passwords (Hachez, n.d.). According to some technical experts, only a combination of smartcard and biometrics can be said to have a higher security level than either of the two (Wilson, 2005; Hachez, n.d.).

The PVN emphasised the difference between biometrics and smartcards and passwords in its case decision. The Tribunal stated that they are not convinced that a smartcard can be an alternative for an accurate identification measure, and there are risks in such measures that do not exist in fingerprint identification.[109] However, they did not mention that biometric identification also raises security risks that do not exist with smartcards and passwords. It is not clear here if the PVN simply assumes that biometrics is more secure than smartcards and passwords. The PVN's decision makes sense if the '*simple*, secure and robust'[110] requirement is considered as a reasonable requirement for the Tysvær municipality. Compared with other identification methods, biometric authentication is simpler and more convenient; simultaneously, it also could satisfy the security and robustness requirement.

In case 2, the discussion of necessity is concerned with the topic of 'consent' and alternatives. First, the PVN agreed that the oil depot of ESSO needs a high level of security considering the consequence of unauthorised people entering into it, and there exists no other unique, accurate identification method that can reach the same level of security. Then, the PVN stated that the system is voluntary, and there exists a real alternative for data subjects who do not want to use fingerprints, which is in accordance with Article 2(7) and Article 12's objective. The PVN pointed out that the general rule is that sole consent cannot be a satisfactory processing basis under Article 8 of the Personal Data Act. The PVN, however, did not exclude the possibility that there can be some special situation where consent can be a sole and valid basis for processing. In this case, where real alternatives are provided and no negative consequences will be rendered to the employees who choose not to use fingerprint, consent will be sufficient for the processing. The PVN admits that '[t]he necessity criteria of Article 12 cannot exclude consent as legal basis of processing'.[111]

Here we have to pose a question: which is more important legally, necessity or consent? Can the use of biometrics be allowed as long as it is consented to? Or can the non-consensual use of biometrics be allowed as long as it is necessary? If we have proved the use of biometrics is necessary, is consent only an added value for the application, as in case 2? I return to the important topic about consent in a later section, while here it is important to clarify the relationship between necessity and consent in Norwegian law when biometrics is concerned.

Necessity remains at the centre of the proportionality debate and is the decisive criterion in terms of Norwegian law. As Tranberg noted in her discussion of the legal framework of processing biometric personal data, only when the purpose specification and proportionality requirements are fulfilled can the processing of such data be built on consent that is voluntary, specific and informed (Tranberg,

2007b). In case 2, the PVN admitted the necessity of using biometrics in ESSO before considering the consent factor. In case 4, although the use of fingerprints for access control occurred on a voluntary basis, the use of biometrics was still prohibited because it was regarded as unnecessary. The Data Inspectorate emphasised in that case that the necessity requirement should be strict.[112] The Inspectorate found that a member card with a PIN code is sufficient to fulfil the security need of a training centre. It pointed out that the cost of spoofing the card overweighs the cost of a member's card; hence, it is not necessary to adopt the more invasive biometric measure. The PVN upheld the Data Inspectorate's decision in this case.

Compared to Article 7(a) of the Data Protection Directive, the requirement of Article 12 of the Norwegian Personal Data Act seems to set quite a high threshold for using biometrics. In the former provision, 'consent' is accepted as an independent legal basis of personal data processing. By contrast, in Article 12 of the Norwegian legislation, the requirement of 'necessity' for unique identification measures is closer to the provision in the Data Protection Directive which requires that special categories of data can be processed when 'the data subject has given his explicit consent to the processing of those data, *except* where the laws of the Member State provide that the prohibition referred to in paragraph 1 may not be lifted by the data subject's giving his consent' (Article 8(2)). Articles 8 and 9 of the Norwegian legislation lay down the basic conditions for processing personal data and sensitive personal data, where consent is accepted as an independent legal basis as in the Data Protection Directive. However, the requirement of Article 12 provides greater restriction for processing biometric data, which indicates the existence of a third group of personal data. This category of data includes 'clear means of identification', which may not fit exactly with the definition of sensitive personal data, but deserves special protection for various reasons. This idea is in accordance with my discussion in Section 3.2.1 about creating a new category of personal data for protecting privacy. Although the PVN emphasised, in its case analysis that they are not excluding 'consent' as a legal basis of personal data processing, the sole fact that the processing of biometrics is based on consent does not lead to lawful processing if the requirement of Article 12 is not fulfilled.

In case 3, Rema 1000 claimed that their objective of using fingerprints for time registration is to prohibit employees from using each other's code or password to help others register their time instead of themselves. The PVN pointed out that this intention unnecessarily renders suspicion on the employees. The introduction of fingerprint time registration is considered a stricter means of controlling the ordinary working time of employees. Such a measure would give a strong sign that the company does not trust its employees using the traditional measures, thus undermining the relationship of trust between employees and employers. In addition, the PVN agreed with the Data Inspectorate that Rema 1000 can use other, less-invasive alternatives such as employee number, ID number in combination with a password for the application. Although the Data Inspectorate agreed that passwords may be shared in some occasions, the security level reached was deemed sufficient in this case. Thus, the deployment of biometrics was not allowed.[113]

In addition to the traditional least drastic means test, the PVN in this case takes into account the influence of such systems on the relationship between employer and employees. The evaluation of the benefits of using the biometric system as an accurate authentication measure and the weakness of it undermining the 'trust' relationship between the employer and employees is another way of using the interest balance test of the proportionality principle. This shows that even in the absence of express language, the Norwegian data protection authorities generally read into the law an implied requirement of a balancing test when proportionality is concerned. This is more or less the same as the ECtHR does, where the proportionality principle is concerned. The risks and consequences of using the fingerprint authentication system are evaluated against the benefits it can bring. This concern is also implied in the other cases when the security need is considered. For instance, training centres and shops are regarded as entities where a lower security level is adequate as the consequences of unauthorised access to a training centre or the incorrect logging of times is far less serious than access to an oil depot or the Tysvær municipality IT system. The balancing test is also apparent in case 4 where the Data Inspectorate mentions the cost of spoofing the card and the cost of a real member's card.

In practice, a data protection authority is often unable or unwilling to resolve all the conflicting policy considerations in the abstract and to reduce them to a single statutory interpretation. In such cases, it may be content to spell out what it regards as the relevant considerations. In the biometric context, security levels, the cost of spoofing or the consequence of human factors can all be 'relevant considerations'. The authority need not use the term balancing in order to convey the idea that an entity should take action on a given problem that, upon giving all the relevant considerations their due weight, seems most desirable. It is not difficult to note what it means when the Data Protection Directive requires that '[p]rocessing is necessary for the purpose of the legitimate interests . . .' (Article 7(f)). A mandate to balance the relevant considerations is probably implicit in nearly all situations. For this reason, the way a data protection authority goes about making its decisions may not depend very much upon whether it is explicitly required by the legislation to 'balance' the relevant considerations when they are aware of what those relevant considerations happen to be in a specific case.

However, the ambiguity of balancing interests adds to the uncertainty of case decisions. The interpretation introduces the uncertainty of what 'unnecessary' generally means and what is necessary in a specific case. For instance, under the same national legislation, the Data Inspectorate held different views about the security needs of the oil depot and Tysvær municipality than the PVN did. The relevant considerations in the specific cases were determined to a large extent by discretion that involved, in turn, social, cultural, legal, economic and technical concerns. The ambiguity allows the data protection authorities to play a greater role in overseeing the balancing of competing considerations. It helps them to achieve the objective of the law by somewhat tangible but flexible means. Provided the authorities have the expertise to make the decisions, this discretion is both unavoidable and reasonable.

(c) Storage

The storage of biometric data is perhaps at the centre of concerns for biometric technology. Should the biometric data be stored? Should the data be stored in a centralised database? How long should they be stored? Both the WP29 and Consultative Committee suggest that the location of storage is a significant factor to decide the 'proportionality' of the biometric application (Article 29 Data Protection Working Party, 2003; The Consultative Committee of Convention 108, 2005). There are four major locations for storing biometric templates: a token or smartcard, a central database on a server and a workstation or directly on the sensing device.

While the storage problem is not the focus of discussion in the above Norwegian case decisions, it is nevertheless relevant to examine where the biometric data would be stored:

Case 1: The biometric data would be centrally kept on an encrypted server, in addition to the storage of biometric data on the sensing device. The biometric data in the central database would only be accessed for the first enrolment; afterwards, authentication is realised by comparing the live biometric template against the template stored in the sensing device.[114]

Case 2: The biometric data would also be stored in a central database that is stored separately from other personal information. The access card is used to link the template in the database to compare it against the live template for authentication.[115]

Case 3: The biometric template would be stored in the local terminal, which would not be linked to the network. The fingerprint is used both for identification and for authentication.[116]

Case 4: The biometric template is centrally stored in a local database, for identification.[117]

When reading through the case decisions about biometrics, it is not readily apparent that the either the Norwegian Data Inspectorate or the PVN followed the suggestion from the WP29 that 'authentication/verification applications which can be carried out without a central storage of biometric data should not implement excessive identification techniques' (Article 29 Data Protection Working Party, 2003: 6). Neither body discussed the central storage issue and its relevance to privacy in their case decisions. In cases 1 and 2, where the use of biometric technology is allowed, the central storage of the biometric data is deployed for authentication, which is technically unnecessary. However, the argument from the Danish Data Inspectorate about the Bornholm traffic case cited in the case decision by the PVN in case 1 is of significant value.[118] It was explicitly noted there that the fingerprint is stored only in the smartcard, which was regarded as an important reason for permitting the use of the biometric application. In another Danish case about access control in a training centre, the Danish Data Inspectorate emphasised that this case is different from the previous Bornholm case in that the biometric template is not stored in a smartcard but in a database, where the data subjects do not have a similar possibility to control their data. Hence, the application at the training centre (case 4) is more privacy-invasive.[119] The Federal German data protection authority has also handed down a favourable decision on the introduction of biometric technology on identity papers in order to prevent their falsification, provided that the data are stored in the microchip of the card rather than in a database for comparison with the owner's fingerprints (Article 29 Data Protection Working Party, 2003).

The fact that the biometric template is stored in a central database poses several risks for privacy and security.[120] It is generally accepted that the risk of reusing biometric information for 'incompatible purposes' and the possibility of protecting the biometric data against accidental or unlawful access, disclosure or alteration would be relatively lower when the information is not centrally stored.[121] The opinion of the WP29 on Council Regulation 2252/2004 argued for the prohibition of a central database of European Union passports and travel documents containing biometric and other data of all EU passport holders, stating:

> There is a risk that the setting up of a centralised database containing personal data and in particular biometric data of all (European) citizens could infringe on the basic principle of proportionality; intensifies the dangers of abuse and function creep; raises possibilities of using biometric identifiers as 'access keys' to various databases, thereby interconnecting data sets.
>
> (Article 29 Data Protection Working Party, 2005: 8)

In addition to the concern about the risks entailed by the use of central databases, it has also been criticised for being ineffective in controlling terrorist attacks. It has been pointed out that many terrorist attacks committed in recent years were perpetrated by people who had been identified as suspicious (Hert et al., 2007). This indicates that the registration of people's information does not actually help prevent attacks. National security and intelligence authorities also criticised the thought that a large central database will assist the preservation of national security.[122] As was highlighted in the London attacks of July 2005, terrorists often live in the country where they intend to strike, which excludes them from the list. Further, the great economic value of such a biometrics central database also causes much concern. Once the biometric information is stored in the database, biometrics could be used very conveniently without the knowledge of the data subjects. Granting private institutions the right to access the database allows the government to establish a paid service; this would be a huge temptation for 'function creep' (Hert et al., 2007).

In contrast, storing biometric information on a portable token such as a smartcard is often welcomed by data protection advocates, though some security concerns still exist.[123] The biometric data are not centrally stored, do not traverse the network, and the user will be able to hold their own information. Some people also promote storing the biometric data separately from other traditional personal data for authentication as in case 2; however, as was discussed in Section 3.2.1.1, this kind of application is less privacy-friendly than the smartcard storage solution.

Yet the smartcard is not perfect. A problem with smartcards is that people often do not know what is actually stored on smartcards and therefore how to access what is there, nor do they know who is going to get the information on the smartcard. A key question is this: who controls the system and is accountable for the subsequent use, disclosure, accuracy and security of the biometric data gathered and distributed via the smartcard? In cases where biometric verification is used widely for multiple applications, it is necessary to address the rights and responsibilities of the card issuers, the data controllers and the cardholders. In addition, another legal question comes up concerning the ownership of the personal data on the card. It has been argued that the data subject only has the right to use the card, but the ownership of the card still belongs to the data controller,[124] which means the latter may have the exclusive legal right to possession of the card and the personal data on the card. This indicates potential risks of abusing this property.[125] Under the data protection principles of Convention 108, and the EU Data Protection Directives, these issues have not been clearly addressed.[126]

Data controllers in the above cases tend to differentiate central storage from local storage, and assume that a local storage of biometric data is more privacy-friendly. However, the Norwegian Data Inspectorate disagrees:

> It is suggested not to differentiate the storage of biometric data in the central database and local database as this may raise the problem of defining the scopes of each. Central storage can for example be a database at national level but also a database with the employer. Storage in a system or device without network linkage will be regarded as local storage. This will however raise cases when the storage is difficult to categorise into either of these definitions. According to the knowledge of the Data Inspectorate, storage of the biometric data by the data controllers is unfortunate, and should be avoided. Therefore it is unnecessary to differentiate between local or central storage.
>
> (Datatilsynet, 2006:15).[127]

The idea of avoiding unnecessary storage is also clear in the suggestion of the WP29 that 'the mechanisms put in place to solve the problems resulting from lost, stolen or damaged cards must be taken into account in this context and those not leading to the storage of biometric data should be promoted. Whenever feasible, the data should be collected once again directly from the data subject' (Article 29 Data Protection Working Party, 2003: 7).

In its proposal for revising Article 12 of the Norwegian Personal Data Act, the Data Inspectorate explicitly suggests that when the biometric data are stored on a smartcard or other similar portable storage device, it should be allowed as long as the requirement in the first paragraph (about objective need and necessity) is fulfilled. However, when central storage or local storage happens, this should be regulated in more detail under the third paragraph (possible additional regulations by the King).[128] This approach has been criticised by Schartum who points out that this kind of central or local storage (not including smart card) is generally inevitable for processing of biometric identification or authentication

(Schartum and Bygrave, 2008). According to my understanding, what the Inspectorate intended to encourage is the storing of a biometric template in the portable token when it is only for authentication, as for identification such storage is unavoidable. But as Schartum pointed out, in practice it seldom happens, even though it is technically possible to do. This is because even when the biometric data are collected for authentication, and they are stored in the smart card held by the data subjects themselves, the data controller usually keeps the database for back up.

EU legislation leaves it up to member states to decide how biometric information will, if at all, be stored nationally. Centralised national databases of biometric data have been established in, for example, the United Kingdom and the United States, and also region-wide in Europe. While encryption is often suggested for centralised databases, the question becomes where would the encryption keys be stored and who would have access to them? If the key is to be stored by a third party for the purpose of backup, more legal parties such as the certificate authority, and other trusted third parties will be involved. Under a key escrow system, a copy of the relevant key would be provided either directly to a law enforcement agency or to a so-called 'trusted third party' that could be required to release the key to government agencies under certain circumstances. It may significantly diminish the attractiveness of encryption to users. Obviously, any involvement by a third party increases the system's vulnerability. Personal information, including biometrics, will be collected and handled during the process, and there may be security and privacy concerns (Greenleaf and Clarke, 1997). Moreover, a key escrow system would also impose significant costs on the use of encryption, especially when such systems are to be used on a global scale.

In addition to questions about the location of storage, the length of storage also raises concerns. In France, the time limit for storing genetic identification material is 40 years for convicted criminals and 25 years in other cases. The time limit is 100 years in England (National Consultative Ethics Committee for Health and Life Sciences, 2007). If such lengthy storage is going to apply to the storage of biometric data,[129] it is likely to be in strong tension with the principle of proportionality and may not be reasonably justified. As noted in the above-mentioned Norwegian cases, the length of storage is unclear, and has not become a focus of discussion in most biometric relevant cases at national levels. The WP29 suggested that biometric data should not be stored longer than necessary, without any further explanation. This introduces the uncertainties of what 'necessary' means, and what is unnecessary when length of storage is concerned in a specific case. This brings us back to the old question of deciding the balance of interests, and the least drastic means test. Whichever way is chosen, one would seem to incur significant risks. If principles are not followed, and in practice no specific guidance is given, it is very likely that the storage of biometric data occurs 'in case it becomes useful', which makes it more subject to abuse.

Case from Sweden[130] The litigation originated in a decision of 13 January 2005 by the Swedish Data Protection Authority – the Data Inspectorate (*Datainspektionen*) – to restrict the use of a fingerprint-based identification system in a secondary school in the Uddevalla municipality.[131] The system was introduced in 1997 to identify those secondary school students who were entitled to eat meals in the school canteens. The system involved scanning some 30 measurement points of one of the fingers of each student. Use of the system was not based on the prior consent of the students or their parents, though the overwhelming majority of these appeared to accept the use of the system. The few students who objected were forced to purchase food coupons, which was a more expensive alternative to the biometrics system.

The Data Inspectorate decided that the system violated the fundamental conditions for processing of personal data, as laid down by section 9(e) and (f) of the Swedish Personal Data Act (*Personuppgiftslagen*). These conditions concern the proportionality requirement of personal data processing.[132] The Inspectorate's decision was finally brought to the Supreme Administrative Court, which upheld the decision of the Administrative Court of Appeals, and decided the biometric system could be allowed based on consent.[133]

The Supreme Administrative Court (*Regeringsrätten*; hereinafter also 'SAC') considered first the issue of whether the biometric system was in breach of section 9. The Court was unanimous in finding that there was no breach. In reaching this conclusion, the Court laid weight on the following factors: (1) relatively few biometric data are registered for each student; (2) the data represent only a partial fingerprint; (3) there are established routines for 'de-registration' ('avregistrering') and for determining access to the data; (4) no evidence has been adduced to suggest that the data are unnecessary for fulfilling the purposes for which they are processed and (5) the fact that the data are biometric does not mean they cannot be seen as adequate and relevant in relation to the goals in question. The second issue considered by the Court concerned the application of section 10, in particular whether implementation of the biometric system required prior consent.

Based on my previous analysis of Norwegian case law and the EU data protection instruments, I will try to briefly analyse the SAC decision about the 'necessity' of the biometric system concerned. The second issue considered by the SAC is dealt with in Section 4.3.2.3 about consent.

The proportionality requirement of personal data processing is basically laid down in section 9 of the Swedish Personal Data Act (*Personuppgiftslagen*), especially sections 9(e) and (f). Section 9(e) stipulates that personal data must be 'adequate and relevant in relation to the purposes of the processing'; section 9(f) requires that 'no more personal data are processed than is necessary having regard to the purposes of the processing'. These requirements are in accordance with the basic meaning of proportionality as 'adequate, relevant and not excessive'.

In applying these provisions, the Swedish Data Inspectorate, who decided the biometric application is an unnecessary interference with personal integrity, adopted the least drastic means test. They pointed out that the control of access to canteen meals could be undertaken in a less intrusive way – i.e. a way that does not involve the scanning, storage and reading of fingerprints. However, this was ignored by the SAC who stated that no evidence had been adduced to suggest that the data are unnecessary for fulfilling the purposes for which they are processed;

the fact that the data are biometric does not mean they cannot be seen as adequate and relevant in relation to the goals in question. This statement shows that the SAC arbitrarily ignored the requirement in section 9(f) which emphasises the non-excessiveness of the data processing system and tried to put emphasis only on the relevance and adequacy criteria, which do not provide the complete meaning of 'necessity'. On the other hand, the SAC also recognised that there exist other less intrusive means for the application when they discussed the necessity of 'consent', but they simply ignored the importance of this fact in their discussion of 'necessity'. However, the 'relevance' and 'adequacy' of the biometric system being used has usually not been the centre of concern when a biometrics-related privacy case is raised.[134] The Data inspectorate, on the other hand, was only concerned about the 'non-excessiveness' of the system and this was central to the litigation. Hence, it is odd that the SAC placed their emphasis on the biometric system's 'relevance' and 'adequacy', and it was also inaccurate to treat, in effect, these criteria as fully constitutive of the test of 'necessity'. It is an approach that does not fit well with either the Swedish Personal Data Act or the Data Protection Directive.

Besides the above reasons, the SAC also listed three other reasons for supporting its view; the first two reasons could be regarded as concerning the allegedly small amount of biometric data being registered. However, as I emphasised before, as long as the biometric data being collected are accurate enough to identify an individual, 30 points does not make so much difference in the legal sense when it is compared with either fewer or more points being used or chosen. For the same reason, it is not legally so different whether it is a partial fingerprint, (technically speaking the biometric template) or the whole fingerprint (raw biometric image). As long as they are regarded as personal data, the collection and storage of such data should be regarded as interference with privacy. This is also the view of the WP29 as highlighted in my previous discussion.

4.3.2.3 Consent

Consent is an essential criterion for fair and lawful processing of personal data, when 'necessary' processing is not allowed. And it is often taken into account when proportionality is concerned. The processing of personal data can only be carried out with the consent of the data subject, except under certain occasions listed in law (Data Protection Directive Article 7(a)).

With regard to the consent of the data subject, Article 2(h) of the Data Protection Directive defines consent as 'any freely given specific and informed indication of his wishes by which the data subject signifies his agreement to personal data relating to him being processed'. Notice is a prerequisite for obtaining valid consent, and users must be informed. There must also be a real alternative for the data subjects that does not undermine their legal rights or cause unreasonable inconvenience. However, these rules are in some cases vague. For example, it is not specified what form an appropriate and effective notice should look like in order to obtain meaningful consent from the data subjects. In addition, to what extent can an alternative be regarded as 'reasonable' and 'real'?

It is also uncertain how the consent requirement of data protection law interacts with the proportionality principle. According to the Data Protection Directive Article 6(1), the proportionality principle (as expressed in that provision) is to be observed regardless of whether the consent is given pursuant to Article 7(a), 8(2)(a) or 26(1)(a). But the Directive also lays down preconditions for data processing in Article 7 for processing of personal data generally and Article 8 for processing of sensitive personal data. According to the wording in Article (7), 'Member States shall provide that personal data may be processed only if: (a) the data subject has unambiguously given his consent; *or* (b)-(f) processing is necessary for . . .'. (emphasis added). If we interpret 'necessary' as a manifestation of the proportionality principle, this expression clearly indicates that either consent or necessity can be the precondition for data processing.[135] However, it has also been argued that the ECtHR could take a slightly different view over the weight of the consent and proportionality in their case law. In that case law (at least that dealing with data protection issues), the proportionality principle (so far) has only been engaged in the absence of consent (Bygrave and Schartum 2009). This makes it uncertain whether and how the ECtHR would restrict data processing consented to by the data subjects but nevertheless not in conformity with a proportionality evaluation. There is no clear guidance from the WP29 or the Consultative Committee about the interaction of proportionality and consent in the biometric context; all these are left to the data protection authorities to decide at the national level.

4.3.2.3.1 THE NORWEGIAN CASES

In case 2, the PVN has affirmed that the alternative provided is real, and mentioned that the information is made to the data subjects in the form of consent about 'where and how the fingerprint will be used, and the security of the system'.[136] The Data Protection Directive lays down a number of information requirements with regard to disclosure. In order to ensure transparency, the data controller must provide his name and address, the purpose of the processing, the recipients of the data and all other information required (Articles 10 and 11). The Directive also requires that the data subjects be given information about their right to access, rectify and delete all information related to them during the processing period (recital 41). It is also important that the user has knowledge of the logic behind the data collection (Article 12), which will avoid decisions resulting from personal profiles without the user knowing the underlying valuation standard. Hence, according to these requirements, the disclosure of ESSO is adequate, and fulfils the legal requirement of 'informed consent'.

Nevertheless, the use of biometrics unavoidably entails some privacy risks, which is why the Data Inspectorate strongly restricts its usage. When the general public is not fully aware of these inevitable risks of using biometrics, and is not clearly informed in the notice from the data controllers, can such consent be regarded as 'informed'? This further raises the question of whether a notice without *adequate information* about the negative consequence of using biometrics should be regarded as *adequate information.* Professor Hert suggested in his report that 'consumer law should make clear that anyone who is asked to voluntarily submit a biometric identifier be fully informed, competent to *understand the impact of actions* and under no threat of harm to agree to such action, which means possible alternatives are provided without too much inconvenience' (Hert, 2005: 27). Here it is unclear if *impact of action* necessarily involves the possible negative risks on privacy. Neither the WP29's working document nor the Consultative Committee's progress report gives a clear answer.

Several studies have shown that people base their decisions about whether or not to participate more upon general beliefs and values than upon evaluation of the specific information provided to them in the consent form (Barr, 2006). However, when the privacy risks of biometrics are not generally well known, the exact wording of consent forms is essential to the realisation of a truly informed consent. In the health law context, it is usually advised that the donors need to understand the possible individual risks and benefits involved when donating samples and filling in questionnaires for biobanks (Brekke and Sirnes, 2006). People are not merely asking 'what is in it for me?' when deciding whether or not to participate. Their orientation is often broader, involving scientific, ethical and social aspects (Levitt and Weldon, 2005). This can be the same case when biometrics is concerned. Biometric technology and the possible technical and social risks are too complex for lay people to understand. This becomes an obstacle for people to understand the full consequences of biometric applications. Consequently, clear information about the risks of using biometric technology will be a very relevant factor to decide whether the consent is based on a truly voluntary basis. Consent obtained on the basis of the data subjects' ignorance and data controllers' misleading information is not in accordance with the legal objective of consent in the data protection laws.

The Data Inspectorate states that consent is of special value as a processing basis for personal data. However, the danger and threat of the new biometric technology is difficult to foresee. The extent to which the biometrics will affect our privacy depends to a large extent on the type of biometrics we choose, the method of collection, the length of storage and the security of the system (Datatilsynet, 2006: 15). This reasonably leads to the following question: to what does the individual give his or her consent? Can it be anything else than agreeing to give consent to an abstract, highly undefined area of application? 'Informed' can be interpreted as a trust relationship, not only between the individuals and the entities, but between the individuals and the unknown future development of the biometrics. It has been noted that a basic problem when informed consent becomes the centre of gravity for ethics is that the gaze is shifted away from possible long-term and underlying societal consequences in favour of an exclusive focus on individual self-determination (Brekke and Sirnes, 2006). In the case of biometrics, because individual users are generally unaware of the possible long-term influence of the technology, it logically follows that they are unaware of what they are consenting to. In this case, one must be quite sceptical of the extent to which 'self-determination' is realised, and to what extent such 'consent' can be regarded as a genuine 'stamp of approval', which could signify more about what is in people's minds than what it actually is supposed to represent.

In addition, while a consent form can be seen as a contract, the two parties are often very unequal in terms of knowledge base and other resources, creating an imbalance between the consenting individual on the one hand, and a major organisation or business on the other. Just as the PVN noted, some special situations can affect the quality of consent.[137] When special relationships such as between customers and enterprises, employer and employees, schools and students are involved, the asymmetry becomes more adverse, leaving data subjects relatively powerless. Therefore, if the imbalance between parties becomes too adverse, true consent becomes difficult, if not impossible. Effective measures should thus be at hand to guarantee that no adverse effect is raised. The Norwegian Data Inspectorate noted that it is still uncertain what kind of policy should be adopted concerning the notice and consent requirements in the biometric context (Datatilsynet, 2006).

Clear, informed notice is very important when consent is required. However, there are several problems with such notice. First, data collectors are usually the ones who write the notice, and thus to require unbiased notice from them is actually quite difficult in practice. Second, data subjects are usually very impatient or careless when reading such notices, and thus render the notice ineffective in protecting their individual rights. A more practical way of getting real consent from individuals may be realised by the introduction of a grading system by the data protection authority. It would also be favourable during the evaluation if the data protection authority used its own expertise and, based on their evaluation of the certain system, graded the system's privacy risk level from A to E. Usually systems that reach F are not approved at all. The data controllers should be required to hand in their privacy policy and privacy impact assessment of the system to the data protection authority. This grade should be reported to individuals when consent is sought. General explanations would be provided by the data protection authority on what each grade indicates and individuals could then make an informed choice. In this way, 'consent' could be regarded as truly informed, and it would be much easier for individuals to understand the risks they are exposed to.

According to Article 6 of the Data Protection Directive, the principle of proportionality must generally be observed regardless of whether or not consent is given pursuant to Article 7(a), 8(2) or 26(1)(a) (The Consultative Committee of Convention 108).

From the final case decisions of the PVN, it may be deduced that consent is more like an added value than an independent legal basis for data processing. This follows from the provisions of the Norwegian Personal Data Act. The quality and quantity of the biometric data being collected must be proportional to the risks associated with the particular data-processing operation for which consent is sought. A similar opinion is shown in the Greek DPA's case decision about the use of fingerprints on identity cards in Greece. In this case, fingerprints were held to be unnecessary for the purpose of verifying identity, and in addition, were held to offend human dignity. The decision further pointed out that the processing remains unlawful even in situations where the data subject has given his consent.[138] As mentioned, biometric data is regarded as a special category of personal data in Article 12. When the data controllers comply with their obligation to notify the Data Inspectorate before processing personal information,[139] the Inspectorate could decide whether a license is necessary for such processing.

In the context of biometrics, it seems that (so far) the Norwegian DPA tend to regard consent as a weaker and less central regulatory mechanism than the proportionality principle, in order to ensure a stronger level of protection to biometric data. Yet, it would be quite dangerous to assume that a stronger level of protection would necessarily result from the application of the proportionality principle. The data subjects' ability to control their personal data might also be jeopardised by over-emphasis on the 'proportionality' principle in contrast to the power of 'consent'. Just as Bygrave and Schartum have pointed out, the relative strength of the proportionality principle depends on a range of factors most notably: how is proportionality formulated? What are the parameters for assessment? Who undertakes the assessment? (Bygrave and Schartum, 2008)

The Data Inspectorate requires that the use of biometrics should be informed, and there can be an additional requirement for consent when the processing includes sensitive personal data (Datatilsynet, 2006). Accordingly, the Norwegian approach to consent and notice may be summarised as follows:

- The covert collection of biometric data is generally prohibited;
- Compulsory collection of biometric data may be allowed when the legal requirements of Articles 12, 11, 8 and 9 are fulfilled, making no distinction between the private and public sectors;
- Consent may be a reasonable legal basis for processing, but if the legal requirements of Article 5, 11 and 12 are not fulfilled, the processing of biometric data is still not allowed;
- Consent is required when the processing of sensitive personal data is involved.

4.3.2.3.2 THE SWEDISH CASE

When the SAC decided whether the implementation of the biometric system required prior consent, they mentioned especially section 10(f) of the Swedish Personal Data Act which states:

> Personal data may be processed only if the registered person has given his/her consent to the processing or if the processing is necessary in order . . . . (f) that a purpose that concerns a legitimate interest of the controller of personal data or of such a third party to whom personal data is provided should be able to be satisfied, if this interest is of greater weight than the interest of the registered person in protection against violation of personal integrity.

This provision basically replicates Article 7 of the Data Protection Directive. This section is evaluated when the processing has already been deemed to pass the 'fundamental requirements for processing of personal data', i.e. section 9. The basic issue in this context was whether the system required prior consent from the data subjects or could be implemented simply on the basis of a weighing up of interests under section 10(f), which is the equivalent of DPD Article 7(f). According to the provision, either 'consent' or 'necessity' could be an independent ground of personal data processing.

Here the SAC was divided: four judges held that consent was required, one judge held that it was not. The dissenting judge essentially followed the reasoning of the Court of Appeal, which decided that a non-consensual implementation of the system could be permitted under section 10(f):

> In balancing the need of the municipality for a flexible and simple system for checking that only those students who have paid a meal fee obtain their school meal against the students' interest in protection from violation of their personal integrity, the Court of Appeal finds that the data processing in issue does not constitute a noteworthy infringement of personal integrity, for which reason the interest of the municipality in the system weighs heavier than the interest of the students.

In contrast, the majority of the SAC noted first that the Personal Data Act builds 'in principle' upon a regime of prior consent and that exemption from such a requirement in this case may pertain only if the municipality's interest in identifying the students weighs more heavily than the students' interest in protection against violation of their personal integrity. On this point, the majority stated:

> Processing of biometric data for the purpose of identification can be regarded by many as sensitive. The particular system in question is based on the existence of data in databases that are accessible by the municipality and these data are stored over a relatively long period. The technical design of the system is accordingly such that the data processing requires taking special consideration of integrity. Against this background and given that the municipality's interest in identifying the students could be fulfilled using other methods, the Supreme Administrative Court finds that the processing may be allowed on the condition that consent is obtained.

It is interesting to note that besides mentioning several privacy relevant factors such as 'length of storage' and 'the existence of the database which the system is based on', the SAC adopted the least drastic means test as a factor to decide whether the municipality's interest weighs more heavily than the students' interest. Although they didn't explicitly state it, according to their evaluation, the municipality's interests weigh less heavily than the students' interests in protection violation of their personal integrity, and therefore fail the 'necessity' requirement in section 10(f) of the Swedish Personal Data Act. Their decision of requiring prior consent actually indicates so. Although I tend to agree with the SAC that prior consent should be sought in this application, the logic of the decision is a bit confusing.

As highlighted in Section 4.3.2.2, the least drastic means test and the balancing test are common measures that are adopted in evaluating 'necessity' issues.

However, these two tests have different contents. Failing or passing either of the tests does not necessarily indicate failing or passing the other. 'Existence of other less invasive alternatives' is a measure to evaluate a data processing system's excessiveness, not to prove that one interest outweighs another. The balancing of interests is an independent parameter for evaluating necessity. Moreover, it is also quite confusing why the SAC did not adopt these two evaluation measures where section 9 is concerned. If they did so, the final decision might have been different.

To decide whether consent is needed, the SAC actually centred its debate on the necessity issue under section 10(f). The Court majority decided finally that the biometric application concerned failed to pass the 'necessity' requirement in that section; therefore, prior consent was needed. On the other hand, they also decided that the biometric applications concerned pass the 'necessity' requirement in section 9. According to this decision, it could be deduced that: (1) the *necessity* requirement in section 9 is different from section 10 and (2) the *necessity* requirement in section 10 is normatively more important than in section 9.

Apart from this discussion, the SAC remained silent on the issue about the current alternative offered for students who refused to give out their fingerprint information. The more expensive food coupons are obviously not a 'real' alternative where 'free' consent should be guaranteed. In addition, whether notice about the risks of using biometrics is given prior to their decision is also unknown.

4.3.3 Biometrics and reasonable expectation of privacy

One can glean from the case law of the ECtHRs that the Court has actually provided some guidance on the application of the law when relevant cases are brought before it. Compared with the comprehensive protection of biometric data provided by the specific data protection instruments, the discussion of 'reasonable expectation of privacy' in public places in the EC(t)HR case law is of significant value.[140] These case decisions are based on the provisions of Article 8(1) of the ECHR concerning the right to privacy and also subject to derogation under Article 8(2).

As discussed in Chapter 3 (Section 3.2), intrusion into physical privacy may occur during information gathering of a person's identity. A breach of physical privacy may therefore be of concern when the collection of biometric information does not have a concrete goal of finding traces, or an object related to crime, but is only for generating identity information of an individual. This kind of collection is not only a breach of individuals' informational privacy but also their physical privacy. Hence, the concept of 'reasonable expectation of privacy' may encompass both kinds of privacy when biometrics is concerned.

In *P.G. and J.H. v. United Kingdom* (2001),[141] a case concerning whether the police may covertly record a conversation between a police officer and a suspect at a police station to obtain an audio sample of the suspect's voice, the applicants argued that the government had infringed upon their private lives because they did not know or have reason to suspect that the police at the station had recorded their conversations. The government countered that private life did not apply because the recordings were made to obtain voice samples not the content of the conversation, and because the 'aural quality of the applicants' voice was . . . a public external feature'.[142] However, the Court rejected the government's view, stating:

> Since there are occasions when people knowingly or intentionally involve themselves in activities which are or may be recorded or reported in a public manner. A person's reasonable expectations as to privacy may be a significant, though not necessarily conclusive factor.[143]

In this case, although the concept of 'reasonable expectation of privacy' is mentioned, the Court in fact did not frame its conclusion within the rubric of reasonable expectation of privacy where the voice sample is concerned. Instead, the Court stressed more whether and how the applicant's voice information was processed. It stated that '[p]rivate life considerations may arise . . . once any systematic or permanent record comes into existence of such material from the public domain.'[144] This focus of discussion on the processing of voice samples shows that even if individuals expect that people may hear and recognise them in the public, they could not foresee that some sort of electronic collection and analysis of their voice occurred, i.e. some sort of data processing is being undertaken. It is an important factor in this case for deciding whether there is reasonable expectation of privacy. The basic idea of this case decision may also apply to biometric features such as face and latent fingerprints in public spaces when they are covertly collected and processed for identification. Another indication of this case is that even when the collection is not of a general surveillance nature, but targets a certain suspect without any judicial procedures, the reasonable expectation of privacy may still exist under the ECHR Article 8(1).[145]

Another relevant case is *Peck v. United Kingdom* (2003).[146] This case dealt with Mr. Peck being filmed by a CCTV camera installed by a municipal authority to oversee a railway station. Peck was filmed some distance from the station, without his knowledge. The CCTV footage of him was later disclosed to the mass media, again without his knowledge, and then televised. The ECtHR held that both the initial CCTV recording of Mr. Peck and the subsequent disclosure to the mass media constituted an interference with his rights under Article 8 of the ECHR, despite the fact that Mr. Peck was filmed in a public area. As in the case *P.G. and J.H*, where the Court principally relied on whether and how personal data were processed,[147] the Court here found an individual's private rights are engaged because 'the relevant moment was viewed to an extent which far exceeded any exposure to a passer-by or to security observation . . . and to a degree surpassing that which the applicant could possibly have foreseen when he walked in [the street]'.[148]

This aspect of the decision appears to widen the scope of what is regarded as private under ECHR Article 8. The court suggests that a distinction can be drawn between different sets of public circumstances, partly on the basis of the extent to which the surveillance concerned is reasonably foreseeable on the part of the subject. As a result, public authorities – at least in Europe – cannot assume that the filming of subjects in a public place will not, under any circumstances, engage privacy rights, even when simple CCTV is being used. Despite the fact that biometric technology, even facial recognition technology, is not exactly the same as traditional CCTV, the use of biometric technology is generally followed with even more systematic and automatic processing of the data collected. Hence, it is reasonable to deduce that the simple fact that biometric characteristics are collected in a public place cannot undermine an individual's right to privacy in that public place, especially when biometrics is concerned.

4.4 Conclusions

This chapter has discussed the major legal problems raised by the use of biometric technology in the European privacy legal framework. To withstand legal and policy challenges, the current initiatives have provided some clues for shaping the legal framework that will govern biometric technology in the future. However, they are far from adequate in providing an effective remedy for dealing with the various legal problems raised by the new technology. The findings of this chapter could be categorised into three groups. The first is how the current legal provisions like the Data protection Directive, ECHR and Convention 108 should be interpreted or understood when biometric technology is discussed (*de lege lata*). These points are as follows:

- Based on the study of case law from the ECtHRs, the derogation of the privacy protection based on Article 8(2) ECHR, Article 3(2) Data Protection Directive, and article 9(2) of Convention 108 can only be applied to the use of biometric technology when there is comprehensive legal regulation of the activities that incur the interference. The regulation must involve administrative procedures designed to ensure the derogation is in accordance with the law, that measures are adopted for protecting the biometric data from unauthorised disclosure and an independent control body has been established for supervision. Second, serious violation of privacy rights can occur when biometric technology is used for surreptitious surveillance and function creep occurs. Third, general surreptitious surveillance shall be prohibited. The derogation is only permissible when no other less-invasive alternatives exist.
- The use of biometric technology implies a restriction on an individual's right to be anonymous. The anonymisation of biometric data is somewhat difficult. The storing of the biometric templates separate from other traditional personal data can only be regarded as a privacy-enhancing approach, but does not amount to anonymous data under the Data Protection Directive.
- When biometric technology is used for surveillance in a public place, it cannot be justified without specific legal evaluation, and the simple fact that the data are obtained in a 'public place' does not mean that an individual's right to privacy is not engaged.

The second group of findings which provides suggestions on special provisions for biometrics (*de lege ferenda*) is mainly concerned with the categorisation of biometric data. Biometric data ought to be categorised as a special kind of sensitive data, even if their health and genetic-related nature are disregarded. Biometric data also have the potential ability to be the unique and universal 'key data' to get all kinds of personal information. Article 12 of the Norwegian Personal Data Act treats biometric data as a special category of personal data, but fails to distinguish the technical term of 'identification' and 'authentication'. Consequently, the provision ought to be reformulated as follows: 'National identity numbers and other accurate means of *identification or authentication* may only be used in the processing when there is an objective need for certain *identification or authentication* and the method is necessary to achieve *such purpose*'.

The third group of findings is a mixture of *de lege lata* and *de lege ferenda*, and concern the understanding of, and specific regulations concerning, the principle of proportionality in the biometric context. The proportionality principle is a decisive factor in the legal review of biometric systems. The evaluation of necessity needs to be determined according to the specific criteria listed in this chapter: the benefits, risks, alternatives, inevitable need, choice of biometrics, storage of template or image, the scope and length of storage and the purpose of biometric application for identification or verification. Moreover, informed consent should include an explanation of the risks of using biometric technology; I suggest a grading system by the data protection authority to evaluate the privacy protection level and to simplify the 'notice' in practice. Consent should not be regarded as an independent legal basis for the deployment of biometrics when the proportionality requirement is not fulfilled.

One substantial indication of necessity is the absence of other (effective) alternatives, and the indispensability of the measure for fulfilling a goal. The case decisions about proportionality actually indicate that necessity is the core of the concern. The least drastic means test and balancing of interests test are basic means of interpreting 'proportionality'. Taking into account the advice from European data protection organisations concerning issues that arise in biometric applications, 'proportionality' may possibly refer to the following concerns:

- There is an inevitable strong need for a *permanent* and *unique* 'identification' or 'verification' method.
- The benefits of using biometric identification outweigh the risks that it may bring.
- No other alternative verification or identification methods (that are as effective as the biometric application) exist to fulfil the task.
- The choice of whether to use biometrics and which biometric measure to choose should be appropriate for the purpose of the applications, which may take into account whether the biometric data chosen could be traceable.
- The distinction between identification and verification is crucial. When only verification is needed, the biometric application system should not be facilitated with a central database that makes identification possible. Biometric verification, generally speaking, is less privacy-invasive because fewer personal data are involved. The need for identification with central storage needs

to be balanced with extra security safeguards on data protection and should be proportional.

- Biometric templates should be stored instead of the biometric image.
- Storage of biometric data should be avoided when unnecessary, and when storage is needed, the least invasive way of storage should be adopted for specific cases.

The first criterion stems from the characteristic that biometric technologies are based on physical or behavioural characteristics that are unique and permanent for identifying individuals. As noted in Chapter 3, the risk of biometrics is such that a unique number would facilitate searches in any database in which it resided. Supposing that the use of biometric information is not strictly restricted, or evaluated for its necessity, the various environments where one might provide biometric information in the public and private sector include banking, medical, public service, retail, training clubs and employment. The prospect of information linkage and collection is extremely problematic. This also logically follows my argument about categorizing biometric data as a third group of personal data and treating it in a special way as Article 12 of the Norwegian Personal Data Act does.

This reasonably leads us to the second criterion: the balancing of benefits and risks. This consideration is based on the balancing test. It has been pointed out by many that the use of biometrics is directly associated with various privacy and security-related concerns (Clarke 2001; Rosenzweig et al., 2004). The evaluation of the risks and benefits to a large extent has to be left to the courts and authorities to decide. It is very difficult, if not impossible, to explicitly create an absolute set of parameters for deciding when one interest shall be considered to outweigh another. Many factors including social, economic, legal and ethical consequences of the risks shall all be taken into account.

The third point is an interpretation of the least drastic means test in the biometric context. Passwords and tokens are traditional authentication methods. Cards or certificates with pictures are widely used traditional identification measures. Just like the use of smartcards, appropriate use of traditional measures may also provide high security levels. There is no absolute better or worse solution between biometrics and other traditional authentication measures when all their weaknesses and benefits are taken into account. As noted in Chapter 3, biometrics' limitations may also raise additional security risks that do not exist in traditional measures. Biometrics is by no means representative of a high security level.

The use of biometrics has often been criticised as overkill relative to the task at hand, and this suggests that we reconsider whether an invasive technology is needed or whether it is sufficient to use a password or token to verify entrance to a training centre instead. Unless the interest of convenience and intimate linkage is of such importance that it outweighs other relevant considerations, the use of biometrics should not be considered proportionate.

The fourth point is based on the consideration that each biometric technology has its strengths and limitations. As noted in Chapter 2, the applicability of a specific biometric technique depends heavily on the requirements of the application domain. For instance, CNIL preferred hand pattern over fingerprints for the biometric application of access by children to a school restaurant (Article 29 Data Protection Working Party, 2003), because hand patterns are not so easily collected from unintentionally left traces. However, the single factor of not leaving traces is not the only factor to consider. No single biometric technology offers a solution to all user requirements. Furthermore, a biometric solution for one requirement may not always be the best approach. Often, analysis of the requirement may reveal that existing solutions are adequate, or may be enhanced by other, non-biometric means (UK Biometrics Working Group, 2002). Faced with little reliable information about biometrics (vendors, products and integrators), it is quite difficult, though important, to give careful thought and consideration before making decisions.

Various factors such as cost, technical issues, general system requirements, target user groups, application environment, operational issues and system administrative concerns need to be taken into consideration.[149] When it comes to the proportionality requirement from a legal perspective, it is important to be aware that certain biometrics engender a greater perception of privacy invasion among the public than others. Also, what legal requirements must be satisfied in governing the collection and storage of the information? For instance, the size of each template (i.e. the information recorded representing a user's biometric features) may be a factor when selecting a particular biometric product for the application. In the Norwegian case decisions discussed above, this is considered relevant for evaluating the possibility of recreating the original biometric image and where the template can be stored. Both of these concerns are relevant to privacy concerns about certain biometric applications. Security and protection of the template data is also an issue. Does the level of risk or the need to protect privacy in the application warrant the encryption of templates and/or the transmission of data (UK Biometrics Working Group, 2002)? How will the solution provide for this? These will be relevant factors to examine in fulfilling the security principle of the Data Protection Directive (Article 17), and will become indirectly relevant for evaluating the 'proportionality' of certain biometric applications. The significance of the fifth and sixth points has been touched upon in the previous sections; it would be tedious to repeat this here.

Notes

* This section is based on the author's article, Liu 2009.

1 See, for example, Article 29 Data Protection Working Party (2003); Commission of the European Communities (2005), The Consultative Committee of Convention 108 (2005). The Consultative Committee established by the Council of Europe's Convention for the protection of individuals with regard to automatic processing of personal data of 1981 Herein after referred to as 'Convention 108 Consultative Committee'.

2 For example, see Kindt (2007).

3 Data Act 1973 (Datalagen, 1973: 289), fully in force 1 July 1974.

4 In 1970, the German state of Hesse enacted the first data protection statute in name. Sweden followed in 1973 with the first national statute. For a current review of the data protection instruments in Europe, European Data Protection Supervisor (2009b).

5 The Council of Europe is an intergovernmental organisation currently encompassing 47 states from the Greater European region. Established in 1949, it seeks to develop common and democratic principles throughout Europe based on the ECHR and other reference texts on the protection of individuals.

6 *Leander v. Sweden*, 26 March 1987, Series A no. 116.

7 The court stated '[b]oth the storing and release of such information, which were coupled with a refusal to allow Mr. Leander an opportunity to refute it, amounted to an interference with his right to respect for private life as guaranteed by Article 8(1)'. Id. §48.

8 *Klass and others v. Germany*, 6 September 1978, Series A no. 28.

9 See also, for example, *Huvig v. France*, 24 April 1990, Series A no. 176-B; *Kruslin v. France*, 24 April 1990, Series A no. 176-A; *Kopp v. Switzerland*, no. 23224/94, ECHR, 1998–II.

10 *Copland v. United Kingdom*, no. 62617/00, ECHR, 2007.

11 *S and Marper v. United Kingdom* [2008] ECHR 30562/04 [Grand Chamber] (4 December 2008).

12 For instance, in *von Hannover v. Germany* (2004) (no. 59320/00, ECHR 2004–VI at § 57), the Court pointed out that, although the goal of Article 8 is essentially that of protecting the individual against arbitrary interference by public authorities, it does not merely compel the state to abstain from such interference: there may be positive obligations on the state inherent in an effective respect for private or family life including the adoption of measures designed to secure respect for private life even in the sphere of the relations of individuals between themselves.

13 As of 29 January 2010 the ratifying states are Albania, Andorra, Austria, Belgium, Bosnia and Herzegovina, Bulgaria, Croatia, Cyprus, Czech Republic, Denmark, Estonia, Finland, France, Georgia, Germany, Greece, Hungary, Iceland, Ireland, Italy, Latvia, Liechtenstein, Lithuania, Luxembourg, Malta, Moldova, Monaco, Montenegro, Netherlands, Norway, Poland, Portugal, Romania, Serbia, Slovakia, Slovenia, Spain, Sweden, Switzerland, the former Yugoslav Republic of Macedonia and the United Kingdom. Three member states have signed the Convention but not yet ratified it: Russia, Turkey and Ukraine, and a further three member states have not yet signed the Convention: Armenia, Azerbaijan, and San Marino.

14 Additional Protocol to the Convention for the Protection of Individuals with regard to Automatic Processing of Personal Data regarding supervisory authorities and transborder data flows (C.E.T.S. No. 181); adopted 23 May 2001.

15 Resolution (73) 22 on the Protection of the Privacy of Individuals vis-à-vis Electronic Data Banks in the Private Sector (adopted 26 September 1973), and Resolution (74) 29 on the Protection of the Privacy of Individuals vis-à-vis Electronic Data Banks in the Public Sector (adopted 24 September 1974).

16 The Directive has been supplemented by two sector-specific Directives for the telecommunications sector. The first of these, Directive 97/66/EC of 15.12.1997 concerning the processing of personal data and the protection of privacy in the telecommunications sector (O.J. L 24, 30 January 1998, p. 1), was repealed and replaced by the second – i.e. Directive 2002/58/EC of 12 July 2002 concerning the processing of personal data and the protection of privacy in the electronic communications sector (Directive on privacy and electronic communications). On 15 March 2006, the European Union formally adopted Directive 2006/24/EC, on 'the retention of data generated or processed in connection with the provision of publicly available electronic communications services or of public communications networks and amending Directive 2002/58/EC'. The Data Retention Directive requires Member States to ensure that communications providers must retain, for a period of between 6 months and 2 years, necessary communications traffic data as specified in the Directive.

17 See, for example, Joined Cases C-465/00, C-138/01, and C-139/01 *Österreichischer Rundfunk and Others* [2003] ECRI-4989; Case C-101/01 *Bodil Lindqvist* [2003] ECR I-129711.

18 The Charter was adopted 07 December 2000 (O.J. C 364, 18 December 2000, p. 1). As of 1 December 2009, the Charter has been given the same legal status of the EU treaties by Article 6(1) of the Treaty of European Union as amended by the Treaty of Lisbon. This new status of the Charter applies to all EU member states except for the United Kingdom and Poland.

19 These guidelines were issued in 1980. They are particularly significant because they represent the first trans-Atlantic agreement relating to privacy and data protection. Like the CoE Convention, the OECD Guidelines aim to harmonise national privacy legislation and provide a framework for facilitating the international flow of data. See OECD Doc. (c 58 final) (1 January 1980). For further discussion of the Guidelines, their international significance, and their relationship to other data protection instruments, see Bygrave (2009), 26–29.

20 See Chapter 4, Kuner (2007).

21 As established in the case *von Hannover v. Germany* (2004), the ECHR may also apply to the private sector indirectly. A state party is under legal obligation to protect persons from interference with their right(s) under Art 8(1) occasioned by other persons or organisations in the private sector.

22 *Lundvall v. Sweden* (1985) Appl 10473/83, 45 DR 121.

23 The European Commission of Human Rights was a special tribunal that existed up until 1998. Individuals who wished to bring a complaint before the ECtHR had first to apply to the Commission. If the Commission found the case to be well-founded, it would launch a case in the Court on the individual's behalf. The entry into force of Protocol 11 to the Convention on 01.11.1998 abolished the Commission, enlarged the Court, and allowed individuals to take cases directly to the latter.

24 *Lundvall v. Sweden* (1985) Appl 10473/83, 45 DR 121, §130.

25 The *Sunday Times v. United Kingdom*, 6 November 1980, Series A no. 38 at §47.

26 Id. At § 49.

27 *Rotaru v. Romania* [GC], no. 28341/95, ECHR 2000–V.

28 *Kruslin v. France*, 24 April 1990, Series A no. 176-A. Another similar case is Huvig v. France, 24 April 1990, Series A no. 176-B. Mr. Huvig's telephone was tapped in the context of investigations into alleged tax fraud by the applicant and his wife. They were convicted. The issues were the same as in the Kruslin case. The applicants complained under Article 8 of the ECHR (right to respect for private life and correspondence) was limited to the alleged failure of the French law to authorise telephone tapping.

29 See further discussion in Bygrave (1998: 265).

30 In *Allan v. The United Kingdom*, no. 48539/99, ECHR 2002–IX, the applicant complained of the use of covert audio and video surveillance within his cell, the prison visiting area and upon a fellow prisoner and of the use of materials gained by these means at his trial. At that time, there existed no statutory system to regulate the use of covert recording devices by the police; the ECtHR held unanimously that there had been violations of Article 8 concerning the use of these devices.

31 The Consultative Committee established by the Council of Europe's Convention for the protection of individuals with regard to automatic processing of personal data of 1981 (2005: 24), The Consultative Committee of Convention 108 (2005: 24).

32 Id.

33 *Klass and others v. Germany*, 6 September 1978, Series A no. 28.

34 *Malone v. United Kingdom*, 2 August 1984, Series A no. 82.

35 See, for example, *Leander v. Sweden*, 26 March 1987, Series A no. 116, at §58, 59.

36 See generally Yourow (1996).

37 Id.

38 Id. At §59.

39 *Klass and others v. Germany*, 6 September 1978, Series A no. 28.

40 Id. At §49.

41 *Z. v. Finland*, no. 22009/93, ECHR 1997–I.

42 Id.
43 *Klass and others v. Germany*, 6 September 1978, Series A no. 28.
44 *Malone v. United Kingdom*, 2 August 1984, Series A no.82.
45 *Klass and others v. Germany*, 6 September 1978, Series A no. 28.§50.
46 For more discussion on this, see van Dijk and van Hoof (2006), Chapter 4.
47 The G-10 law is an important federal law in Germany. It imposes limitations on the secrecy of certain communications. The G-10 law was amended in 2001 to require that service providers give law enforcement the means to monitor data as well as voice lines.
48 *Klass and others v. Germany*, 6 September 1978, Series A no. 28 §51.
49 *S and Marper v. United Kingdom* [2008] ECHR 30562/04 [Grand Chamber] (4 December 2008).
50 The Scottish system allows DNA retention of unconvicted persons only in the case of adults charged with violent or sexual offences for a maximum of 3 years.
51 *S and Marper v. United Kingdom* [2008] ECHR 30562/04 [Grand Chamber] (4 December 2008) §70–77.
52 Ibid. at §78–86.
53 *S and Marper v. United Kingdom* [2008] ECHR 30562/04 [Grand Chamber] (4 December 2008) at pp. 24–29.
54 This is in accordance with my previous discussion on the evaluation about derogation of the article 8(1) in Section 4.3.1.2.
55 *S and Marper v. United Kingdom* [2008] ECHR 30562/04 [Grand Chamber] (4 December 2008) at §99.
56 *S and Marper v. United Kingdom* [2008] ECHR 30562/04 [Grand Chamber] (4 December 2008) at §104.
57 *S and Marper v. United Kingdom* [2008] ECHR 30562/04 [Grand Chamber] (4 December 2008) at §102.
58 *S and Marper v. United Kingdom* [2008] ECHR 30562/04 [Grand Chamber] (4 December 2008) at §102.
59 Such as the UK school, and Swedish school mentioned in previous sections.
60 I return to this discussion in Chapter 8.
61 Home Office Press Office (2009).
62 Id. Emphasis added.
63 See the Commission's commentary with respect to its amended proposal for a data protection Directive of 15.10.1992(COM (92)422final-SYN287), p. 9. See also its commentary with respect to the original Directive proposal of 24.9.1990 (COM (90)314final-SYN287) p. 15.
64 Id.
65 Grijpink (2001: 154–160). The scenario is that a visitor to the museum uses a biometric template and a pin code selected by him to authenticate. The visitor's template is captured when he enters the museum and a pin code is chosen by himself that links to the template. When he leaves the museum, he will need to re-enter the same biometrics and the code he selected to match against the saved template, if a match is found, his biometric data will be deleted. In the end, if someone's template remains in the computer, it indicates he has not left the museum. The system is described as offline and anonymous.
66 More or less the same, within the boundary of errors.
67 The Consultative Committee established by the Council of Europe's Convention for the protection of individuals with regard to automatic processing of personal data of 1981 (2005).
68 *R. v. Department of Health, ex parte Source Informatics Ltd* (2000) [2000] 1 All ER 786.
69 Id. At § 45.
70 For more discussion about the anonymisation of personal data.
71 'personal data – Definition', 14 December 2000.

72 *Z. v. Finland*, no. 22009/93, ECHR 1997–I, at §96.
73 Id.
74 This is in light of the exceptions listed out in Article 8 of Data Protection Directive.
75 For more discussion, see Chapter 3, Section 3.2.1.1.
76 This is a logical assumption by the author.
77 See, for example, Adler (2004).
78 This is based on personal knowledge, and it is quite possible that some researcher has already begun to do some relevant research.
79 The Consultative Committee of Convention 108 (2005: 19–20).
80 For a comprehensive overview of health information that can be discovered in raw data, see Kindt and Muller (2007).
81 PVN, 2005/12, Klage på vedtak om begrensninger i adgangen til kameraovervåking av bussenes publikumsområder. Personvernnemndas avgjørelse av 2. august 2006 (Jon Bing, Gro Hillestad Thune, Hanne Bjurstrøm, Tom Bolstad, Jostein Halgunset, Tore Hauglie, Leikny Øgrim).
82 For another interesting case in Norway about the blood sample as personal information, see further Bygrave (2003).
83 I wonder if the categorisation of personal data or sensitive personal data may not make too much difference for the public sector, especially when it is used for 'national security' or 'public purpose,' as the privacy restrictions can be derogated.
84 *Perkey v. Dept. of Motor Vehicles*, 42 Cal. 3d 185 (1986) at 187–188. For further discussion of the case, see Chapter 5 about biometric regulations in the United States.
85 Information Commissioner's Office (2007: 15). The Survey was conducted by telephone. All interviews were conducted in house by SMSR's WHO telephone interviewing team. The total sample was 1066 interviews.
86 This has been mentioned in the previous section discussing the derogation of the privacy right.
87 Emphasis added by the author.
88 *Heinz Huber v. Germany*, C-524/06 of 3 April 2008, § 82 (similar wording also appears in §62).
89 See further, Article 29 Data Protection Working Party (2003: 6) and The Consultative Committee of Convention 108 (2005: 18); European Data Protection Supervisor (EDPS) (2006a: 2).
90 I return to this further later in this section.
91 For instance, fingerprints are quite easily left unintentionally, while complete hand geometry is supposed to be more difficult to be left unintentionally.
92 See, for example, CINL Opinion N° 04-018 of 8 April 2004, on the request for an opinion by the hospital of Hyeres relating to the employment of a fingerprint verification application for the management of employees' time and attendance.
93 Deliberation 02-070 of 15 October 2002 (School Refectory).
94 This section is based on the author's paper Liu (2009b).
95 According to the Personal Data Act of 2000, the Data Inspectorate has been given authority to issue legally binding decisions. These decisions may be appealed to another administrative body, a quasi-judicial appeals board – the Data Protection Tribunal. The findings of these two bodies are not as authoritative as a court's finding with respect to data protection matters; however, the court would accord considerable weight to such findings. See Bygrave (2002: 18) and references cited therein.
96 This is a translation taken from the website of the Data Inspectorate. The original text is in Norwegian and is as follows: 'Fødselsnummer og andre entydige identifikasjonsmidler kan bare nyttes i behandlingen når det er saklig behov for sikker identifisering og metoden er nødvendig for å oppnå slik identifisering'.
97 PVN, 2006/07.
98 PVN, 2006/10.
99 PVN, 2006/11.

100 PVN, 2006/08, see also similar decisions and reasoning for Oslo training centre (PVN-2006–9), where Datatilsynet's decision is also upheld.
101 See, Appendix 1, Personal Data Act of Norway (2000).
102 Odelstingsproposisjon no. 92 (1998–1999), Om lov om behandling av personopplysninger (personopplysningsloven), p. 114 (in Norwegian).
103 My translation.
104 For further information, see discussion in Chapter 2.
105 E-mail consultation with Patrik Bours (Associate Professor in Information Security, Norwegian Information Security Lab, Gjøvik University College).
106 Oxford English Dictionary online, retrieved from Oxford Reference Online Premium database.
107 This builds expressly on the passage from the preparatory works to the Personal Data Act, cited earlier in this section.
108 For example, it appears that some manufacturers do not even try to protect communications between the measure device and the central unit, thus leaving the system completely open to anyone having access to that line.
109 PVN, 2006/07 p. 13.
110 PVN, 2006/07 p. 12 (emphasis added).
111 PVN, 2006/10 p. 9.
112 PVN, 2006/08 pp. 2–3.
113 It is interesting to note the thumbprint authentication system for registering working time has been used in an Australian Chain store Woolworths for years now, while it is prohibited in Norway.
114 PVN, 2006/07 p. 4.
115 PVN, 2006/10 p. 5.
116 PVN, 2006/11 p. 6.
117 PVN, 2006/08 p. 5.
118 PVN, 2006/07 p. 9 Danish Data Inspectorate Decision 4.3.2003.
119 PVN, 2006/09, p. 4 Danish Data Inspectorate Decision of 26.11.2004 case 2004-219-0208.
120 For discussion about the security risks, see Tuyls and Goseling (2004: 158–170.
121 There has been much in-depth analysis about the risks of central storage, from both a legal and technical perspective; for an example, see Article 29 Data Protection Working Party (2003: 6).
122 Interview with officials of the Dutch Agency for intelligence and security (AVID), NRC Handelsblad, 'The Public will never love us', 14 August 2006, p. 2.
123 See Svigal (1994: 107–114) for a description of the opportunities for unauthorised access to and use of smartcards such as radiation tapping and tampering with reader units.
124 One example is the Privium Smart Card, which is owned by the Schiphol Group. It is ambiguous who owns the data contents of the card. See Schiphol Group (2007: 1 and 2).
125 The idea of treating personal information as property for protection will be debated in the next section.
126 For a more thorough discussion about smartcards, see Neuwirt (2001).
127 My translation.
128 Paragraph 3 of Article 12: 'The King may by regulations prescribe further rules regarding the use of national identity numbers and other clear means of identification'.
129 There is a clear difference between genetic information and biometric data, but there are also some similarities.
130 The case summary is based on an unpublished working paper by Lee Bygrave, dated 20 January 2010.
131 Case number 1601–2004.
132 The Inspectorate took the same line in several other cases involving the application

of similar types of biometric applications to control provision of meals to school students in other municipalities. For example, decision of 09.03.2005 in case number 1976–2004 dealing with the use of a palm-scanning system in a secondary school in Värnamo municipality; decision of 15.06.2004 in case number 42–2004 dealing with a fingerprint system in a secondary school in Vetlanda municipality. For an overview of these decisions, see also Tranberg (2007b).

133 Case number 6588–05; decision of 16.12.2008.

134 This is quite clear from the previous Norwegian case discussion and the other mentioned cases in other European Countries and the United States.

135 For more discussion about consent and proportionality in general, see Bygrave and Schartum (2009).

136 PVN, 2006/10 p. 4.

137 PVN, 2006 /10 p. 4.

138 Decision 510/17/15–05–2000.

139 Article 31 of the Personal Data Act requires that: The controller shall notify the Data Inspectorate before a) processing personal data by automatic means; b) establishing a manual personal data filing system which contains sensitive personal data.

140 This is somewhat similar to the concept in the Fourth Amendment of the US Constitution (cf. Chapter 5, Section 3.2.1.2 for more discussion), while the legal basis and argument differ.

141 *P.G. and J.H. v. United Kingdom*, no. 44787/98, ECHR, 2001–IX.

142 Id. at §54.

143 Id. at §57.

144 Id.

145 For general discussion about reasonable expectation and privacy and the case law in ECHR, see Arostegui (2005).

146 *Peck v. United Kingdom*, no.44647/98, ECHR, 2003–I.

147 *P.G. and J.H. v. United Kingdom*, no. 44787/98, ECHR, 2001–IX.

148 *Peck v. United Kingdom*, no.44647/98, ECHR, 2003–I.

149 For more discussion about selecting biometric products, see generally, UK Biometrics Working Group (2002).

Reference

Adler, A. (2004) Images can be regenerated from quantised biometric match score data. In *IEEE* (ed.) *Proceedings of Canadian Conference of Electrical and Computer Engineering*, Niagara Falls: IEEE, 469–72

Article 29 Data Protection Working Party (2003). *Working Documents on Biometrics (12168/02/EN WP 80)*. Available at HTTP; <http://ec.europa.eu/justice_home/fsj/privacy/docs/wpdocs/2003/wp80_en.pdf>.

Article 29 Data Protection Working Party (2005). *Opinion 3/2005 on Implementing the Council Regulation (EC) No 2252/2004 Of 13 December 2004 on Standards for Security Features and Biometrics in Passports and Travel Documents Issued by Member States (1710/05/EN-revWP 112 04/09/12)*. Available at http://ec.europa.eu/justice_home/fsj/privacy/docs/wpdocs/2005/wp112_en.pdf.

Article 29 Data Protection Working Party (2007). *Opinion 4/2007 on the concept of personal data (01248/07/EN WP 136)*. Available at http://ec.europa.eu/justice_home/fsj/privacy/docs/wpdocs/2007/wp136_en.pdf.

Barr, M. (2006) I'm not really read up on genetics: biobanks and the social context of informed consent, *Biosocieties*, 1(1): 251–62.

Bing, J. (1972) Classification of personal information with respect to the sensitivity aspect, in K. Lenk and L.J. Blaock (eds.), *Proceedings of the First International Oslo Symposium on Data Banks and Society*, Oslo: Universitetforlaget, 98: 141

Bing, J. (2008) Sensitive personal data: An introduction, in V. María, A. Pérez, and P. Pablo (eds.) *Challenges of Privacy and Data Protection Law*, Brüssel: Bruylant, 91–208.

Brekke, O.A. and Sirnes, T. (2006) Population biobanks: the ethical gravity of informed consent, *BioSocieties* 1: 385–98.

Bunyan, T. (2004) *Slide into Authoritarianism?* Available at http://www.statewatch.org/news/2004/sep/03uk-dna-database.htm.

Bygrave, L.A. (1998) Data protection pursuant to the right to privacy in human rights treaties, *International Journal of Law and Information Technology*, 6: 247–84.

—— (2002) *Data Protection Law: Approaching its Rationale, Logic and Limits*, The Hague: Kluwer Law International.

—— (2003, September) The Body As Data? Reflections on the Relationship of Data Privacy Law With the Human Body, paper presented at conference on 'The Body as Data', Federation Centre, Melbourne, Australia.

—— (2004) Privacy protection in a global context – A comparative overview, *Scandinavian Studies in Law*, 47: 319–48.

—— (2009) International agreements to protect personal data', in J.B. Rule and G. Greenleaf (eds.) *Global Privacy Protection*, Cheltenham UK and Northampton, MA: Edward Elgar, 15–49.

Bygrave, L.A. and Schartum, D.W. (2009). Consent, proportionality and collective power, in S. Gutwirth, P. J. A. de Hert and Y. Poullet (eds.), *Reinventing Data Protection?* Heidelberg: Springer, 157–75.

Bromba, M. (2006) *On the Reconstruction of Biometric Raw Data from Template Data.* Available at http://www.bromba.com/knowhow/temppriv.htm.

Cate, F.H. (1999) The changing face of privacy protection in the European Union and the United States', *Indiana Law Review*, 33: 173–232.

Clarke, R. (2001) Biometrics and Privacy. Available at http://www.rogerclarke.com/DV/Biometrics.html.

Commission of the European Communities (2005) *Communication to the council and the European parliament on improved effectiveness, enhanced operability and synergies among European parliament on improved effectives, enhanced operability and synergies among European databases in the area of justice and home affairs (COM (2005) 597 final)*, Brussels, Belgium.

Craig, P. and Burca, G.D. (2008) *EU Law*, Oxford: Oxford University Press.

Datatilsynet (2006) *Notat Fra Datatilsynet – Forslag Til Revisjon Av Personopplysningslovens § 12. og Ny Bestemmelse Om Bruk Av Biometriske Data.* Available at http://www.datatilsynet.no/upload/Dokumenter/saker/2006/Revisjon_12_biometri.pdf.

Ellies, E. (1999) *The Principle of Proportionality in the Laws of Europe*, Oxford: Hart Publishing.

European Data Protection Supervisor (2006a). *Opinion of the European data protection supervisor on the proposal for a regulation of the European parliament and the council amending the common consular instructions on visas for diplomatic missions and consular posts in relation to the introduction of biometrics including provisions on the organisation of the reception and processing of visa applications (COM (2006) 269 final –2006/0088 (COD) (2006/C 321/14))*. Available at http://www.edps.europa.eu/EDPSWEB/webdav/shared/Documents/Consultation/Opinions/2006/06-10-27_CCI_EN.pdf.

European Data Protection Supervisor (2009b). *Data protection legislation.* Available at http://www.edps.europa.eu/EDPSWEB/edps/Home/EDPS/Dataprotection/Legislation.

FIDIS (2007) *Biometrics in Identity Management.* Available at http://www.fidis.net/resources/deliverables/hightechid/int-d37001/.

Flesland, A. (2006) *Begrenset Adgang til Bruk av Biometriske Kjennetegn.* Available at http://www.datatilsynet.no/templates/Page____1342.aspx.

GeneWatchUK (2009) *Home office DNA consultation.* Available at http://www.genewatch.org/sub-564539.

Greenleaf, G. and Clarke, R. (1997) *Privacy Implications of Digital Signature.* Available at http://www.rogerclarke.com/DV/DigSig.html.

Grijpink, J. (2001) Privacy law: biometric and privacy, *Computer Law and Security Report*, 17(3): 154–60.

—— (2005) Two barriers to realizing the benefits of biometrics, *Computer Law & Security Report*, 21(2): 138–145.

Harris, D., O'Boyle, M., Bates, E. and Buckley, C. (2009) *Law of the European Convention of Human Rights* (2nd edn), Oxford: Oxford University Press.

Hert, P.J.A. de (2005) Biometrics: Legal Issues and Implications', background Paper for the Institute of Prospective Technological Studies, DG JRC – Sevilla. Available at http://www.statewatch.org/news/2005/apr/jrc-biometrics-paul-de-hert.pdf.

Hert, P.J.A. de, Schreurs W. and Brouwer E. (2007) Machine-readable identity documents with biometric data in the EU, *Kessing Journal of Documents and Identity*, 22, 23–6.

Hert, P.J.A. de and Gutwirth, S. (2009) Data protection in the case law of Strasberg and Luxemburg: Constitutionlisation in action, in P.A.J. de. Hert and S. Gutwirth (eds.), *Reinventing Data Protection?* Heidelberg: Springer, 1–45.

Home Office Press Office (2009). *Government announces proposals for DNA retention.* Available at http://press.homeoffice.gov.uk/press-releases/proposals-dna-retention.

Hustinx P.J. (2005) *Data Protection in the European Union.* Available at http://www.edps.europa.eu/EDPSWEB/webdav/site/mySite/shared/Documents/EDPS/Publications/Speeches/2005/05-04-21_Data_Protection_EN.pdf.

Information Commissioner's Office (2007) *The use of Biometrics in Schools.* Available at http://www.ico.gov.uk/upload/documents/library/data_protection/detailed_specialist_guides/fingerprinting_final_view.pdf.

Kindt, E. (2007) Biometric applications and the data protection legislation, *Datenschutz und Datensicherheit*, 31 (3), 166–70.

Kindt, E. and Dumortier, J. (2005) Summary of Legal Data Protection Requirements for the Processing of Biometric Data (Based on D.4.2.of BioSec), *BioSec Consortium.* Available at https://www.law.kuleuven.be/icri/publications/765EBFBioSecgeneralsummaryD4%202v1%200%20Copy.pdf?where (no longer accessible).

Kindt, E. and Muller, L. (2007) *D.E.10. Biometrics in Identity Management, FIDIS.* Available at http://www.fidis.net/uploads/media/fidis-wp3del3.10.biometrics_in_identity_management.pdf.

Kuner, C. (2007). *European Data Protection Law: Corporate Regulation and Compliance*, Oxford: Oxford University Press.

Levin, A. and Nicholson, M.J. (2005) Privacy law in the United States, the EU and Canada: The Allure of the middle ground, *University of Ottawa Law & Technology Journal*, 2(2): 357–95.

Levitt, M. and Weldon, S. (2005) A well placed trust? Public perceptions of the governance of DNA databases, *Critical Public Health*, 15(4): 311–21.

Liu, Y. (2009) The Principle of Proportionality in Biometrics: Case Studies from Norway, *Computer Law and Security Report*, 2009, 25(3), 237–250.

McCullagh, K. (2007, January) Data Sensitivity: Resolving the Conundrum. Paper presented at 22nd BILETA Conference. University of Hertfordshire.

Mountfield, H. (2001) The concept of lawful interference with fundamental rights, in J. Jowell and J. Cooper (eds.), *Understanding Human Rights Principles*, Oxford and Portland: Hart Publishing, 5–31.

Prins, C. (1998) Biometric technology law, making our body identify for us: legal implications of biometric technologies, *Computer Law & Security Report*, 14(3): 159–65.

National Consultative Ethics Committee for Health and Life Sciences (2007) *Biometrics, Identifying Data and Human Rights*. Available at http://www.libertysecurity.org/IMG/pdf_avis098.pdf.

Neuwirt, K. (2001) *Council of Europe Report on the Protection of Personal Data With Regard to the Use of Smart Cards*. Available at http://www.coe.int/t/e/legal_affairs/legal_co-operation/data_protection/documents/reports%20and%20studies%20by%20experts/1Report_Neuwirt_smartcards_2001.pdf.

Rosenzweig, P., Kochems, A., and Schwartz, A. (2004) Biometric technologies: Security, legal, and policy implications. *Legal Memorandum*, 12 (June). Available at http://www.heritage.org/Research/HomelandSecurity/upload/65326_65321.pdf.

Schartum, D.W. and Bygrave, L.A. (2008) *Utredning om fødselsnummer, fingeravtrykk og annen bruk av biometri i forbindelse med lov om behandling av personopplysninger §12*. Available at http://www.regjeringen.no/pages/2213921/G-0406.pdf.

Schiphol Group (2007) *General Terms and Conditions Privium*. Available at http://www.schiphol.nl/media/portal/privium/pdf/pdf_files/alg_voorw_uk_v1_m56577569830689445.pdf.

Simitis, S. (1999) *Revisiting sensitive data*. Available at http://www.coe.int/t/e/legal_affairs/legal_co-operation/data_protection/documents/reports%20and%20studies%20by%20experts/ 1Report_Simitis_1999.pdf.

Svigal, J. (1994) Smart cards-a security assessment, *Computer and Security*, 13: 107–114.

Struve, M.G. (1967) The less-restrictive-alternative principle and economic due process, *Harvard Law Review*, 80(7): 1463–88.

The Consultative Committee of Convention 108 (2005) *Progress report on the application of the principles of Convention 108 to the Collection and Processing of Biometric Data*. Available at http://ec.europa.eu/research/science-society/pdf/3-t-pd-2005-biom_e_en.pdf.

Tranberg, C.B. (2007a) Necessary processing of personal data, *Legal Privacy and Security Issues in Information Technology*, 2: 347–60.

—— (2007b). Persondata og biometri i Skandinavien, *Lov & Data, 90* (June), 1–5.

Tridimas, T. (1999) *The General Principles of EC Law*, Oxford: Oxford University Press.

Tuyls, P. and Goseling, J. (2004, May) Capacity and examples of template protecting biometric authentication systems, in D. Maltoni and A.K. Jain (eds.) *Biometric Authentication*, Heidelberg: Springer, 158–70.

UK Biometrics Working Group (2002). *Use of Biometrics on Identification and Authentication and Advice on Selection of Products*. Available at http://www.cesg.gov.uk/site/ast/biometrics/media/BiometricsAdvice.pdf.

van Dijk, P. and van Hoof, G.J.H. (2006). *Theory and Practice of the European Convention on Human Rights*, Mortsel: Intersentia.

Walden, I. (2003) Anonymising personal data under European law, in J.E. Nicoll, J.M. Prins, and D. Hague, *Digital Anonymity and the Law: Tensions and dimensions*, The Hague: T.M.C. Asser Press, 147–59.

Wilson, S. (2005). *Lockstep Submission to Senate Privacy Inquiry*. Available at http://www.aph.gov.au/senate/committee/legcon_ctte/privacy/submissions/sub11.pdf.

Yourow, H.C. (1996). *The Margin of Appreciation Doctrine in the Dynamics of the European Court of Human Rights Jurisprudence*, London: Martinus Nijhoff Publishers, Kluwer Press.

5 Biometrics and privacy protection in the United States

5.1 Introduction

The United States government is continually developing and exploring reliable methods to ensure the country's security, and increasingly, it has turned to biometric technology. The implementation of the United States Visitor and Immigrant Status Indicator Technology program,[1] for instance, involves large-scale use of biometric technology at US airports. Additionally, the US Patriot Act,[2] which was passed soon after the September 11 attacks, encouraged agencies to spend funds on biometric authentication systems. The list of biometrics applications in the public and private sectors is almost endless.[3]

The United States has approached regulation of the use of personal information in another manner than Europe and Australia. Despite various patches of legislation concerning privacy at the federal level, it is very difficult to find systematic and comprehensive privacy protection legislation in the United States that is comparable to the European legislative regime or, for that matter, the Australian regime. Generally speaking, the United States has fairly weak laws in terms of data protection and those laws have been developed in a piecemeal, ad hoc fashion.

In the United States, different sets of privacy laws apply to various government sectors. While the Constitution applies generally to all branches of the public sector, some branches are more specifically governed by State Constitutions and sets of codified laws, whereas federal statutes govern federal government agencies. No single privacy statute governs private entities or individuals in the United States. Instead, different private industries are governed by sector-specific privacy statutes.[4] Because of the diversity and spread of privacy laws in the United States, the study in this chapter pretends in no way to be exhaustive; this chapter focuses only on aspects of the main pieces of legislation that are considered most relevant and that are most widely referred to when biometric technology and privacy are discussed. Specifically, in its discussion on privacy legislation in the public and private sectors, this chapter evaluates the Constitution, the Federal Privacy Act of 1974 and the Gramm–Leach–Bliley Act of 1999 Privacy rights recognised in tort law, as well as the proposed Personal Data Privacy and Security Act of 2009, which covers both private and public sectors, are also discussed and explored for their impact on bio-privacy issues. The analysis focuses on the substantive law aspect of federal legislation rather than enforcement issues.

Like the analysis in Part II of this book, this chapter primarily focuses on information privacy and physical privacy. These concerns are explored through relevant case law and legal sources at the federal level. Section 5.2 of this chapter gives a brief overview of the relevant sources of protection for biometric information, whereas Section 5.3 investigates both the effectiveness and inadequacy of privacy legislation with regard to biometric information. This discussion is divided into small sections that specifically focus on different pieces of legislation related to privacy in the biometric context. Section 5.4 deals with specific cases about biometrics and biometric privacy legislations at state level. The chapter concludes with a summary of the findings and some modest recommendations for more effective privacy protection for biometric technology.

5.2 Legal sources of privacy protection

5.2.1 Privacy protection under the US Constitution

There is no explicit right to privacy guaranteed in the United States Constitution. However, based on several provisions in the Bill of Rights, the Supreme Court has held that the right to privacy is one of the most fundamental constitutional rights.[5] The Court has generally held that the First, Third, Fourth, Fifth and Ninth Amendments to the Bill of Rights implicitly protect privacy.[6] The closest the Court has come to identifying a right to information privacy is in *Whalen v. Roe* (1977).[7] However, the US Constitution only provides protection against unreasonable intrusion on privacy by the government, not against intrusion by private commercial entities.[8]

5.2.2 Federal statutory privacy protection

Several federal statutes are designed to protect privacy. However, most of them only concern the public sector's intrusion into privacy.[9] The Privacy Act of 1974[10] requires federal government agencies to follow certain procedures in the collection and disclosure of the personal information that these agencies collect. The Privacy Act can generally be characterised as an omnibus 'code of fair information practices'[11] that attempts to regulate the collection, maintenance, use and dissemination of personal information by federal executive branch agencies.

The recently introduced Data Privacy and Security Act of 2009,[12] which is intended to assist consumers in better protecting the privacy of their personal information in the face of recurrent data security breaches across the country, would cover both government and commercial entities if it can be passed. Title II of the act is a formal regulatory framework for data brokers and Title III deals with Privacy and Security of Personally Identifiable information.

A patchwork of federal laws also covers some specific categories of personal information.[13] Several statutes specifically regulate the banking industry, for instance.[14] Currently, the most wide-reaching federal legislation in the United States to prevent financial institutions from buying and selling personal information without the individual's permission is the Gramm–Leach–Bliley Act of 1999,[15] also known as the Financial Modernisation Act of 1999. The Act regulates institutions and businesses that are engaged in banking, insurance, stocks and bonds, financial advice and investment. The Gramm–Leach–Bliley Act allows states to formulate protections that exceed federal law. Title V of the Act, which deals specifically with privacy, develops precautions to ensure the security and confidentiality of customer records and information to protect against any anticipated threats or hazards to the security or integrity of such records. It also aims to protect against unauthorised access to or use of such records or information that could result in substantial harm or inconvenience to any customer.

5.2.3 Actions under common law in torts

Various legal actions are also available under the common law (Prosser, 1960). The development of privacy rights under the common law began with the oft-quoted 1890 law review article by Samuel and Louis Brandeis, which characterised privacy as the 'right to be let alone' (Warren and Brandeis, 1890).

Dean William Prosser further developed the right to privacy by classifying the invasion of privacy into four distinct torts. These four categories are as follows: (1) misappropriation; (2) unreasonable intrusion upon the plaintiff's seclusion or solitude; (3) public disclosure of private facts and (4) false light in the public eye. These distinctions were later adopted in the Restatement of Torts (Second).[16] The four categories of torts each have their own detailed characteristics. An individual can file a claim related to the use of biometric information under any one of these categories. At that point, the details of what does or does not qualify as the measure of injury would become the focus of the discussion.[17]

5.3 Countering privacy risks of biometric technology through privacy law

As discussed in Part II, the main privacy concerns about biometric technology centre on possible intrusion into informational privacy[18] and physical privacy. Privacy law in the United States provides a potentially wide range of sources that support an individual's claim to informational and physical privacy. However, it is neither clear exactly how these sources will apply to address the possible privacy violations related to biometric technology, nor to what extent the resulting protection is adequate.

*5.3.1 US Constitution**

5.3.1.1 Informational privacy

As previously noted, the privacy protection offered by the relevant constitutional provisions extends only to unwarranted governmental invasions of privacy. This means a large portion of commercial biometric applications fall outside the scope of these provisions. Additionally, even when governmental use of biometric technology is the primary concern, constitutional protection is still quite limited.

The US Supreme Court addressed the issue of the right to informational privacy in *Whalen v. Roe* (1977).[19] The case concerned whether a New York statute for processing personal medical data in a centralised computer database constituted invasion of a patient's privacy. Although the court upheld the statute in the end, it recognised an individual's right to have his or her personal information remain private:

> We are not unaware of the threat to privacy implicit in the accumulation of vast amounts of personal information in computerised data banks or other massive government files. The . . . supervision of public health . . . require[s] the orderly preservation of great quantities of information, much of which is personal in character and potentially embarrassing or harmful if disclosed. The right to collect and use such data for public purposes is typically accompanied by a concomitant statutory or regulatory duty to avoid unwarranted disclosures. Recognising that in some circumstances that duty arguably has its roots in the Constitution, nevertheless New York's statutory scheme, and its implementing administrative procedures, evidence a proper concern with, and protection of, the individual's interest in privacy. . . .[20]

The leading opinion seems to indicate that prior Supreme Court decisions recognised the individual's interest in avoiding disclosure of his or her personal matters. Justice Brennan expressed his concern about central databases, and admitted they vastly increased the potential for abuse of information.[21] He suggested that there is a core right to informational privacy.[22] Although he upheld the New York statute, Justice Brennan remarked that different facts might have made a prima facie case of a constitutional violation: 'broad dissemination by state officials of such information . . . would clearly implicate constitutionally protected privacy rights and would presumably be justified only by compelling state interests'.[23] This legal reasoning may be of significant value to biometric technology, especially when the *trade off* between privacy and security is concerned.[24] Consciousness of privacy risks of a central database is particularly relevant to the controversial debate about central biometric databases. However, Brennan's view was criticised by Justice Stewart in a dissenting opinion. Stewart disavowed the notion that there is a constitutional informational privacy right and dismissed the cases Brennan cited in support of such a proposition: 'Whatever the *ratio decidendi* of Griswold, it does not recognise a general interest in freedom from disclosure of private information'.[25]

Various understandings of informational privacy appear in the justices' legal reasoning, and it is thus very difficult to conclude that the court has established a clear constitutional safeguard for informational privacy in the face of emerging biometric technologies. Neither can we predict the Supreme Court's likely approach to the constitutionality of a biometric database in the future.

There is, nevertheless, other case law related to government-mandated information requirements that touch on biometric technology. In *Thom v. New York*

Stock Exchange (1969),[26] the US District Court for the Southern District of New York upheld a state law requiring employees in the securities industry to provide fingerprints. This decision was upheld on appeal and the Supreme Court denied a review.[27] The District Court noted that fingerprints do not convey the type of information that would make their collection or dissemination constitutionally suspect: 'Possession of an individual's fingerprints does not create an atmosphere of general surveillance or indicate that they will be used for an impermissible purpose. Fingerprints provide a simple means of identification; no more'.[28] The Supreme Court also rejected the plaintiff's request for a less invasive means of identification on the grounds that there was no real invasion of privacy to begin with.[29] The Federal Court has similarly upheld the constitutionality of several other laws requiring biometric information. In *Newport v. Iacobucci* (1986)[30], the Court of Appeals for the Sixth Circuit upheld the constitutionality of a municipal ordinance that required employees to register with the Police Department and be fingerprinted, and noted that it was a valid implementation of the city's police power. In *Brown v. Brannon* (1975),[31] the US District Court for the Middle District of North Carolina upheld a similar municipal ordinance that required the fingerprinting of business license applicants. That decision was also affirmed on appeal. Interestingly, none of the courts considered the legal notion of 'informational privacy' in their decisions.

Instead, discussion of the constitutionality of collecting biometric information usually centres on interpretations of the Fourth Amendment, which concerns 'unreasonable searches and seizures'. Moreover, although the issue of a constitutional right to informational privacy was touched upon in *Whalen*, lower courts often ignore this review and refuse to read *Whalen* as settling the issue.[32] Indeed, in its decision in *Brown*, the court did not make a single approving reference to Justice Brennan's opinion regarding the possible constitutional interest 'in avoiding disclosure of personal matters'.[33] In addition, court decisions usually focus only on the constitutionality of *mandatory* collection of biometric information, which is only a small part of the information privacy concerns related to biometric technology.

5.3.1.2 The Fourth and Fifth Amendments to the US Constitution

At the outset, it should be noted that biometric technology is used by both government and private entities. Government use of biometrics is already constrained by the Fourth and Fifth Amendments to protect citizens' physical privacy. However, given the potential privacy concerns related to biometric technology use, these amendments might also be used to protect informational privacy.

5.3.1.2.1 THE FOURTH AMENDMENT

The Fourth Amendment, which governs searches and seizures conducted by government agents, ensures that 'the right of people to be secure in their persons, houses, papers and effects, against unreasonable searches and seizures, shall not be violated'. To apply the Fourth Amendment to biometric technology use, several issues must first be considered. Does the use of biometric technology constitute a 'search' or 'seizure' subject to Fourth Amendment protections? If so, is this search/ seizure reasonable? Does it fall under the scope of the administrative search doctrine,[34] and can therefore be excluded from the privacy protection argument?

5.3.1.2.2 SEARCH

In traditional jurisprudence, violations of the Fourth Amendment are found as a result of physical trespass.[35] Actual physical intrusion into the body, such as the taking of blood samples, breathalyzer testing and urine analysis, is ordinarily considered a 'search'.[36] For instance, in *Vernoniz School District 47J v. Acton* (1995),[37] the Supreme Court held that state-compelled collection and testing of urine constituted a 'search' under the Fourth Amendment. In *Skinner v. Railway Labour Executives Association*, the Supreme Court further ruled on when the collection of physical evidence should be considered a search. The Court held that 'although the collection and testing of urine under the regulations do not entail any intrusion into the body, they nevertheless constitute searches, since they intrude upon expectations of privacy as to medical information and the act of urination that society has long recognised as reasonable'.[38] The basis of the Supreme Court's holding that 'the collection and subsequent analysis of the requisite biological samples must be deemed Fourth Amendment searches' indicates that a search occurs when (1) there is intrusion into the physical body, and/or (2) collected evidence has the potential to reveal private medical facts about an individual. Thus, even when the manner of collection is not intrusive, it may still be considered a search if the information collected can reveal sensitive personal facts. A similar approach is evident in *Kyllo v. United States* (2001),[39] in which the Supreme Court held that the use of thermal imaging technology to detect the amount of heat radiating from a house was a search, even though the device could not penetrate the walls of the house.[40] Although *Kyllo* dealt with searching a person's house and not the person's body, the house in *Kyllo* is analogous to the body.[41]

This understanding of a search indicates that the collection and use of biometric information could also be considered a search under the Fourth Amendment. Although the collection of biometric data like fingerprints and retina and iris images does not require physical intrusion into a person's body, the information they contain can reveal private facts. Like the use of breath, blood and urine samples discussed in *Skinner*, biometric information could be regarded as physical evidence. Thus, the concerns addressed in *Skinner* could be relevant to biometrics.

Nevertheless, to date, the Supreme Court has held that fingerprinting is not a search under the Fourth Amendment.[42] In *Davis v. Mississippi* (1969), the defendant was detained by police and required to provide a fingerprint sample.[43] The Court held that because the detention was not authorised, the fingerprint evidence obtained during the detention was inadmissible.[44] However, the Court stated that 'fingerprinting involves none of the probing into an individual's life and thoughts that marks an interrogation or search'.[45]

Yet it is important to note that the collection and use of biometric information does not absolutely resemble 'fingerprinting', especially when biometric data are used for tracking or acquiring additional information about people. Depending on the level of integration after the collection and matching process, a biometric system could disclose as little as a person's name, or the entirety of cross-referenced data available in any linked database. This could potentially include any criminal information, tax information, credit information, health information and vehicle registration information. The data could also record an audit trail, which can disclose the time at which the match was made, the location of the biometric device making the match, etc.

The use of electronic tracking devices, which may be relevant to biometric systems, has been met with mixed decisions depending on where the devices are placed.[46] In *United States v. Knotts* (1983),[47] the Supreme Court ruled that there was no search when the government attached a beeper to a chemical container that was being transported in a car because 'scientific enhancement of this sort raised no constitutional issues which visual surveillance would also raise'.[48] However, the Supreme Court made the opposite ruling in *United States v. Karo* (1984), where a beeper was used to ascertain that a can of ether was located inside a house.[49] Despite their differences, neither decision directly discussed whether such tracking devices constitute physical intrusion. Instead, the Supreme Court was more concerned about the place where the tracking device was used.

However, both cases are relevant to possible biometric privacy issues. Indeed, commentators have focused their discussion about both cases on whether privacy was being threatened. For instance, regarding *Karo*, Tomkovicz has suggested that the installation and monitoring of an electronic tracking beacon to follow an individual or object indisputably involves a purposeful governmental quest for information and thus possesses the initial essential characteristic of a Fourth Amendment search (Tomkovicz, 2003). Although the Supreme Court tends to determine the 'search' nature of a tracking device by its location, critics have contended that the essence of a 'search' under the Fourth Amendment is whether that activity threatens legitimate needs for privacy. This idea is further supported by *Maryland v. Macon*, in which the Supreme Court established that a Fourth Amendment search 'occurs when an expectation of privacy that society is prepared to consider reasonable is infringed'.[50]

Although there are different views among the courts and commentators about how to decide whether privacy is threatened,[51] it is clear that, at least in some cases, tracking devices can constitute a 'search' if they threaten privacy, and thus the use of biometric technology could similarly be regarded if it is used for tracking. However, a more important concern is how to decide if such tracking threatens privacy; in other words, do individuals have a 'reasonable' expectation of privacy about their biometric information?

5.3.1.2.3 REASONABLENESS

A search must generally be conducted under a judicial search warrant, and thus a warrantless search is *per se* unreasonable.[52] However, sometimes the Supreme Court has held that 'searches conducted without grounds for suspicion of particular individuals have been upheld [by the courts] in "certain limited circumstances"'[53]. A vast majority of government-mandated biometric applications are performed under non-criminal circumstances. For example, biometric technology can be used to monitor airline passengers, drivers' license applicants, police officers and governmental employees. These certain limited circumstances include 'brief stops for questioning or observation at a fixed Border Patrol checkpoint, or at a sobriety checkpoint, and administrative inspections in "closely regulated" businesses'.[54] It would be doubtful if these limited circumstances can be expanded to cover the circumstances in which biometric applications are used.

To determine when a 'search' has occurred for Fourth Amendment purposes, courts have utilised the principles first enunciated in *Katz v. United States* (1967).[55] In *Katz*, the Supreme Court held that a Fourth Amendment search does not occur unless 'the individual manifested a subjective expectation of privacy in the object of the challenged search', and 'society [is] willing to recognise that expectation as reasonable'.[56] This test may be productively used for the discussion of the 'reasonableness' of biometric applications.

To what extent do individuals have a subjective expectation to not be scanned by a biometric device when they appear in public? To a large extent, the answer depends on the individual, and because of the 'subjective' nature of this expectation, it is very difficult for a third person to determine. Additionally, it is not particularly meaningful to discuss whether a plaintiff has a 'subjective' expectation of privacy since if they did not, they would not bring the case to court. Hence, the more important *Katz* principle is the following one: will a subjective expectation of privacy be accepted by the court?

There are some fairly good reasons why an individual might have a subjective expectation that his biometric information should not be collected. Before discussing these reasons, however, it will be helpful to quickly overview the types of information that the court has ruled are not subject to expectations of privacy. First, most people have no subjective expectation that they will not be noticed or watched by others when they are in public. Hence, in the *United States v. Dionisio* (1973),[57] the *Katz* decision was cited as standing for the proposition that:

> [the] physical characteristics of a person's voice, its tone and manner, as opposed to the content of a specific conversation, are constantly exposed to the public. Like a man's facial characteristics or handwriting, his voice is repeatedly produced for others to hear. No person can have a reasonable expectation that others will not know the sound of his voice, any more than he can reasonably expect that his face will be a mystery to the world.[58]

The *Dionisio* court came to this conclusion by interpreting the *Katz* court's finding that the Fourth Amendment provides no protection for what 'a person knowingly exposes to the public, even in his own home or office'. Some supporters of biometric surveillance cite the above as standing for the proposition that biometric surveillance does not constitute an 'unreasonable search' under Fourth Amendment analysis (Woodward, 2001). However, this argument is problematic because it assumes that biometric information includes only those characteristics that are easily noticed, such as face, voice and gait. Yet other kinds of biometric information, such as fingerprints, palm prints, iris and retina patterns, are not easily obtained without close detection. Thus, biometric information of this kind is arguably non-public information.

Legal commentator Susan McCoy has discussed some of the privacy implications of facial-recognition technology when it is incorporated into video surveillance systems aimed at public places, and it is worth considering her views here. Her argument demonstrates all of the major issues that commentators use to support the constitutionality of facial recognition technology. However, she departs from other commentators in her argument that people have a reduced expectation of privacy in public settings, and, additionally, that no individual can reasonably expect to maintain privacy in a public forum.[59] Based on an analysis of the US Supreme Court decisions in *Katz*[60] and *Kyllo*,[61] McCoy goes on to list several reasons why facial-recognition technology does not violate privacy:

- Facial recognition is implemented in an open field, and there cannot be a reasonable expectation of privacy in public places.[62]
- Video surveillance is not a search regulated by the Fourth Amendment because it captures exactly what the naked eye beholds: 'What a person knowingly exposes to the public . . . is not a subject of Fourth Amendment protection'.[63]
- Facial recognition technology only identifies criminals who are filed in the system's database and does not automatically store images of ordinary citizens who pass by its line of sight. Biometric technology was designed to locate and identify criminals, not innocent people.
- Facial recognition technology does not violate privacy rights because it merely makes a procedure currently used by law enforcement more efficient; it employs the same procedures as fingerprinting. If fingerprinting does not violate the constitution, then neither should facial recognition technology.

(McCoy, 2002)

Each of these four arguments is very interesting. At the same time, they expose a misunderstanding of the technologies concerned, as well as a controversial understanding of the legal doctrine of 'reasonable expectation of privacy'. There is no simple answer to whether facial recognition violates the right to privacy, because it can be applied in various ways, some of which are more privacy-friendly (or privacy-invasive) than others. However, it is generally the case that a covert facial-recognition system is inherently privacy-invasive. Below, I respond to McCoy's arguments one by one, in the order they are listed above.

Argument 1: Certainly, it is true that people generally have less reasonable expectations of privacy in public places. Yet, as is also emphasised in leading case law, notions of privacy remain tied to the individual rather than certain categories of space.[64] Although the Supreme Court in the *Kyllo* decision invalidated the disputed search because it occurred in and around a residential home, it does not follow that there is no privacy or reasonable expectation of privacy in existence at other places; it only means that the protection of privacy in respect of those other places may be more controversial or difficult to uphold but not impossible. Whether the use of facial-recognition systems or other biometric technology in public places violates privacy depends on specific applications.

Argument 2: McCoy seems to equate facial-recognition technology with video surveillance technology in general, at least in respect to the privacy interests at stake. It is important to remember, although, that the two technologies are not fully commensurate. Traditional standard video surveillance may not equal the facial-recognition technology we are talking about here. Traditional video surveillance, beginning with simple closed-circuit television monitoring (CCTV), is simply a means to observe at a distance what happens or happened in a certain area. With the development of video cassette recorders, the observations were able to be preserved on tape. Later, digital computer technology was used for capturing, storing and analysing more detailed images. Only the latest generation of video surveillance tools includes automated facial recognition. The software for facial recognition captures an image of a person's face, generates a template of that image and then compares the template against other templates saved in a database. In so doing, the software makes it possible for law enforcement officials to proactively target and monitor persons of interest. Theoretically, the software can be used for both authentication (one-to-one) and identification (one-to-many) matches, but, in practice, it is usually used for screening (negative identification).[65] Generally, unless a special privacy-friendly system is applied, the templates stored in the database are also linked with other personal information.

In a facial-recognition scheme, people are checked against a database one by one; they are, as a point of departure, all under suspicion, and for no apparent reason. By contrast, in traditional standard video surveillance, people are typically only subjected to attention and identification when they actually commit some crime (or carry out other non-conformist actions). Criminals generally do not attract special attention under such systems if they do not do something overtly illegal. For example, video surveillance of a shop will usually not pick out a person unless they steal goods; otherwise they remain anonymous and are not checked against some database. Whether this kind of surveillance is problematic is outside the scope of this book, but it is clear that the biometric facial-recognition system is more privacy-invasive than traditional standard video surveillance.

McCoy further claims that one knowingly exposes one's face to the public. This claim is true to some extent, yet it is also true that facial-recognition technology greatly expands the exposure of one's face and enables access to considerably more information than would be available to the ordinary public view; the technology reveals, in other words, more than just a face. Furthermore, data subjects generally have no ability to refuse such scanning. Large-scale use of facial recognition does not look for someone in particular; rather, it looks at anyone, suspicious or not, who happens to wander past. Even if individuals expect to be watched by law enforcement officers, they do not generally expect to be automatically checked against a particular database and then monitored if a positive match is made. But this is what happens with facial-recognition systems: 'To the extent that the database tracks the location of faces it successfully scans, it operates as a homing device on a person's movements' (Brogan, 2002).

Argument 3: Although one might expect that only an individual whose template happens to match the saved data of a criminal is recorded and monitored, this is a naïve misconception of biometric technology. Facial recognition can be used both for authentication and identification, depending on the kind of application adopted. There are various ways the facial recognition system can be applied. For verification purposes, all individuals' facial images are collected. This is the case, for instance, with the United States Visitor and Immigrant Status Indicator Technology program: visitors' facial images are collected in advance, and saved in a huge database with their other personal data. The facial recognition can be used in various ways, and the level of privacy invasiveness depends on how it is applied. There is no absolute guarantee that innocent people's biometric templates will not be stored or matched for verifying their identity and tracking their movements. McCoy's argument only makes some sense when facial recognition is used solely for screening; however, even in this case there is high possibility of FAR when the current state of the technology is used in the public space, due to the poor quality of the technology. This means a lot of innocent people's personal information will be generated, even when facial recognition technology is not being misused.

Argument 4: McCoy also claims that facial recognition technology is 'merely making a procedure currently used by law enforcement more efficient' and draws a comparison to both fingerprint matching in forensic science and the police officer standing in a crowd with a stack of mug shots. However, the efficiency increase she notes actually changes the effect of the activity on privacy. In the traditional manual settings described above, there is a certain delimited time, place, purpose, and reason for the checking process. The police officer stands at a certain place checking the people who pass because it is a place where a certain criminal or criminals may appear, and it is only these persons who are the aim of the surveillance. The latent fingerprint is matched against the database of criminals because it is collected at a crime scene. However, the automated facial recognition system has far more comprehensive surveillance capabilities: it may be operated in many places, around the clock, and for a broad range of purposes. While the present limitations of such systems mean that they are currently not aimed at finding one or several particular criminals out of thousands of people passing by, such capabilities may be implemented in the future – thus making for an even more privacy-invasive society.

Moreover, quantitative differences could become qualitative differences. Even if a certain technology is constitutional when used within certain contexts, it might raise constitutional concerns when it is applied on a larger scale. In *United States v. Place* (1983),[66] the Supreme Court held that that a sniff by a police dog specially trained to detect narcotics is not a 'search' pursuant to the Fourth Amendment. However, Justice Brennan noted that widespread and indiscriminate use of 'dog-sniffing' practices might raise constitutional problems:

> Before excluding a class of surveillance techniques from the reach of the fourth Amendment, therefore, we must be certain that none of the techniques so

> excluded threatens the areas of personal security and privacy that the Amendment is intended to protect . . . It is certainly true that a surveillance technique that identifies only the presence or absence of contraband is less intrusive than a technique that reveals the precise nature of an item regardless of whether it is contraband. But by seizing upon this distinction alone to conclude that the first type of technique, as a general matter, is not a search, the Court has foreclosed any consideration of the circumstances under which the technique is used, and may very well have paved the way for technology to override the limits of law in the area of criminal investigation . . . If a device were developed that could detect from the outside of a building the presence of cocaine inside, there would be no constitutional obstacle to the police cruising through a residential neighbourhood and using the device to identify all homes in which the drug is present . . . hence at some point in the future, if the Court stands by the theory it has adopted today, search warrants, probable cause, and even "reasonable suspicion" may very well become notions of the past.[67]

Government collection of everyone's biometric information to check it against a database indicates a similar assumption that everyone is guilty until proven otherwise. The scale problem is further aggravated by the need to create databases that can handle vast amounts of data using very few resources.

5.3.1.2.4 ADMINISTRATIVE SEARCH DOCTRINE

Administrative searches are searches done to further administrative goals, such as security, rather than to secure evidence of a crime. It is one exception to the warrant requirement for a search under the Fourth Amendment. The theory behind this doctrine 'is that searches conducted as part of a general regulatory scheme in furtherance of an administrative purpose, rather than as part of a criminal investigation to secure evidence of a crime, may be permissible under the Fourth Amendment although not supported by a showing of probable cause directed to a particular place or person to be searched'.[68] Searches of this kind may be allowed under the Fourth Amendment if (i) the government can establish a compelling need to conduct the search, and (ii) the search will serve that need, and (iii) that the decision to search a particular person is not left to the discretion of a field officer. These criteria form the basic elements of the administrative search doctrine.

The judiciary applies the doctrine to ensure that government searches avoid arbitrary treatment of the citizens being searched.[69] Administrative searches must also be conducted in a 'reasonable' manner. The presence of a compelling governmental interest is required to justify an administrative search. The 'reasonableness' of a search is determined by balancing the individual's privacy interest against the government's interest in promoting public safety.[70] In the *United States v. Edwards* (1974),[71] the Fifth Circuit Court determined the reasonableness of an administrative airport search by 'balancing the need for a search against the offensiveness of the intrusion'.[72] The court determined that 'when the risk is the jeopardy to hundreds of human lives . . . inherent in the pirating or blowing up of a large airplane, the danger alone meets the test of reasonableness, so long as the search is conducted in good faith and with reasonable scope and the passenger has been given advance notice'.[73] In the *United States v. Davis* (1973),[74] the Ninth Circuit reaffirmed the administrative search doctrine. The Court recognised that reasonableness still remained the critical question regarding the constitutionality of the administrative search.[75] Under the test of reasonableness, the court held that the compelling governmental interest in maintaining air safety justifies the limited privacy intrusion required to satisfy the administrative need.[76]

Apart from the 'reasonableness' evaluation, 'public interest' is usually the justification for biometric searches as well. For example, the administrative search doctrine may apply at airports or border control where the security of the public is considered of great importance. However, as noted above, even under such justification, the 'search' still needs to be limited. Recognising the high-level privacy interests in biometric data, courts may focus on how government agencies collect and handle the dissemination of biometric data, and what kinds of safeguards are in place to prevent its unauthorised collection and use. Furthermore, according to *Michigan v. Clifford* (1984),[77] government officials are not permitted to use an administrative search to investigate criminal activity. If the purpose of the search is to uncover evidence of criminal conduct, government officials must obtain a valid search warrant.[78] Hence, it is important to find out if the gathering of the identification information is a mere pretext to help law enforcement gain evidence of criminal violations.[79] This has been affirmed by the Supreme Court in *Kyllo v. United States* (2001)[80]. In this case, the Supreme Court acknowledged the importance of the issue raised by sophisticated surveillance technology by stating, '[i]t would be foolish to contend that the degree of privacy secured to citizens by the Fourth Amendment has been entirely unaffected by the advance of technology . . . The question we confront today is what limits there are upon this power of technology to shrink the realm of guaranteed privacy'.[81]

5.3.1.2.5 CONSENSUAL ENCOUNTER

The Fourth Amendment dictates an individual's rights against *search* and *seizure* both in public and at home. If the collection of biometric information could be regarded as a kind of 'seizure', itself within the meaning of the Fourth Amendment, then it is also possible to justify the intervention if the target subject is free to decline such intervention when it happens.

In *Terry v. Ohio* (1968), the Supreme Court established that a stop is a 'seizure' under the Fourth Amendment: 'It must be recognised that whenever a police officer accosts an individual and restrains his freedom to walk away, he has "seized" that person'.[82] In the context of biometrics, if the individual is stopped by the police and required to hand in his biometric information, which provides law enforcement with information about the individual that could be quite detailed, this arguably could be regarded as an *encounter*. This 'encounter' is one kind of 'stop' according to *Terry*.[83]

In *United States v. Mendenhall* (1980),[84] which involves an encounter between Mendenhall and Drug Enforcement Administration agents at the Detroit

Metropolitan Airport, the court held that the initial encounter between the government and the defendant was consensual and did not escalate to the level of seizure.[85] In determining the moment at which an individual is intruded upon for the purposes of the Fourth Amendment, the court found that '[a]s long as the person to whom questions are put remains free to disregard the questions and walk away, there has been no intrusion upon that person's liberty or privacy as would under the Constitution require some particularised and objective justification'.[86]

This requirement of being able to 'walk away' is further developed in *Florida v. Bostick* (1991),[87] which concerns whether a defendant was restrained because police questioning took place on a bus.[88] The court sustained the basic idea of *Mendenhall* and concluded that:

> Bostick's freedom of movement was restricted by a factor independent of police conduct – i.e. by his being a passenger on a bus. Accordingly, the 'free to leave' analysis on which Bostick relies is inapplicable. In such a situation, the appropriate inquiry is whether a reasonable person would feel free to decline the officer's requests or otherwise terminate the encounter.[89]

Hence, where the Fourth Amendment is concerned, the court needs to assess whether the individual had the ability to choose to refuse such invention. Put simply, the justification of the 'stop' hinges on the 'consent' of the individual. If the individual is notified and given the possibility to accept or refuse such inquiry, there is no Fourth Amendment violation. This idea is in accordance with most other privacy legislation, such as the European Data Protection Directive 95/46/EC and the Australian Privacy Act 1988.[90] The slight difference is that in US case analysis there is no clear requirement of 'notice', which means the government is not required to tell individuals about their right to decline; as long as the individual is not prohibited from asserting his right to terminate the search, the intervention may be justified. The same analysis applies for biometric technology. The government could seek to justify its intervention into individual privacy by initiating a voluntary program. However, in most occasions when biometric identification or authentication is used, individuals are not given any choice. This renders this justification irrelevant to most biometric applications.

5.3.1.2.6 THE FIFTH AMENDMENT

Another concern about the use of biometric technology is that, by using an individual's physical characteristics to identify him or her, an individual is forced to be a witness against him or herself, thereby putatively violating the Fifth Amendment. This provision requires that 'no person . . . be compelled in any criminal case to be a witness against himself'.[91] This privilege against self-incrimination has historically functioned to protect a 'natural individual from compulsory incrimination through his own testimony or personal records'.[92] The Fifth Amendment, in conjunction with the exclusionary rule,[93] prevents the government from gathering evidence in violation of the Fifth Amendment or using that evidence at trial.[94]

Instances of biometric technology that are used for administrative purposes, such as for driver's licences or national ID cards, may not fall directly under the scope of the Fifth Amendment. However, in *McCarthy v. Arndstein* (1924),[95] the Supreme Court extended the right against self-incrimination to non-criminal procedures that may lead to criminal prosecution. Nonetheless, both federal and state courts have held that the right against self-incrimination does not extend to forced subjection to fingerprinting, photographing[96] or physical measurements, nor to forced writing,[97] speaking, standing, walking or gesturing for identification purposes.[98] In *Schmerber v. California* (1966), the Supreme Court held that only evidence that is 'testimonial or communicative [in] nature' falls within the umbrella of protection offered by the Fifth Amendment.[99] Since an individual's forced participation in line-up is not considered testimonial or communicative,[100] it might follow that a physical characteristic collected from an individual is also not testimonial or communicative in nature. This conclusion is obviously based on the premise that biometric technology is not intrusive and involves no physical penetration.

5.3.1.3 Anonymity

As highlighted in Chapter 3, to the extent that identifiable biometric information is accessible, an individual's basic privacy rights are eroded. Thus, if individuals can claim a constitutional right to biometric anonymity, the use of biometric technology may to some extent be restricted. While there may be no expectation of privacy in public places, there may still be an expectation of anonymity. When biometric technology is concerned with identification,[101] anonymity may be a valuable argument. Some constitutional principles, while not explicitly recognising a right to anonymity, provide groundwork for it.

Anonymity is recognised as important to the exercise of other constitutional rights than just privacy (Kent and Millett, 2003). The discussion about the right to anonymity often argues for a First Amendment right to be free of the inhibiting effects of surveillance in public unless the government can proffer some justification for it (Slobogin, 2002). The Supreme Court has declared that, notwithstanding a significant government justification, a person who writes a pamphlet or is involved in collecting signatures for a petition cannot be required to reveal his or her name.[102] It has also held that membership lists of organisations need not be revealed.[103] In *McIntyre v. Ohio Elections Commission* (1995), Justice Stevens stated:

> [T]he decision in favour of anonymity [is] motivated by fear of economic or official retaliation, by concern about social ostracism or merely by a desire to preserve as much of one's privacy as possible . . . [it] is an aspect of freedom of speech protected by the First Amendment.[104]

When individuals' biometric data, such as a fingerprint or facial image, is collected by governmental agencies, and a certain individual is actually engaging in expressive conduct, they may have a basis for claiming violation of anonymity under the First Amendment. When the use of biometric technology infringes First Amendment values, the government should have to justify the presence of the biometric application on meaningful law enforcement grounds. However, this only provides constitutional protection for expressive conduct,[105] a category that has been narrowly defined as a dichotomy between actions regarded as 'communicative' and those regarded as 'pure conduct'.[106] A First Amendment challenge to biometric surveillance thus cannot be successful unless the government also intrudes upon expressive conduct. For example, to compile and store private information of law-abiding citizens for non-investigative purposes that could qualify as 'expressive conduct'.

Yet another way to prove the value of anonymity in terms of First Amendment is to argue that it is essential to a broad range of non-speech activities that are implicitly connected to speech (Nguyen, 2002). Under this conception, many activities, such as going to the store or purchasing certain goods, could all be regarded as a part of speech. Thus biometric technology that interfered with those actions could potentially be regarded as a First Amendment violation. But to the author's knowledge there is not yet any Supreme Court decision that is directly based on such an interpretation of expressive conduct.

The right to anonymity is also related to the right to movement, which is derived from the Due Process Clause.[107] The right to movement was defended in *State v. Baumann* (1995),[108] in which the Wisconsin Court of Appeals upheld an order that permanently enjoined 32 individuals from photographing or videotaping people entering and leaving an abortion clinic under circumstances that exhibited 'intent to harass, intimidate or interfere with any person seeking access to or departing from such a facility'.[109] Does the use of biometric applications, such as facial recognition at various public places, similarly affect the right to movement? Some commentators claim it certainly does. They point out that such technology may cause people to avoid going to specific places, for example, or cause people to feel the need to walk instead of run (Slobogin, 2002). However, the extent of this claim is yet to be evaluated[110].

5.3.2 Federal privacy statutes

In assessing privacy statutes that affect biometric technology, I seek to discover the following: (1) Do the principles of the statutes effectively address the informational privacy concerns posed by the biometric technology mentioned in Part II and (2) are any privacy concerns posed by biometric technology currently unaddressed?

5.3.2.1 The Privacy Act of 1974

The Privacy Act of 1974 establishes a code of fair information practices that govern personally identifiable information about individuals that is maintained in systems of records by federal agencies. Nevertheless, despite its name, the Act is quite limited in scope. It only regulates the federal government's collection, maintenance, use and dissemination of personal information of US citizens and legal resident aliens. It does not apply to the use of such information by state and local governments, nor does it apply to private individuals or private entities.

Significantly, the Act only applies to 'records' maintained within a 'system of records'; hence, the definition of 'record' and system of records is critical to any discussion of whether biometric information would be subject to this statute. The Act defines a 'record as:

> any item, collection, or grouping of information about an individual that is maintained by an agency, including, but not limited to, his education, financial transactions, medical history, and criminal or employment history and that contains his name, or the identifying number, symbol, or other identifying particular assigned to the individual, such as a finger or voice print or a photograph.[111]

To qualify as a Privacy Act 'record', the information must identify an individual.[112] This requirement is similar to the personal data requirement in the EU Directive and fits well with the biometric information in general.[113] The use of terms 'finger or voice print or photograph' leaves little doubt that biometric information presumably falls within the scope of the Act. Additionally, the 'record' needs to be kept in a 'system of records',[114] which is described as an 'indexing or retrieval capability using identifying particulars that is built into the system'. Furthermore, the agency must 'in fact, retrieve records about individuals by reference to some personal identifier'.[115] The Department of Justice points out that 'a major criticism of the Privacy Act is that it can easily be circumvented by not filing records in name retrieved formats'.[116] With regards to biometric records, government agencies could similarly circumvent the legislation by building up a facial recognition records database that retrieves by date and place of collection instead of by name or biometric identifier[117] or by not keeping the biometric information in a central database. The Privacy Act is also criticised for its exceptions for law enforcement and for its easily abused 'routine use' exception.[118] These exceptions render the Act ineffective to a large extent[119] because it is relatively easy to qualify for either exception, especially since virtually all government agencies can be said to be involved in some type of law enforcement.[120]

The Privacy Act embodies eight principles, as summarised by the Privacy Protection Study Commission:[121]

1. There shall be no personal data record-keeping system whose very existence is secret and there shall be a policy of openness about an organisation's personal data record-keeping policies, practices, and systems. (The Openness Principle)
2. An individual about whom information is maintained by a record-keeping organisation in individually identifiable form shall have a right to see and copy that information. (The Individual Access Principle)
3. An individual about whom information is maintained by a record-keeping organisation shall have a right to correct or amend the substance of that information. (The Individual Participation Principle)

4. There shall be limits on the types of information an organisation may collect about an individual, as well as certain requirements with respect to the manner in which it collects such information. (The Collection Limitation Principle)
5. There shall be limits on the internal uses of information about an individual within a record-keeping organisation. (The Use Limitation Principle)
6. There shall be limits on the external disclosure of information about an individual a record-keeping organisation may make. (The Disclosure Limitation Principle)
7. A record-keeping organisation shall bear an affirmative responsibility for establishing reasonable and proper information management policies and practices which assure that its collection, maintenance, use, and dissemination of information about an individual is necessary and lawful and the information itself is current and accurate. (The Information Management Principle)
8. A record-keeping organisation shall be accountable for its personal data record-keeping policies, practices and systems. (The Accountability Principle)

The first three principles seek to ensure the transparency of federal agencies' data use by providing individuals with the right to access and correct their records. These interests are significant in general, but not very relevant to biometric information in particular. Because of the sophisticated technical nature of biometric information, it would be impractical for most individuals to determine whether their stored biometric information is accurate or correct without specific assisting measures provided by the data controller.

The fourth principle, which seeks to establish general rules for collecting personal records, may be relevant to tackle the problem of *unauthorised collection* of biometric information. In furtherance of this principle (and, arguably, the Openness Principle), the statute requires federal agencies to 'collect information to the greatest extent practicable directly from the subject individual when the information may result in adverse determinations about an individual's rights, benefits, and privileges under Federal programs'.[122] Through this provision, 'notice' is required for the collection of personal information, which may exclude 'covert collection' of individuals' biometric information. It also establishes that the 'notice' should inform individuals about the collecting authority, the purpose of collection, the possible routine use and the effect on the individual if the information is not provided.[123] This specific notice requirement has been described as 'most promising' concerning transparency (Schwartz and Reidenberg, 1996). However, despite its promising language, the Act is generally viewed as having little impact in securing information privacy. According to the 1977 Privacy Protection Study Commission, the Act only 'resulted in a modest amount of revision and reduction of data collection forms, and consequently a modest reduction in data collection itself'. A similar view is also indicated in the book concerning data privacy law in the United States by Schwartz and Reidenberg (1996).

Let us look in more detail at the reasons for the ineffectiveness of the legislation. Under the Privacy Act, an individual cannot 'control' their personal information by mere 'notice' because 'consent' is not required for first-time collection although written consent is required for disclosure of the personal records that have been collected already.[124]

Moreover, even when collection happens on a so-called voluntary basis, the 'notice' of the 'effect of not providing this information' may in fact compel the individual to 'consent'. Beyond getting 'notice', individuals cannot order an agency to change its practice. Biometric technology has often been used as a flag for enhancing public security or fighting fraud, which provides a good excuse for the unrestricted use of biometric information. This fact contributes to the imbalance of power between individuals and federal agencies – that is, individuals are concerned that they may be disadvantaged by not getting a certain kind of social benefit or that they may have to bear legal consequences if they don't 'consent'. For instance, in *Perkey v. Department of Motor Vehicles* (1986),[125] the state court required individuals to provide fingerprints as a condition for obtaining a driver's license. Although individuals were free to refuse, doing so resulted in the loss of an important social benefit. Clearly then, consent and notice are relatively meaningless if alternatives are not also provided.[126]

The condition of directly collecting information only when 'practicable' and only '*when* the information may result in adverse determinations . . .' does not exist in the Australian federal Privacy Act or in the EU's Data Protection Directive. The provision seems to be problematic for two reasons. First, it is unclear who is entitled to decide whether such information may result in adverse determinations. Where biometric technology is concerned, privacy advocates and federal agencies are very likely to have different views about the consequence of the technology. The adoption of biometrics usually enhances the surveillance power of federal agencies; it thus becomes difficult for them to objectively decide when information is necessary. Second, the use of the criterion 'practicable' may lead to abuse and the collection of information from third parties. Without clear interpretation, inconvenience, cost of time or money and lowered efficiency are all likely to be equated with impracticability. The gathering of biometric information from the private sector that agencies cannot legally collect on their own without oversight and the appropriate protections has already become a trend among government agencies.[127] The recent decision to award ChoicePoint an Internal Revenue Service contract highlights this tension.[128] As a digitalised template, biometric information can be easily copied and transmitted. Hence, collecting from third parties would be much more efficient and cost-effective than to ask the individual and go through the 'notice' process.

The requirement to 'collect only information that is relevant and necessary to accomplish a lawful purpose' helps restrict the general collection of personal records. However, it is unlikely to be an adequate safeguard against the *unnecessary collection* of biometric records[129], because it seemed that courts were unwilling to extend strict scrutiny to biometrics regulations. For instance, in *United States v. Doe* (1972)[130], the Second Circular Court compelled fingerprinting of voice or handwriting exemplars ordered by a grand jury; in *Brown v. Brannon* (1975)[131] the Supreme Court supported a police officer taking fingerprints in connection with an application for a business license. These decisions indicate that biometric intrusions are not regarded by courts as infringing on fundamental interests. As a result, the 'unnecessary collection' of biometric technology is still possible. However, it is also relevant to note once again that the exceptions of 'routine use', 'law enforcement purpose' and 'data matching' are also responsible for this failure, as they can make 'unnecessary' collection appear 'necessary'.

The Privacy Act also attempts to regulate the processing of personal information by federal agencies in §552a (b). Limitations on the internal and external use and disclosure of personal information are imposed on federal agencies, and the accountability of information management is stipulated. Yet it is unclear whether these regulations could effectively prohibit the 'function creep' of biometric information. The act requires an agency to obtain an individual's written consent before disclosing a record about him or her to any of its employees other than those who have a need for the record in the performance of their duties (§552a(b)(1))[132]. Nevertheless, the act has been criticised because the words 'need' and 'duty' are open to interpretation, which limits the effectiveness of this restriction.[133] Moreover, the restriction of 'external use' is subject to ten exceptions. These include the quite problematic 'routine use', 'officers and employees' needs and duty' and 'law enforcement use', which may, to a large extent, undermine the effectiveness of such disclosure restrictions.[134]

One of the most important aspects of the Privacy Act is that it restricts the sharing of information between government agencies. It does this by limiting 'matching programs', which it defines as the computerised comparison of databases in order to determine the status, rights or benefits of individuals within those systems of records. The Privacy Act prohibits agencies from running matching programs on systems of records unless there is a written arrangement between the agencies.[135] This regulation may have a significant impact on the use limitation of biometric data if effectively implemented.

However, it has been demonstrated that individuals are generally unaware of how federal agencies match their personal information (Schwartz and Reidenberg, 1996). Furthermore, data matches are often considered simply a 'routine use' of data (US Congress Office of Technology Assessment, 1986), a factor that has been criticised as representing another limitation on the Privacy Act's ability to create transparency in data processing (Schwartz and Reidenberg, 1996). As a result, government agencies are able to skirt the consent requirement for the use of biometric information for purposes other than the initial one. This represents a significant limitation on the Privacy Act's implementation of a 'disclosure limitation' principle.

The principles of information management and accountability related to security issues can also be relevant for the protection of biometric information. In §552a (e)(10), the Act requires federal agencies to:

> Establish appropriate administrative, technical, and physical safeguards to insure the security and confidentiality of records and to protect against any anticipated threats or hazards to their security or integrity which could result

in substantial harm, embarrassment, inconvenience or unfairness to any individual on whom information is maintained.

It is unclear what kind of security standard would be adequate where biometric technology is concerned. As was demonstrated in Part II, the security problem of biometric information has raised much debate. Compared with other personal records, biometric information needs to be given more attention. Should encryption be used when biometric data are stored? Should central storage of biometric information be strictly limited? Should biometric data be anonymised when only authentication is used? Who should be responsible for keeping the storage system secure and accurate? In practice, the broad language of the Privacy Act fails to convey a precise conception of the requisite security measures.

5.3.2.2 Personal Data Privacy and Security Act of 2009

In July 2009, the Personal Data Privacy and Security Act was introduced into the US Senate. The Bill is similar to Bills introduced in 2005 and 2007, and requires government and commercial entities to ensure that the personal data they collect are protected by adequate security. As of now, the PDPSA has not been formally passed, but it was last placed on the Senate Legislative Calendar on 5 November 2009. It is uncertain whether the Bill will be passed, but it would be very interesting to examine to what extent this proposal could improve the current privacy protection level in the United States, especially where biometric data are concerned.

Financial institutions and health care service providers are excluded from coverage under the PDPSA because they are covered by other relevant legislation.[136] Biometric information is explicitly included in the Bill's definition of 'sensitive personally identifiable data', which includes 'any information or compilation of information in electronic and digital form that includes . . . unique biometric data such as fingerprint, voice print, a retinal or iris image or other unique physical representation . . .'[137]

To protect the privacy and security of personal data, the Personal Data Privacy and Security Act adopts several similar measures to the Privacy Act of 1974.[138] To avoid repetition, I will only comment on the new measures. The Bill seeks to ensure the following (Open Congress, 2009):

1. Americans will be provided with notice when they have been harmed (Section 3 (12)).
2. Unauthorised access to personally identifiable information will be criminalised (Section 101).
3. Directs the US Sentencing Commission to review and amend, if appropriate, federal sentencing guidelines for persons convicted of using fraud to access, or to misuse, digitised or electronic personally identifiable information, including sentencing guidelines for identity theft (Section 103).
4. Requires interstate data brokers (defined as business entities which, for monetary fees or dues, regularly engage in the practice of collecting, transmitting

or providing access to sensitive personally identifiable information on more than 5000 individuals to non-affiliated third parties on an interstate basis) to implement several transparency measures for the data subjects (Section 101). Civil penalties are imposed on data brokers for violation of the transparency requirements (Section 102).

5. Companies that have databases with personal information on more than 10,000 Americans must establish and implement data privacy and security programs. Exempts certain financial institutions, covered entities under the Health Insurance Portability and Accountability Act of 1996, and public records from such requirements (Section 301).
6. Notice to law enforcement, consumers and credit reporting agencies must be given when digitised sensitive personal information has been compromised (Section 311). Exempts National Security, Law Enforcement (Section 312a), Safe Harbour (Section 312 b) and Financial Fraud Prevention Exemption (Section 312 c).
7. The government's use of personal data is restricted by requiring (1) evaluation of the privacy and security practices, (2) audit of the information security and privacy practices, and (3) privacy impact assessments on their use of commercial databases (Sections 401–403).
8. Tough monetary penalties for various security and privacy breaches.[139]

On its face, the PDPSA constitutes a significant addition to extant data protection rules in the United States If it is passed, audits, privacy impact assessment and the evaluation of privacy and security practices will be compulsory. Organisations will need to devote resources and personnel to manage the program requirements. Considering that businesses can be held civilly liable for security violations under the PDPSA, there would appear to be a real incentive for getting these assessments correct. This seems to be promising for biometric technology because it will provide a solid legal basis for assessing the risks associated with biometric technologies and their deployment. With the compulsory adoption of such measures, the deployment of biometric technology is likely to occur in a more privacy-friendly manner.

In addition, the PDPSA recognises encryption as a vital component of any data security program (Section 302) and explicitly recognises encryption as an effective technical safeguard during the use, transmission, storage and disposal of personal electronic records. Thus, organisations that use encryption as a method to protect stored biometric information will find it much easier to demonstrate the effectiveness of their data privacy and security programs than organisations that do not employ encryption technologies.

When compromise and harm occur, the notice requirement ought to significantly aid the transparency of data collection and storage processes. However, the PDPSA includes some potentially troubling exemptions. For example, Section 312 b(1)(A) specifies that 'an agency or business entity will be exempt from the notice requirements' if 'a risk assessment concludes that there is no significant risk that a security breach has resulted in, or will result in, harm to the individuals whose sensitive personally identifiable information was subject to the security breach'. This exemption could become a huge loophole in privacy protection in the biometric context because there is no clear regulation about who should conduct the relevant risk assessment. Provided the risk assessment is conducted by the agency or business entity that collects the biometric information, it would be very difficult to guarantee the fairness of such risk assessment.

According to Section 311(a), '[t]he agency and business that engaged in interstate commerce, that uses, accesses, transmits, stores, disposes of or collects sensitive personally identifiable information shall *following the discovery of a security breach* of such information, notify any resident of the United States whose sensitive personally identifiable information has been, or is reasonably believed to have been, accessed, or acquired' (emphasis added). However, how can the organisations provide notice when the biometric data are compromised? The general concerns about the security risks of biometric technology are as follows.

1. Once it is compromised, it is very likely to be compromised forever because there is no fall-back system to create a new finger or new face (at least at the current stage of technology).[140]
2. Generally, unless a systematic security breach occurs with respect to the biometric information database without the individuals' cooperation, it would be difficult for the commercial entities to discover a single compromise because people usually have a firm belief that biometric information is indisputable. If this happens, the victim might be in a very difficult situation to prove otherwise.

Moreover, although the PDPSA gives individuals the right to access their personal identifiable information profile and correct inaccurate information, it remains silent on how the data subject can control the flows of personal identifiable information. Like the Privacy Act of 1974, there is no provision on the manner of collection, nor any limitations concerning the use of personal identifiable information after collection. This should be considered a serious loophole in the Bill because when biometric information is collected by commercial entities, an individual is given no opportunity to receive notice about the collection, the purpose of the collection or the way it may be used after collection. Additionally, the individual is not entitled to any notice about how or when biometric information will be shared. It reasonably follows that the individual has no right to consent to or deny any such collection, usage or sharing. This loophole is a serious privacy risk because it allows for precisely the kind of privacy concerns about biometrics that must be avoided. Unauthorised collection, unnecessary collection and function creep of biometric information are unrestricted under the Privacy Act. This renders the security and privacy management regulations of the Data Privacy Act weakened, at best, given that personal identifiable information subjects are not entitled to any right to control their information flows. Instead, the sharing, usage and collection of biometrics are likely to be determined and controlled by commercial entities. To compound the problem, commercial entities can sell, share, lease or otherwise trade individuals' profiles with other companies, thus increasing the number of privacy invasions.

Furthermore, although the PDPSA criminalises the unauthorised access of personally identifiable information – which may benefit enforcement – it does not provide a precise definition of the term 'unauthorised access'. Since data subjects lack the right to receive notice of or give consent to the collection or usage of data on themselves, commercial entities are likely to be the ones to give 'authorisation'. The PDPSA provides no specific guidance on the restrictions to accessing personally identifiable information and 'unauthorised' is likely to be defined by data controllers in ways that assure them (and not the data subjects) the greatest benefit.

5.3.2.3 Gramm–Leach–Bliley Act of 1999

The Financial Modernisation Act of 1999, also known as the 'Gramm–Leach–Bliley Act', includes provisions that protect consumers' personal financial information held by financial institutions. As noted earlier in the book, the financial industry was among the first in the private sector to use biometric technology. In order to reduce fraud, many banks have turned to biometrics-based authentication schemes.[141]

Title V of the Gramm–Leach–Bliley Act has been described as the most extensive federal privacy legislation in US history (Fischer and Camper, 1999). It imposes an on-going duty for every financial institution to safeguard the privacy of its customers' non-public personal information. 'Non-public personal information' is composed of two categories: (1) personally identifiable financial information that is not publicly available information and (2) lists, descriptions or other groupings of consumers that are derived from the use of personally identifiable financial information.[142] 'Personally identifiable financial information' is defined as any information that a financial institution generates about a consumer pursuant to the issuance of a financial product or service. It has been interpreted as including such information as name, telephone number credit history and income (Federal Deposit Insurance Corporation, 2001). Accordingly, biometric information, if being collected for identification purpose will be categorised as 'personally identifiable financial information'.

Title V contains a set of general notice requirements and restrictions on information sharing:

1. A financial institution is not allowed to disclose an account number, access number or access code for a credit card or a consumer's deposit account to any non-affiliated third party for use in telemarketing, direct mail marketing or other marketing through electronic mail (§502(d)(e)); however, disclosure is allowed to consumer reporting agencies[143] (id.).
2. Financial institutions must provide 'clear and conspicuous' disclosure of privacy policies to consumers at the beginning of a customer relationship and 'not less than annually' thereafter (§503(a)).

3. Unless a disclosure is pursuant to an exception described in the Act, a financial institution may not disclose to a non-affiliated third party any 'non-public personal information' (§502(a)).

The latter type of information means personally identifiable financial information that is (1) provided by a consumer to a financial institution, (2) results from any transaction with the consumer or any service performed for the consumer, or (3) is otherwise obtained by the financial institution (§509). Thus, the privacy provisions do not apply to non-consumer customers and accounts.

Nevertheless, despite the progress for consumer privacy represented by Gramm–Leach–Bliley Act, many privacy advocates are disappointed.[144] This disappointment stems from two complaints. First, Gramm–Leach–Bliley Act requires consumers to make the effort to opt-out of information sharing. Information is shared with non-affiliates and other parties unless,[145] after receiving notice, the consumer takes action and tenders an objection. In practice, when consumers are not appropriately informed about the 'opt out' policy, their ignorance will be regarded as a silent commitment to sharing. In an opt-out regime, effective privacy notices are contrary to industry interests. The opt-out framework actually creates a financial incentive for the creation of confusing privacy notices. Companies are versant in how to best phrase and send opt-out notices to maximise customer confusion and minimise the chance that customers will read the notices.[146] Vague, illusive and unhelpful notices can frustrate and confuse consumers and the right to opt-out is almost always at the end of the privacy statement. This kind of notice usually keeps silent on how information is used and how much the financial institution profits from the use of this information.[147] In addition, consumers generally have insufficient knowledge of biometric data and its potential risk to privacy and security; hence, without clear and informed notice, it is very likely that consumers will keep silent, which is a defect in the opt-out regulatory standard.

Furthermore, individuals also face a multitude of potential risks through unrestricted and undisclosed information-sharing of their biometric information among affiliates and non-affiliates under the Gramm–Leach–Bliley Act. For instance, the privacy rule does not impose limitations on information sharing with affiliated parties (§502(d)).[148] The rule also contains exceptions to allow transfers of non-public personal information to unaffiliated parties to process and service a consumer's transaction, and to facilitate other business transactions (§502(b)(2)).[149] The natural result of this collection and sharing of personal data is individual profiling. A customer profile can be developed by a financial affiliate of the company and sold or shared with an affiliate that does not fall within the broad definition of financial institution. The combined databases of distinct banking, insurance and securities companies may contain a vast amount of medical, financial and other sensitive information. This enables the 'function creep' of biometric information. When appended with information easily obtainable from outside sources, a comprehensive profile of each individual customer of the financial institution can be compiled with a simple biometric identifier. Such affiliate sharing between financial and non-financial companies may be a great threat to a customer's privacy rights, and a customer of the bank or other financial institution has no ability to knowledgeably consent to the sharing of information, no right to know what or when information is shared and with what company, and no right to stop the onward flow of information.

Moreover, the Gramm–Leach–Bliley Act also allows an exception for information disclosed to a non-affiliated third party in connection with a joint marketing agreement (§502(b) (2)). If the disclosure of the customer's personal information, which may include biometric information, fits within this exception, the customer is given no opportunity to opt out, nor is the individual entitled to any notice about the type of information shared. This becomes a great loophole for privacy protection, especially considering the wide scope encompassed by the term 'financial institutions', which is defined as 'any institution the business of which is engaging in financial activities as described in Section 4(k) of the Bank Holding Company Act of 1956'.[150] (Section 509(3) (a)). Customers have no means of assessing the 'third party' that share their biometric information, and the use of such information by the third party then becomes difficult to restrict according to the Gramm–Leach–Bliley Act. In short, the individual has very limited ability to control the flow or content of their biometric information under the Gramm–Leach–Bliley Act.

Additionally, the Gramm–Leach–Bliley Act draws a distinction between 'consumers' and 'customers' for the purpose of notice requirements. A 'consumer' is defined as an individual who obtains a financial product or service from a financial institution to be used primarily for personal, family or household purposes (§509(9)). A consumer becomes a customer for a financial institution only upon establishing a continuing relationship with the institution.[151] Thus a customer's relationship is defined as an ongoing relationship between a consumer and a financial institution, whereby the institution supplies one or more financial products or services to the consumer that are utilised predominantly for personal, family or household purposes (§509(11)). For instance, opening a cheque account, borrowing money or receiving investment advice constitute transactions that would qualify a consumer to become a customer of a financial institution (Cochran and Chavarria, 2000). However, isolated actions, such as cashing a cheque or engaging in an ATM transaction, are not sufficient to establish a customer relationship (Kempler and Woody, 2000). The Gramm–Leach–Bliley Act only places restrictions and requires written policies for information in a company's files on its own customers (§503). This is quite problematic for scenarios in which biometric information of non-customers is collected. For instance, in the case *Messing v. Bank of America* (2003),[152] the bank required that non-account holders provide thumbprint signatures in order to cash cheques. The non-customer's biometric information was collected by the bank, and this was regarded as 'reasonable' according to the court decision. Financial institutions can also get biometric information of non-customers from other companies or institutions via information sharing – as allegedly occurred in the case Charter Pacific.[153] Since information-sharing restrictions are weakened to a large extent by the various exceptions in the Gramm–Leach–Bliley Act, this kind of sharing will not be prohibited. Financial institutions that receive biometric information from non-customers will be able to use it without much restriction. As a result, the various privacy concerns about biometric information are unlikely to be solved.

One of the security concerns about the use of biometric information is permanent identity theft. Although biometric technology has always been considered a tool to prevent identity fraud, it is not perfectly secure, and once it is compromised, it is very difficult to repair at its current stage.[154] Hence, measures taken to prevent biometric information being hacked into are considered an essential part of the whole picture. The Gramm–Leach–Bliley Act requires financial institutions to implement security measures to protect consumer information, but it does not specify what measures are to be taken nor what the penalty would be for failure to comply with the statute, although it has been claimed that bank and regulatory agencies have issued joint guidelines for safeguarding confidential consumer information (Schiller, 2003). However, the guidelines are not mandatory; financial institutions need only implement methods suggested by the guidelines if they choose to do so. Additionally, the Gramm–Leach–Bliley Act does not provide for a private right of action for consumers to directly sue financial institutions for violation of the statute (Schiller, 2003). Hence, consumers have no explicit private course of action under the Gramm–Leach–Bliley Act, unless the Federal Trade Commission strictly enforces its mandatory 'Safeguarding Rule', which requires financial institutions under its jurisdiction to safeguard customer information.[155] If this is not the case then the security measures are unlikely to be implemented effectively.

5.3.3 Common law of torts

Currently, the common law offers privacy remedies for four main torts. To suggest that the existing scheme of privacy torts would be an effective tool for addressing the complex problems of biometric technology raised in the private domain is foolhardy. However, in the absence of other effective legal solutions, it may be a possible alternative and hence it is worth exploring in more detail.

5.3.3.1 Tort of intrusion upon seclusion

This tort provides a solution against someone who 'intrudes physically or otherwise' upon the plaintiff's 'solitude or seclusion' or 'private affairs or concerns' in a way that would be 'highly offensive to a reasonable person'.[156] The tort may be relevant to the use of biometric technology when the latter involves surveillance methods to collect people's biometric information where they have a reasonable expectation of privacy.[157] The tort does not address issues after the collection. However, unless the data subject can prove the data collector surreptitiously acquired the biometric information, the data subject cannot show that there is trespass or intrusion.[158] This indicates that it may be difficult for data subjects to seek a solution for unauthorised use of their biometric information, provided that information is collected legally. For instance, in *Humpher v. First Interstate Bank* (1985), the court found no intrusion when a doctor revealed information on a confidential adoption form in his possession because he did not 'pry into any personal facts that he did not [already] know'.[159]

Superficially, however, this tort may at least be an argument for preventing 'unauthorised collection' of biometric information, provided it passes the 'reasonable expectation of privacy' test, which requires subjective evidence that a data subject expects privacy and that this expectation of privacy is objectively considered reasonable by the court. It was previously noted that only expectations in a 'private space' or a 'private seclusion that the plaintiff has known about his person' is recognised.[160] Would it be difficult for an individual to prove that he or she has a subjective or objective expectation of privacy regarding his or her biometric information, such as face image or fingerprint? It will be recalled that the Court of Appeals of Maryland accepted compulsory collection of biometric information by private organisations as a reasonable means of identification in the other states;[161] this tort of intrusion may only be valid when the collection manner is covert. However, even when it is not notified, facial recognition in shops or banks may still be considered 'reasonable' because these are not private places. People who fiercely guard their biometric information will be hard pressed to establish a reasonable expectation of privacy under this tort.

However, it may be possible to argue privacy intrusion if the data collector can be shown to have used biometric information in combination with other personal information such as medical records, bank account information, transaction history, etc. This is because even if we do not have a reasonable expectation to keep our faces private, we have a reasonable expectation that this information is not used to create a comprehensive profile of our private matters. This expectation was upheld in *Weld v. CVS Pharmacy, Inc.* (1999).[162] Weld sued CVS for selling customers' personal information (names, addresses and birthdates) to a direct marketer. CVS argued that the Massachusetts Privacy Statute was not violated by the disclosure of this information. However, the Massachusetts Superior Court disagreed, noting that the plaintiff had complained about the use of that information 'in conjunction with the systematic searching of customer prescription records'.[163] This indicates that data mining and profiling of personal information could amount to violation of privacy. Although the tort claim was not the focus in this case, the decision indicates that unauthorised use of biometric information through data mining and profiling could come under the scope of the tort. For instance, when biometric fingerprints collected for authentication in medical services are sold or disclosed to other unauthorised parties by the hospital for data mining, relevant medical history, medical situations and other personal information are compromised. In such scenarios, data subjects may be able to prove the violation of a 'reasonable expectation of privacy' as required by the tort.

5.3.3.2 Tort of public disclosure of private facts

This tort requires that the plaintiff's private information not be publicised. Such information must 'not [be] of legitimate concern to the public' and its publication 'would be highly offensive to a reasonable person'.[164] The tort can provide a legal solution for mental harm such as shame, humiliation and so on, caused by the publication of intimate and embarrassing information. One problem with this tort is that it has been severely weakened by limitations under the First Amendment in the US Constitution – particularly limitations on the definition of private information and what is of legitimate concern to the public.[165] However, these limitations do not apply to individuals who are not public figures (Ludington, 2006).

Nevertheless, data subjects who want to use this tort to redress the misuse of biometric information may face a significant hurdle in proving that data traders have publicised their biometric information. While data collectors usually sell or disclose information to companies or government agencies, they do not broadcast the information. According to the Restatement, publicity means that a private matter is made public when it is communicated 'to the public at large or to so many persons that the matter must be regarded as substantially certain to become one of public knowledge'.[166] Often the sale or disclosure of biometric information does not clearly fall into the scope of the Restatement's definition of publicity as information communicated to 'the public at large'.[167] However, it may still be possible to redress the disclosure or sale of biometric information when the defendant's security has been so inadequate that the information has effectively been publicised.

Another problem with applying the tort to biometrics concerns the extent to which the publicity of biometric information may be regarded as offensive and embarrassing. Typically, biometric information is quite neutral information, especially when it is transformed into biometric templates, which are only numerical digits. This generally fails the highly offensive test. It is more likely that the publication of biometric information could be considered highly offensive when it obviously discloses certain health problems or disabilities, such as a palm print of only four fingers or a facial image with an obvious scar. Despite this possibility, plaintiffs may still struggle to show that the publication of such biometric information is highly offensive in the present surveillance society.

5.3.3.3 Tort of false light

This tort provides a solution against someone who publicises a matter concerning an individual and in so doing portrays the latter in a way that is both false and objectionable.[168] This intends to protect the individual's dignity and their right to be portrayed as they are and to avoid being portrayed in an offensively false manner.[169] For instance, in the case *Braun v. Flynt* (1984),[170] the Fifth Circular Court held that the publication of a photo of the plaintiff with 'Ralph, the Diving Pig' among pornographic photographs created actionable, disparaging suggestions about the plaintiff's character.

Again, in circumstances when data subjects want to use false light to redress the misuse of biometric information, they may face similar hurdles as in previous torts. However, there are a few courts that have moved away from the Restatement's publicity test and have instead applied a 'particular public' or 'special relationship' test for deciding the existence of the tort of false light. For instance, the District of Columbia Circuit Court of Appeals deviated substantially from the traditional publicity requirement in *McSurely v. McClellan* (1985).[171] The court held that the publicity requirement was met when embarrassing information about the plaintiff reached only one person.[172] While the private facts tort ordinarily applies only to communications reaching the public at large, the *McSurely* court stated that the publication requirement under the false light tort could be satisfied by a disclosure to a small number of people when a 'special relationship' exists between the plaintiff and the people receiving the information.[173] This decision indicates a possible flexibility in applying the 'publicity' test, which may be useful for protection against misuse of biometric information.

If the disclosure or sale of biometric information passes the 'publicity' test, the tort may be useful for addressing the sale of inaccurate biometric information that result in profiling of the data subject in an objectionable way, such as falsely disclosing that data subjects have certain diseases or transferring inaccurate biometric information, which, for instance, cause false rejection of an eligible data subject for getting certain services or benefits or entering certain places. It is important to note that this tort requires that the resulting portrayal of the data subject be entirely a misuse of information. However, the most difficult hurdle remains the 'publicity' definition, which renders this provision less meaningful for protection against misuse of biometric information.

5.3.3.4 Tort of misappropriation

This provides a solution against a person who misappropriates the name or likeness of another without the latter's permission and uses it for their own benefit, usually, but not necessarily, in a commercial context.[174] It has been noted that this tort seeks to protect the commercial interest in one's identity,[175] and compensates for mental distress.[176] The Restatement notes that the tort does not protect a person's name, as such, but rather protects the 'value or benefits' of that name, such as the individual's 'prestige, social or commercial standing'.[177] A typical case is illustrated in *Fanelle v. Lojack corp.* (2000),[178] where the plaintiff's photograph was used without permission and associated with car theft in an anti-theft advertisement. It is noteworthy that while the commercial use of the photo was emphasised in this case, it is not necessary for the data to be used in a commercial way for there to be a successful claim; cases have also been successful in non-commercial contexts.[179] Furthermore, the plaintiff doesn't necessarily need to be a celebrity to prove the 'value' of their identity, as illustrated in the above case.

Nevertheless, it has been emphasised that if the use of the individual's identity is considered newsworthy, is protected by the First Amendment, and is not commercial, it may be difficult for the individual to bring appropriation claims (Gorman, 2004). Despite this, the misappropriation tort may still be the most promising for addressing the misuse of biometric information. Understood broadly, the rights protected by the tort are infringed whenever biometric information is used commercially without consent or in some cases notice and either causes mental distress or damages the value of an individual's identity. Tort law could thus be used to redress unauthorised use of biometric information such as function creep, linking, tracking and profiling. The fear of permanent identity theft could also be a valid ground for mental distress.

Additionally, most common law countries interpret the traditional phrases 'name or likeness' to include any aspect of the plaintiff's personality. In *Abdul-Jabbar v. General Motors Corp.* (1996),[180] the court found that 'name and likeness' is not limited to present use and allows an athlete to pursue action against the appropriation of his former name. In *Waits v. Frito-lay, Inc.* (1992),[181] the court upheld a claim against a snack manufacturer for appropriating the voice of a well-known singer in a radio commercial. According to the court, any information that can identify a person may be accepted. Biometric information, therefore, can be regarded as a sufficiently accurate identifier of individuals and could be included under 'name and likeness'.

Despite this, the fact remains that individuals using privacy tort claims have never managed to hold data traders[182] liable for misusing the personal information of data subjects. However, some recent cases show that the appropriation tort may be possible to accommodate informational privacy claims. In *Remsburg v. Docusearch, Inc.* (2003),[183] the Supreme Court of New Hampshire dismissed an appropriation claim against a private investigator who sold information about a woman to her stalker who later killed her. The decision is based on the Restatement view that:

> Tortious liability for appropriation of a name or likeness is intended to protect the value of an individual's notoriety or skill. Thus, the Restatement notes, in order that there may be liability under the rule stated in this Section, the defendant must have appropriated to his own use or benefit the reputation, prestige, social or commercial standing, public interest or other values of the plaintiff's name or likeness. The misappropriation tort does not protect one's name *per se*; rather it protects the value associated with that name.[184]

Misappropriation is not actionable if a person's name or likeness is published for a 'purpose other than taking advantage of [the person's] reputation, prestige or other value'.[185] Misappropriation occurs most often when a person's name or likeness is used to advertise a product or when someone impersonates a person for gain. A person who sells personal information sells the information for the value of the information itself, not to take advantage of the person's reputation or prestige. In other words, the benefit derived from the sale in no way relates to the social or commercial standing of the person whose information is sold.[186] This reasoning may suggest that if a data trader compiles and sells a list of personal information not just for the information itself but for the reason of taking advantage of an individual's 'reputation, prestige or other value' that stems from the personal information, (for example, for compiling the profile of an individual's medicinal history, spending habits, travel frequencies, and so on) it would be theoretically liable, at least in the state of New Hampshire.

Promising *dicta* aside, obstacles still exist for using tort claims to address the misuse of biometric information. These include obstacles such as the 'publicity' requirement, the 'highly offensive' requirement and the requirement of selling information to 'profit from [the individual's] social standing or prestige'. All these make it difficult to bring a privacy claim in tort when biometric information is being used. Thus, tort law in its current form may not be able to sufficiently address privacy concerns raised by biometric technology.

5.4 Bio-privacy case law and regulations

5.4.1 Case law concerning biometrics

This section explores how biometrics-related cases have been dealt with in the US legal regime. All pertinent court rulings found by the author about the legality of biometric identifiers have upheld the application of such identifiers when it is justifiable on the grounds of public safety.[187] Constitution-based privacy rights have not been supported by the courts for people who are required to prove their identity with a fingerprint, although the government may have to show that it has a reasonable public purpose for taking a fingerprint. However, most cases about biometrics only focus on whether the compulsory collection of biometric information is reasonable or necessary. The judicial decisions about the use of biometric technology, therefore, only touch upon the tip of the iceberg. The privacy concerns about the use of biometric technology not only focus on whether to adopt this technology, but the restrictions on use of the technology.

5.4.1.1 Perkey v. Department of Motor Vehicles

Among the various cases regarding biometric technology use by governmental agencies, *Perkey v. Department of Motor Vehicles* (1986)[188] is unique. This case is worth mentioning because the California Supreme Court's decision went beyond merely deciding whether the collection of fingerprints *per se* is legal or not. In this case, one of the petitioner's arguments was that the mandatory fingerprint requirements for obtaining the new driver's license violated her right to privacy. The court examined the procedure to determine whether it involved the kind of intrusive invasion of bodily integrity that has been found either to violate 'due process' rights guaranteed by the Fifth and Fourteenth Amendments or to contravene the Fourth Amendment's proscription against 'unreasonable' searches.[189] Based on the precedents, the court concluded that fingerprinting alone did not improperly infringe upon an individual's right of privacy. The reasons are, in summary, as follows.

1. Fingerprinting does not implicate the Fourth Amendment since it 'involves none of the probing into an individual's private life and thoughts that marks an interrogation or search'.[190]
2. The physical process by which fingerprints are taken does not require 'penetration beyond the body's surface'. Therefore, it does not readily offend those principles of dignity and privacy which are fundamental to the US notion of due process.[191]

3. In light of the unobtrusive nature of the fingerprinting process, the courts in California have permitted the use of fingerprinting in many non-criminal contexts.[192]

However, concerning the petitioner's claim that the collection and dissemination of fingerprints for identification use by various governmental agencies or private entities violates her right of privacy as protected by Article 1, Section 1 of the California Constitution; the court concluded that the DMV's dissemination of fingerprint data to third parties for purposes unrelated to motor vehicle safety was unlawful under several provisions of the Civil and Vehicle Codes.[193] The relevant legislation mentioned was the Information Practice Act 1977.[194] The Act states that the right to privacy is threatened by 'the indiscriminate collection, maintenance, and dissemination of personal information and the lack of effective laws and legal remedies'.[195] The court stated the following:

> Although fingerprints, as such, are not specifically listed in the definition of 'personal information', there can be no doubt that the Act was intended to provide protection for precisely this type of record. Fingerprints fall within the category of information 'that identifies or describes an individual.' Moreover, a fingerprint is a much more private matter than a name or address and its disclosure can lead to much greater intrusions on privacy than the simple disclosure of an individual's name.[196]

The court explicitly recognised fingerprints as uniquely accurate and immutable forms of personal identification and claimed that 'fingerprint can serve as a "key" to sensitive and potentially dangerous information about a person [. . .] With current technology, an individual who provides a fingerprint for use in authenticating a personal identification system is ceding the ability to preserve a measure of anonymity in the sensitive areas of his or her life'.[197] Furthermore, the court found that '[f]ingerprints have been described as "sensitive" information in contrast to less sensitive identification information such as a name or a date of birth'.[198] The court then pointed out that although the fingerprints are collected by the department primarily for departmental purposes, a wide variety of agencies and private parties regularly use the files to identify persons from fingerprints. They concluded that dissemination of fingerprints for a purpose other than that originally intended is a violation of the applicant's right to privacy under the state constitution.

Thus, it is clear that the collection of physically non-intrusive biometric data for a specific legislated agency purpose may be accepted by US courts. However, unlike previous precedents, which stated that a fingerprint is merely an additional means of identifying the applicant, similar to age, sex, weight, height, hair and eye colour,[199] the court admitted that (1) a fingerprint is a relatively sensitive kind of personal information, (2) the collection of fingerprints affects anonymity, and (3) the misuse of a fingerprint violates individual privacy. The court's opinion is in accordance with my previous discussion about the Fourth Amendment and facial recognition in the sense that while individuals may not have a reasonable expectation of the privacy of their biometric information *per se*, they may well have a reasonable expectation of privacy with respect to how that biometric information will be used after its collection.

5.4.1.2 Messing v. Bank of America

In 2002, the Maryland Court of Special Appeals in *Messing v. Bank of America*[200] became the first court in the United States to decide the issue of mandatory collection of biometric means of identification in the context of commercial transactions. This case concerns a plaintiff who sued the Bank of America because he was required to provide a thumbprint before cashing a cheque at the bank. The plaintiff was not a regular customer of the bank. The court decided although not unanimously, that 'the Bank's request of a thumbprint upon the instrument constitutes a request for "reasonable identification" within the meaning of Maryland Code, Commercial Law Article, §3–501(b) (2)'.[201] Before this case, many similar cases that involved mandatory commercial use of biometric data were also supported by the supreme courts of other states.[202] Although, to my knowledge, there is not yet a case that has been brought to the federal court, it is generally held that the collection of biometric data in the private sector is reasonable in terms of scope and necessity. However, there are also dissenting views. For example, in the case *Messing v. the Bank of America* (2003), the Chief Judge of the Appeals court dissented:

> Today, honest citizens attempting to cope in this world are constantly being required to show or give drivers' license, photo identification cards, social security numbers, the last four digits of social security numbers, mothers' 'maiden names', 16 digit account numbers etc. Now the majority takes the position that it is "reasonable" for banks and other establishments to require, in addition, thumbprints and fingerprints. Enough is enough. The most reasonable thing in this case was petitioner's 'irritation with the bank of America's Thumbprint Signature program'.[203]

The majority of the Court of Appeals nonetheless found a thumbprint a 'reasonable identification' requirement set forth in federal banking regulations and state commercial transactions law for the following reasons.

1. A thumbprint is specified as one of a number of means of signature and authentication in the Maryland Uniform Commercial Code (UCC) (§1–201(39)).
2. Other biometric identification means have been upheld in non-criminal circumstances; furthermore, such identification means are reasonable and *necessary*[204] to combat growing cheque fraud.
3. The thumbprint is not unreasonably inconvenient and the inkless method used was even less intrusive than traditional fingerprinting, which has been repeatedly upheld as an 'unobtrusive' form of identification.
4. The fact that a thumbprint does not permit immediate identification but only assists in remedial prosecution is irrelevant, since both functions are valid purposes of a 'reasonable identification' requirement.[205]

The first two reasons were given in response to the petitioner's argument that the use of biometrics was a violation of privacy. The third reason is also relevant in the sense that it discussed the intrusiveness of the inkless 'thumbprint'. The fourth reason is in response to the petitioner's arguments that the thumbprint was not actually used for identification at the time it was provided, since the petitioner was not a usual customer of the Bank of America, and the bank actually had no storage of his thumbprint beforehand. Considering the relevance of privacy issues here, we will just focus on the first three reasons from the court.

The legal basis for the first reason states that '"Signed" includes any symbol executed or adopted by a party with present intention to authenticate writing'.[206] The court is correct to agree that this broad definition can possibly include any kind of biometric signatures. However, it is significant to point out that this provision can only indicate that the UCC accepts biometric signatures as an *adequate* means of providing authentication. There is no indication that it also amounts to a *necessary* means for authentication in the private context and accordingly mandatory collection could be regarded as reasonable. As we mentioned, the problem of using biometrics in the private context is not whether it is strong enough to be used as identification, but whether it is disproportionate to the task at hand.

The legal basis for the second reason is that there have been precedents from other courts that upheld biometrics as a reasonable identification method. It is true that many other cases have approved use in a number of other non-criminal situations, such as *Thom v. New York Stock Exchange* (1969),[207] *People v. Stuller* (1970)[208] and so on. Nevertheless, each case concerned biometric information in the public domain and the court decisions are generally based on discussions of whether there are Fourth or Fifth Amendment violations of the US Constitution. Consequentially, using these precedents as grounds for a final decision about this case in the private domain, which applies different legal resources, does not seem to be very persuasive.

The third reason is based on the subjective judgement that the inkless thumbprint is less intrusive than a thumbprint using ink. It is critical to note that an image that becomes digitally stored in the computer is actually much more easily preserved and copied than an image on paper. In consequence, the inkless thumbprint is actually at more risk of misuse than an image on paper.

The reasons listed by the court thus do not appear persuasive enough from a legal point of view. This is further evident in the dissenting views for *Messing v. Bank of America*, as stated above.

Nevertheless, the final decision of the court is generally in accordance with most similar court decisions about the collection of biometric information by government agencies. Many similar cases that involved mandatory commercial collection of biometric data have also been supported by the supreme courts of other states.[209] Although there has not yet been a case brought to the federal court, US courts are generally of the opinion that the collection of biometric information *per se* should not be restricted in any sense. Both commercial entities and government agencies can thus make the collection of the biometric information mandatory. However, subsequent misuse of the biometric information will still be prohibited, as the biometric information is regarded as 'sensitive personal information' by the California Supreme Court.[210]

This leads to two paradoxes. First, biometric promoters claim that the submission of biometric information is usually on a voluntary basis; hence, although privacy protection in the private domain is very limited in the United States, it may not result in much privacy concern (Woodward, 1997). This statement may be understood in reverse to indicate that mandatory collection of biometric information in the private domain would cause such concern. In the United States, most federal statutes and the constitution are only targeted at the privacy protection of cases relevant to government agencies. The mandatory collection of biometric information in the private domain may raise even more concerns about the subsequent misuse of biometric information because of the lack of legal remedies. Hence, the court's support of mandatory collection of biometric information in the private domain actually raises many potential privacy threats to individuals.

Second, in Europe, when personal information is categorised as 'sensitive personal information', it indicates that the collection of such information should be more restricted.[211] Once biometric information is collected, it becomes more difficult to restrict misuse, even if the misuse is prohibited by law. Hence, it is of great importance to restrict the collection of biometric information and set up a strict threshold for the collection of such data in order to protect privacy more effectively. However, in the United States, the courts generally hold the view that the collection of biometric information should be viewed as similar to the collection of other ordinary personal information such as names and addresses; on the other hand, the courts admit that biometric information is a kind of sensitive information, unlike ordinary personal information, which needs to be protected. Hence, the general supporting view of collecting biometric information in various situations may greatly undermine the protection of an individual's privacy concerning the biometric information in practice, especially in the private domain, where fewer legal remedies exist.

5.4.2 Privacy legislations concerning biometrics

Some states in the United States have worked on guidelines and provisions for the privacy of biometric information.[212] Among these are several bills that attempt to systematically regulate the use of biometrics.[213] For instance, in 2002, the New Jersey Senate introduced the Biometric Identifier Privacy Act (BIPA), which aims to provide a series of safeguards for biometric identification.[214] In 1999, the California Senate debated a bill to limit the private sector's use of biometrics.[215] A clash of views on privacy and biometrics erupted when the bill was proposed. The bill was amended by the committee; however, as it banned compulsory collection of biometric information by commercial entities, financial institutions opposed the amended bill because they use biometrics to verify non-customer payees and the bill would preclude facial scanning in automated cheque cashing services.

5.4.2.1 *Overview*

The Biometric Identifier Privacy Act would prohibit a person or government agency that possesses biometric data to sell, lease, transfer or disclose such information unless one of the following criteria have been met:[216]

- The individual to whom the biometric data belongs consents to the sale or transfer.
- The sale or transfer completes a financial transaction requested or authorised by the individual.
- The sale or transfer is required or permitted by federal or state law.
- The sale or transfer is made for law enforcement purposes.
- In addition, the bill requires the entity storing the biometric information to protect that data in a manner similar to other confidential information and provides for civil penalties for violation. Finally, it regulates a specific penalty for the breach of the sections in the bill.

SB 71 in California would regulate and restrict the use of biometric identifiers such as fingerprints, voice prints, and retinal scans to prevent their use in identifying theft or discrimination. It would prohibit compulsory collection of biometric information by an entity other than law enforcement or governmental agencies. For those who obtain consent, the bill would allow the biometric identifier to be used only to verify identity, and not to discriminate against any individual or community. The bill prohibits any person from providing biometric identifiers to third parties unless the transmission was encrypted and the encryption code is not shared with a third person. The bill would allow biometric verification or identification to be used as a condition of a commercial transaction if a notice of the policy was posted or if the advance consent of the individual was obtained.[217]

5.4.2.2 *Analysis*

Although the purposes of the two bills are to protect privacy and restrict misuse of the biometric information, individuals would still face a multitude of risks due to several obvious loopholes in the bills. To avoid repetition, the definition of the *biometric identifier* will be dealt with later in Chapter 6. Here, I will just highlight three main inadequacies of the two bills.

5.4.2.2.1 CONSENT

Consent is required as a prerequisite for the collecting and commercial use of biometric information in both bills, which is favourable for the protection of biometric information. However, the bills do not clarify how the consent should be obtained. Should it be an informed consent with clear notice of the risks for submitting the biometric information? Should it be opt-in consent or opt-out consent? Without real informed consent, it is unlikely an individual's privacy right to their biometric information would be effectively protected.

Where an individual has surrendered his personal information to the private sector, generally the Supreme Court has not found a constitution-based privacy right. For example, in *Smith v. Marylan* (1979),[218] the court noted that it 'consistently has held that a person has no legitimate expectation of privacy in information he voluntarily turns over to third parties'.[219] In another case, *United States v. Miller* (1976),[220] which involved a bootlegger's private financial records that were turned over to US Treasury agents pursuant to a grand jury subpoena, the court found that Miller had no expectation of privacy in his records, reasoning that 'the depositor takes the risk, in revealing his affairs to another that the information will be conveyed by that person to the government'.[221] The records, therefore, could not be considered confidential communications because they had been voluntarily conveyed to the bank in the ordinary course of business. Hence, once biometric information is submitted with consent by individuals in the primary market, even if it is captured only once, the biometric identifier could easily be replicated, copied and otherwise shared among countless public and private sector databases.

This kind of sharing of biometric information in the secondary market may, to some extent be restricted by the BIPA, which prohibits 'sale, lease or disclosure' to another 'person' except under limited occasions; and by the SB71, which states that 'except as otherwise provided, the person shall not sell, transmit, exchange or otherwise provide to third parties biometric identifiers or data containing biometric identifiers in its possession'. However, the ambiguity of 'another person' and 'otherwise provided' may permit the misuse of biometric information on some occasions. For example, if 'otherwise provided' can be explained as publication of a simple notice or policy, the 'person' holding the biometric information can avoid getting the individual's consent for the disclosure or sharing of biometric information.

5.4.2.2.2 UNAUTHORISED COLLECTION AND USE

In the BIPA, the collection of a biometric identifier is restricted in Article 3, which states, 'Notwithstanding any other provision of law to the contrary, no person shall obtain a biometric identifier of an individual, for the purpose of commercial advantage, without authorisation of the individual'. This provision emphasises the restriction of unauthorised collection of biometric identifiers for commercial use, but also leaves a significant loophole to other kinds of unauthorised collection as long as it is not for commercial advantage. The privacy rules in the bill do not impose explicit limitations on using the biometric information for purposes other than those consented by the data subjects at collection; thus, 'function creep' is still possible. Although in the BIPA, 'a person who possesses a biometric identifier of an individual shall not sell, lease or otherwise disclose the biometric identifier to another person', 'a person' is not explicitly prohibited from using the biometric identifier for purposes other than commercial benefit. For instance, the data controller can first collect the fingerprint for access control verification, and afterwards use it for profiling and surveillance. Since he is not gaining direct commercial advantage from it, the prohibition in the bill cannot apply to him.

Instead of stating that the usage of biometric identifiers should always be in accordance with the consented purpose, SB71 states that 'the collection and use of a biometric identifier shall be used solely to verify identity'. This statement is problematic in two ways: (1) as biometric technology is usually used for either 'verification' or 'identification', the term 'verify identity' is not precise enough from a technical point of view because it is likely to be mixed with the sole 'verification' process; (2) although the regulation intends to restrict the use of the biometric identifier only to 'verify identity', without a reference to ascertain context, this restriction is almost meaningless in practice because the basic function of the biometric identifier is to 'verify identity'. When it is used for profiling personal records, it acts as a key to 'verify identity'; when it is used as a surveillance method, it is also used to 'verify identity'. This actually permits a data controller to use the biometric identifier for any circumstances as long as it is being used to 'verify identity'.

5.4.2.2.3 ENCRYPTION

Pursuant to Article 3 of SB71, 'biometric identifiers may be transmitted electronically through or to third parties solely to verify their identity, if the transmission is for no other purpose, and the transmission is encrypted. Encryption codes may not be shared with any third parties'. This provision seems to afford greater privacy protection to the biometric identifier by requiring encryption for transmission, but the confusing wording of the provision makes it problematic in practice. According to the content of this provision, the purpose of the transmission must be solely to verify identity, and the criteria for transmission are that the transmission is encrypted and the encryption codes are not shared.

In essence, this only concerns security during the 'transmission process', and does not concern 'biometric identifiers' as such. From a technical point of view, encrypting the transmission process is only to guarantee the security of the transmission. As long as the 'encryption code' is shared with a third party, this measure does not prevent the third party from getting the unencrypted biometric identifier stored in the database. Although the regulation provides for the security of the transmission process, it does not guarantee the confidentiality of the biometric identifier that is not in the transmission process. Rather, this provision allows the data controller to transfer the biometric identifier to third parties without the authorisation of the data subject. As discussed above, the premise of using the biometric identifier solely for 'verifying identity' without any context restrictions cannot amount to any restrictions on the use of biometric identifiers in several circumstances.

5.5 Conclusions

This chapter has examined the right to privacy issues regarding both physical and informational privacy, as well as the property right of biometric information. Despite the comprehensive privacy regulations underlying United States privacy law, the analysis above illustrates the inadequacies of existing legislation with regard to the protection of biometric information. At the federal level, considering the substantial doubts on the strength and scope of the constitutional right of information privacy discussion in *Whalen*, it would be advisable to seek legislative protection for informational privacy about biometric technology rather than rely on the supposed constitutional right to informational privacy. The privacy statutes are not likely to restrict the collection of biometric information in any sense. Although limited privacy protections are available in the privacy statutes to prohibit the subsequent misuse of biometric information, significant loopholes still exist. Furthermore, most of these regulations are only targeted at government agencies. Other than several sector-specific laws and the recent Personal Data Privacy and Security Act of 2009, the implementation of biometric applications in private programs might be to a large extent exempted from constitutional guidelines and most private statutes.

Nevertheless, the prospect that biometric information may receive additional potential protection from the constitution and tort law has also been noted as possible. The Supreme Court's analysis in the *Whalen* decision provides valuable lessons for the informational privacy concerns in the biometric context. Court reasoning in biometric cases at the state level is also to some extent problematic, and the specific biometric regulations are far from mature. It is likely that the privacy proposal concerning the use of biometric technology in general will continue to fail in court, and strict bio-privacy legislation may not be supported by legislators. In states where bio-privacy legislation has been introduced, the lobbying groups whose interests will be affected have successfully defeated the proposals. However, the courts' recognition of the sensitivity of biometric information and the prohibition of unauthorised dissemination in *Perkey* may be expected as a promising precedent for the privacy concerns related to biometrics. Moreover, although there still seems to be a long way to go for bio-privacy legislation to succeed, the impact of the bio-privacy legislative movement has not been negligible. Biometric regulation is necessary to prevent unauthorised collections and transfers of biometric data, although comprehensive regulatory oversight, or the complete ban of biometrics in the private sector, is premature.

Notes

* This section is based on the author's article, Liu 2011.

1 On 28.10.2003, the Department of Homeland Security unveiled the United States Visitors and Immigrant Status Indicator Technology. Beginning 12 Jan 2004, visitors from selected countries that enter the United States at various ports of entry are photographed and fingerprinted by customs officials. See Electronic Privacy Information Centre (2003).

2 USA PATRIOT Act, H.R. 3162.

3 For instance, more than 90% of the nation's nuclear facilities use hand geometry readers to validate entry (Huddart, 2001), Disney World uses a two finger geometry system to verify its season pass holders (Higgins, 2003) and a thumbprint or eye scan is now being used instead of a card or password to access funds from ATMs. See Business Wire (2002).

4 For a more comprehensive overview, see Schwartz and Reidenberg (1996).

5 See, for example, *Olmstead v. US*, 277 U.S. 438 (1928) at 478 (Brandeis, J. dissenting: 'the makers of our constitution . . . sought to protect Americans in their beliefs, their thoughts, their emotions and their sensations. They conferred, as against the Government, the right to be let alone – the most comprehensive of rights and the right most valued by civilised men'). See also *Griswold v. Connecticut*, 381 U.S. 479 (1965) at 494 (Goldberg, J., concurring: 'The right of privacy is a fundamental personal right, emanating from the totality of the constitutional scheme under which we live'. Quoting *Poe v. Ullman*, 367 U.S. 497 (1961) at 521 (Douglas, J., dissents).

6 *Griswold v. Connecticut*, 381 U.S. 479 (1965) at 484.

7 *Whalen v. Roe*, 429 U.S. 589 (1977).

8 See, for example, *Bowers v. Hardwick*, 478 U.S. 186 (1986); *Burton v. Wilmington Parking Authority*, 365 U.S. 715 (1961); *Roe v. Wade*, 410 U.S. 113 (1973); *Jackson v. Metropolitan Edison Co*, 419 U.S. 345 (1974).

9 See, for example, the Tax Reform Act of 1976, 26 U.S.C.§6103 (1976), which provides that the Internal Revenue Service should limit the disclosure of an individual's tax information. See Driver's Privacy Protection Act, 18 U.S.C.§§2721–2725 (1999).

10 5 U.S.C.§552(a) (1999).

11 The Code of Fair Information Practices was the central contribution of the HEW (Health, Education, Welfare) Advisory Committee on Automated Data Systems in 1973. It supplied the intellectual and statutory framework for the Privacy Act of 1974 and served as a model for privacy legislation in the United States.

12 S.B. 1490; introduced in Congress on 22 July 2009. The Bill was approved by the Senate Judiciary Committee on 05 Nov 2009 and is now before the full Senate for consideration.

13 For example, Standards for Privacy of Individually Identifiable Health Information, 45 CFR Parts 160 and 164, promulgated under the Health Insurance Portability and Accountability Act of 1996, Pub. L. no. 104–191, (1996); the Family Educational Rights and Privacy Act, Pub. L. no. 93–380, (1974); the Children's Online Privacy Protection Act, 15 U.S.C.§§6501–6506 (1998).

14 See, for example, The Right to Financial Privacy Act of 1978, 12 U.S.C.§3401 (1978). It protects certain customer financial records from disclosure. The Fair Credit Reporting Act, 15 U.S.C.§1681 (1999), also applies to financial institutions, and regulates the disclosure of consumer credit information.

15 Gramm–Leach–Bliley Financial Services Modernisation Act, Pub. L. no. 106–102 (1999).

16 Restatement (Second) of Torts, §652(A) (1977).

17 Return to this in Section 5.3.3.

18 See Part II of the book. As noted there, the possible intrusion into informational privacy in respect of biometrics revolves mainly about the following: unauthorised collection of biometric information; unnecessary collection of extra personal information; unauthorised use of the collected biometric information; loss of anonymity and security loopholes in biometric technology that may raise privacy concerns. These are very important parameters for the analysis further below.

19 *Whalen v. Roe*, 429 U.S. 589 (1977).

20 *Whalen v. Roe*, 429 U.S. 589 (1977), at 605.

21 Id. At 607.

22 Id. At 607.

23 Id. At 606 (Brennan, J., concurring).

24 See Part II of the book.

25 *Whalen v. Roe*, 429 U.S. 589 (1977), at 609 (Stewart, J., dissenting).

26 *Thom v. New York Stock Exchange*, 306 F Supp 1002 (S.D.N.Y. 1969), affirmed, 425 F.2d 1074 (2d Cir. 1970).

27 398 U.S. 905 (1970).

28 Id., at 1011.
29 Id., at 1010.
30 *Newport v. Iacobucci*, 479 U.S. 92 (1986).
31 *Brown v. Brannon*, 399 F Supp. 133 (M.D.N.C 1975).
32 See, for example, *Doe v. Wigginton*, 21 F.3d 733 (6th Cir. 1994), which held that 'the Constitution does not encompass a general right to nondisclosure of private information', at 740 quoting *J.P. v. DeSanti*, 653 F.2d 1080 (6th Cir. 1981) at 1090.
33 *Whalen v. Roe*, 429 U.S. 589 (1977), at 599–600.
34 The so-called administrative search doctrine is explained further below.
35 *Olmstead v. United States* 277 U.S. 438 (1929).
36 See *Skinner v. Railway Labour Executives Ass'n*, 489 U.S. 602 (1989) at 616.
37 *Vernoniz School District 47J v. Acton*, 515 U.S. 646 (1995).
38 *Skinner v. Railway Labour Executives Ass'n*, 489 U.S. 602 (1989) at 616–618.
39 *Kyllo v. United States*, 533 U.S. 27 (2001).
40 Id. At 34.
41 Id. At 34.
42 *Davis v. Mississippi*, 394 U.S. 721 (1969).
43 Id. at 723.
44 Id. at 728.
45 Id. at 727.
46 I return to the factor of 'place' in the following section.
47 *United States v. Knotts*, 450 U.S. 276 (1983).
48 Id. at 282.
49 *United States v. Karo*, 468 U.S. 705 (1984) at 714–718.
50 *Maryland v. Macon*, 472 U.S. 463, (1985) at 469 (citing *United States v. Jacobsen*, 466 U.S. 109, (1984) at 113).
51 I return to this discussion in the following section.
52 *Vernonia School Dist.47J v. Acton*, 515 U.S. 646 (1995) at 653–654.
53 *Chandler v. Miller*, 520 U.S. 305 (1997) at 308.
54 Id.
55 *Katz v United States*, 389 U.S. 347 (1967).
56 *California v. Ciraolo*, 476 U.S. 207 (1986) at 211.
57 *United States v. Dionisio*, 410 U.S. 1 (1973).
58 Id. At 771.
59 See, for example, McCoy (2002) and references cited therein.
60 *United States v. Katz*, 389 U.S. 347 (1967).
61 *Kyllo v. United States*, 533 U.S. 27 (2001).
62 See Id. At 33.
63 *United States v. Katz*, 389 U.S. 347 (1967), at 351.
64 See, for example, *Kyllo v. United States*, 533 U.S. 27 (2001), at 32–33. Compare also the decision of the European Court of Human Rights in *Peck v. UK*, no.44647/98, ECHR, 2003-I. The latter judgment is elaborated in Chapter 4 Section 4.3.
65 See further, for example, Bolle et al. (2003), p. 26 screening establishes that a person is not on some watch list of 'interesting' people. See also Chapter 2 Section 2.3, especially footnote 42.
66 *United States v. Place*, 462 U.S. 696 (1983).
67 Id. at 138.
68 *United States v. Davis*, 482 F.2d 893 (9th Cir. 1973) at 908.
69 See *Skinner v. Railway Labor Executives Ass'n*, 489 U.S. 602 (1989) at 613–614.
70 *United States v. Albarado*, 495 F.2d 799 (2nd Cir. 1974) at 804.
71 *United States v. Edwards*, 498 F.2d 496 (5th Cir.1974).
72 Id., at 500.
73 Id.
74 *United States v. Davis*, 482 F.2d 893 (9th cir.1973).

75 *United States v. $124,570 US Currency*, 873 F.2d 1240 (9th Cir.1989).
76 Id., at1243.
77 *Michigan v. Clifford*, 464 U.S. 287 (1984) at 294–295.
78 Id.
79 *Skinner v. Railway Labor Executives Ass'n*, 489 U.S. 602(1989) at 616.
80 *Kyllo v. United States*, 533 U.S. 27 (2001).
81 Id. At 33–34.
82 *Terry v. Ohio*, 392 U.S. 1 (1968) at 16.
83 Id. At 21–22, in a sliding scale of interaction between the individual and the police, we have consensual encounter, the investigative stop, the quasi-arrest and the custodial arrest.
84 *United States v. Mendenhall*, 446 U.S. 544 (1980).
85 Id. At 555.
86 Id. At 554.
87 *Florida v. Bostick*, 501 U.S. 429 (1991).
88 Id. At 431.
89 Id. At 436.
90 For further discussion see Chapters 4 (Europe) and 6 (Australia).
91 U.S. Const., Amend. V.
92 *Andresen v. Maryland*, 427 U.S. 463, 470–71 (1976) (Blackmun, J., Quoting *United States v. White*, 322 U.S. 694 (1944) at 701 at 470–471.
93 *Wong Sun v. United States*, 371 U.S. 471, 484 (1963) at 484 (Brennan, J., The exclusionary rule which prevents evidence seized in violation of the Fourth Amendment from being introduced in court, 'extends as well to the indirect as the direct products' of unconstitutional conduct.).
94 *United States v. Verdugo-Urquidez*, 494 U.S. 259 (1990) at 264. (Rehnquist, C.J., 'Although conduct by law enforcement officials prior to trial may ultimately impair that right, constitutional violation occurs only at trial'.)
95 *Mc Carthy v. Arndstein*, 266 U.S. 34 (1924).
96 *New York v. Quarles*, 467 U.S. 649 (1984) at 671.
97 *Gibert v. California*, 388 U.S. 263 (1976).
98 *United States v. Chibbaro*, 361 F.2d 365 (3rd Cir. 1966).
99 *Schmerber v. California*, 384 U.S. 757 (1966) at 761
100 *United States v. Wade*, 388 U.S. 218 at 222 ('we have no doubt that compelling the accused merely to exhibit his person for observation . . . involves no compulsion of the accused to give evidence having testimonial significance'.)
101 When biometrics is used for authentication, anonymity is not so relevant because the technology is used to ensure the individual is who he claims to be.
102 *Talley v. California*, 362 U.S. 60 (1960) at 64 ('persecuted groups and sects from time to time throughout history have been able to criticise oppressive practices and laws either anonymously or not at all'.)
103 *NAACP v. Alabama ex. Rel Patterson*, 357 U.S. 449 (1958) at 462.
104 *McIntyre v. Ohio Elections Commission*, 514 U.S. 334 (1995) at 341–342.
105 Expressive conduct is also referred to as symbolic speech and involves communicative conduct that is the behavioural equivalent of speech. The conduct itself is the idea or message. Some expressive conduct is the equivalent of speech and is protected by the First Amendment. I will not go into too much detail about it here, as it is not the focus of the discussion.
106 This dichotomy excludes a large variety of actions (car theft, for example) from the kindly protections of the First Amendment.
107 It is the principle that the government must respect all of the legal rights that are owed to a person according to the law. The Fifth Amendment guarantee of due process is applicable only to actions of the federal government.

108 *State v. Baumann*, 191 Wis. 2d 824 (1995).
109 Id., at 2.
110 Since the focus of my book is on privacy, it would be excessive to go into more case discussion about the right to movement and its relevance to biometrics.
111 5 U.S.C.§552a(a)(4) (1994).
112 See, *Reuber v. United States*, 829 F.2d 133 (D.C. Cir. 1987) at 142, (letter reprimanding individual sent to and disclosed by agency was a 'record' because it clearly identified the individual by name and address),
113 For more discussion see Chapter 4.
114 A concept of 'System of Records' is regulated in the Privacy Act as 'a group of any records under the control of any agency from which information is retrieved by the name of the individual or by some identifying number, symbol, or other identifying particular assigned to the individual'.
115 OMB Guidelines, 40 Fed. Reg. 28, 948, 28, 952 (1975).
116 United States Department of Justice (1996) citing the Privacy Commission Report at 503–504 and no.7, at 556.
117 This is the case even though they may contain the same kind of information and may still be used in the same way that is covered by the act as 'system of records'. See further, The Privacy Act of 1974.
118 According to 5 U.S.C.§552a (a)(7), the term 'routine use' means, with respect to the disclosure of a record, the use of such record for a purpose which is compatible with the purpose for which it was collected. 5 U.S.C.§552a (b) routine use is a condition for disclosure of personal information.
119 The Privacy Act has been heavily criticised for its scope, or lack thereof, in general and many question whether the Privacy Act grants effective right to individuals. See, for example, Schwartz (1995) criticised the Privacy Act regarding the 'routine use' exemption, computer matching and transparency of data use. This chapter focuses solely on the discussion about how this Act can tackle the information privacy challenges posed by biometric technology.
120 For more general discussion about the deficiency of these exemptions of the Privacy Act, see Schwartz and Reidenberg (1996): 94–103.
121 Privacy Protection Study Commission (1977) chapter 13. These eight principles result from commission analysis, not a specific Congressional statement. I will use it for simplifying the analysis of the basic regulations of the act.
122 5 U.S.C.§552a (e) (2).
123 5 U.S.C.§552a (e) (3).
124 5 U.S.C.§552(a)(b) 'No agency shall disclose any record which is contained in a system of records by any means of communication to any person, or to another agency, except pursuant to a written request by, or with the prior written consent of, the individual to whom the record pertains . . . except . . .'.
125 *Perkey v. Department of Motor Vehicles*, 42 Cal.3d 185 (1986).
126 For more discussion about free and informed consent, see Section 3 of Chapter 6.
127 In June 2001, EPIC filed a series of requests under the Freedom of Information Act seeking access to government records regarding companies that sell personal information to the government. See further Hoofnagle (2004).
128 The Internal Revenue Service announced in Oct 2000 that it has awarded ChoicePoint Inc. (NYSE: CPS) a major new agreement to provide on-line public records for nearly all of its office locations over the next five years. ChoicePoint is a company that sells a wide array of information to the government. It is also a company that has attracted the attention of privacy advocates. At Privacy International's Big Brother Award ceremony held in Cambridge, Massachusetts in March 2001, ChoicePoint received the 'Greatest Corporate Invader' award 'for massive selling of records, accurate and inaccurate, to cops, direct marketers and election officials'. At Privacy International's Big Brother Award ceremony held in Seattle, Washington in April 2005,

ChoicePoint received the 'Lifetime Menace Award' for its continued efforts to build dossiers on individuals.

129 Unnecessary collection refers both to occasions when biometric technology is adopted over other adequate authentication measures, as well as the potential disclosure of excessive information because of the collection of biometric information.

130 *United States v. Doe*, 457 F.2d 895 (2nd Cir. 1972) at 898.

131 *Brown v. Brannon*, 399 F Supp 133 (M.D.N.C. 1975) at 138.

132 'Conditions of Disclosure. No agency shall disclose any record which is contained in a system of records by any means of communication to any person, or to another agency, except pursuant to a written request by, or with the prior written consent of, the individual to whom the record pertains, unless disclosure of the record would be (1) to those officers and employees of the agency which maintains the record who have a need for the record in the performance of their duties'.

133 See Privacy Protection Study Commission (1977) for further analysis of this restriction.

134 See generally Privacy Protection Study Commission (1977) for more analysis.

135 5 U.S.C.§552(a) (supp.1995). This agreement must be given to the Committee on Governmental Affairs of the Senate and the Committee on Government Operations of the House, and should be available to the public.

136 Gramm–Leach–Bliley Act and Health Insurance Portability and Accountability Act.

137 Section 3(12) Personal Data Privacy and Security Act of 2009.

138 Personal Data Privacy and Security Act of 2009, s. 1490, 111th.

139 This enforcement measure will be excluded from discussion here, because the focus of the book is the substantive level.

140 For further analysis of this issue, see Chapter 3, Section 3.4.

141 See, for example, Evers (2006). For instance, customers without a bank account wishing to cash a payroll cheque are often required to provide a fingerprint or thumbprint. Voice recognition software can be used in the banking industry to control access to account information. The trend of using biometric identifiers for identification at banks has also led to their use at many Automatic Teller Machines.

142 GLBA§509(4)(A)–(C); 12C.F.R.§332.3(n)(1) (2001).

143 Consumer reporting agency is an entity that collects and disseminates information about an individual's creditworthiness.

144 See, for example, Electronic Privacy Information Centre (2005).

145 However financial institutions are prohibited from disclosing, other than to a consumer reporting agency, access codes or account numbers to any non-affiliated third party for use in telemarketing, direct mail marketing or other marketing through electronic mail.

146 See, for example, in *Ting v. AT&T* 319 F.3d 1126 (9th Cir. 2003), Darcy Ting, individually and on behalf of all others similarly situated, and Consumer Action, a non-profit membership organisation, both as private attorneys general, brought suit against AT&T, alleging that AT&T's Consumer Services Agreement violates California's Consumer Legal Remedies Act and that state's Unfair Practices Act by barring customers from, among other things, pursuing claims against AT&T on a class wide basis.

147 There are numerous examples that can be easily found on the internet. It would be inappropriate for me to name some specific examples here due to banks' concern for their prestige.

148 GLB§509(6) defines 'affiliate' as 'any company that controls, is controlled by, or is under common control with another company'.

149 GLB§502(b)(2) 'This subsection shall not prevent a financial institution from providing non-public personal information to a non-affiliated third party to perform services for or functions on behalf of the financial institution, including marketing of the financial institution's own products or services, or financial products or services offered pursuant to joint agreements between two or more financial institutions that comply with the

requirements imposed by the regulations prescribed under Section 6804 of this title, if the financial institution fully discloses the providing of such information and enters into a contractual agreement with the third party that requires the third party to maintain the confidentiality of such information'.

150 Section 4(k) of Bank Holding Company Act is very long, and describes many activities that are financial in nature, for further information see Bank Holding Company Act of 1956.

151 According to the definition of 'customer relationship' the term 'time of establishing a customer relationship' shall be defined by the regulations prescribed under Section 504, and shall, in the case of a financial institution engaged in extending credit directly to consumers to finance purchases of goods or services, mean the time of establishing the credit relationship with the consumer. (GLBA §509(11)).

152 *Messing v. the Bank of America N.D.*, 373 Md. 672, 821 A.2d 22 (2003). The case is dealt with in more detail in Section 5.5.1.2.

153 *ABA v. Lockyer* Nos. 04-16334 and 04-16560 (9th Cir. June 20, 2005). The Electronic Privacy Information Centre (EPIC) describes the case as follows: 'Charter Pacific Bank sold its database containing 3.6 million valid credit card account numbers to a convicted felon who then fraudulently billed the accounts for access to Internet pornography sites that victims had never visited. In fact, approximately 45% of the victims did not even own a computer. Charter Pacific did not develop the database from its own customers' information. Instead, it compiled the information from credit card holders who had purchased goods and services from merchants that had accounts at Charter Pacific. The unrestricted sharing of this information resulted in over $44 million of unauthorised charges'. See Electronic Privacy Information Centre (2005).

154 For further discussion, refer to Chapter 3.

155 This new rule was issued in May 2002 pursuant to the GLBA (Section 501(b)) and applies to, among others, mortgage lenders and mortgage brokers. 'The Safeguards Rule sets forth standards for developing, implementing, and maintaining reasonable administrative, technical and physical safeguards to protect the security, confidentiality and integrity of customer information handled by all financial institutions over which the Federal Trade Commission has jurisdiction' (Benoit and Lovoy, 2003). The rule does not contain a specific penalty for non-compliance. However, non-compliance could lead to enforcement proceedings and/or lawsuits by the FTC or customers.

156 Restatement (second) of Torts§652B.

157 See, for example, *Schuchart v. La Taberna del Alabardero, inc.*, 365 F.3d 33 (D.C. Cir. 2004) at 36, which notes that liability for intrusion upon seclusion lies when a defendant 'offensively pries into a plaintiff's zone of privacy' including peeping through windows; *Doe 2 v. Associate Press*, 331 F.3d 417 (4th Cir. 2003) at 422, which held that liability for the tort of intrusion requires actions such as watching or spying in an area in which a person expects to be free from surveillance.

158 See, *Wolf v. Regardie*, 553 A.2d 1213 (D.C.1989) at 1217–1218, listing examples of intrusion including spying, reading another's mail, eavesdropping and trespassing.

159 *Humpher v. First Interstate Bank of Oregon*, 696 P.2d 527 (Or.1985) at 532–533.

160 Restatement (second) of Torts§652B cmt. C.

161 See, *Messing v. Bank of America N.D.*, 373 Md. 672, 821 A.2d 22 (2003).

162 *Weld v. CVS Pharmacy, Inc.*, 11 Mass L Rep 21 (1999).

163 Id.

164 See, *Taylor v. NationsBank N. A.*, 78 A.2d 893 (Md.Ct.Spec.App.1999) at 897, which states that the defendant's disclosure must be sufficiently offensive to carry 'overtones of mental distress' for the plaintiff to recover for unreasonable publicity.

165 See, for example, *Cox Broad Corp. v. Cohn*, 420 U.S. 469 (1975). Based on First Amendment grounds, the Supreme Court invalidated a state statute that established liability for the broadcast of a rape victim's name, when the broadcaster obtained the name from public records. After this case, if the information contained in a public record is

subject to disclosure under freedom of information or sunshine laws, it will generally not be considered private. In addition, persons who are public figures, voluntarily or otherwise, generally cannot sue for the publication of truthful information, even if the information published exceeds the scope of the figure's notoriety.

166 Restatement (second) of Torts§652D.

167 Restatement (second) of Torts§652D cmt. a. The Restatement explicitly distinguishes the publicity requirement from 'publication' as used in the defamation context, which requires disclosure only to a third party.

168 Restatement (second) of Torts§652E; it is yet undecided whether the defendant must act with reckless disregard for the truth or whether they can be liable for mere negligence.

169 Id. §652E cmt. False light is distinct from defamation, which protects the individual's interest in maintaining a good reputation. False light thus protects a personal interest – what one thinks of oneself, while defamation protects a relational interest – what others think of one. Keaton et al. (1984) §117.

170 *Braun v. Flynt*, 726 F.2d 245 (5th Cir.1984).

171 *McSurely v. McClellan* 753 F.2d 88 (D.C. Cir.1985).

172 Id., at 113.

173 Id., at 112.

174 Restatement (second) of Torts§652C cmt. B.

175 See McCarthy (2007) §1:26.

176 *Taylor v. City of Demopolis* (2005), unreported, noted that appropriation tort compensates 'damage to human dignity' (quoting *Allison v. Vintage Sports Plaques*,136 F.3d 1443 (11th cir.1998) at 1446).

177 Restatement (second) of Torts§652C cmt. C.

178 *Fanelle v. Lojack corp.*, 79 F.sup.2d 558 (E.D.Pa.2000).

179 See, for example, *Hinish v. Meier & Frank co.*, 113 p.2d 438 (Or.1941) at 439, 448, the plaintiff brought an appropriation claim against the defendant who signed the plaintiff's name on a telegram to the governor.

180 *Abdul-Jabbar v. Gen. Motors Corp.*, 85 F.3d 407 (9th Cir.1996) at 414–415.

181 *Waits v. Firto-lay, Inc.*, 978 F.2d 1098 (9th Cir.1992).

182 Here it refers to the data controllers who trade individual personal information for their own benefit without the data subject's authorisation.

183 *Remsburg v. Docusearch, Inc.* 816 A.2d 1001 (N.H.2003) at 1010.

184 Restatement (Second) of Torts§652 C Comment at 381.

185 *Matthews v. Wozencraft*, 15 F.3d 432 (5th Cir.1994) at 437.

186 Id.

187 See, for example, *Thom v. New York Stock Exchange*, 306 F Supp 1002 (S.D.N.Y. 1969), *People v. Stuller*, 10 Cal.App.3d 582 (1970); A California appellate case upheld a municipal ordinance requiring bartenders to submit fingerprints to the local police department.

188 *Perkey v. Dept. of Motor Vehicles*, 42 Cal. 3d 185 (1986) at 187–188.

189 Id.

190 See *United States v. Sechrist*, 640 F.2d 81 (7th Cir. 1981) at 86; *Davis v. Mississippi*, 394 U.S. 721 (1969); *United States v. Dionisio*, 410 U.S. 1 (1973) at 15.

191 See *Rochin v. California*, 342 U.S. 165 (1952).

192 See, for example, *Miller v. Murphy*, 143 Cal.App.3d 337 (1983) at 344–346. *People v. Stuller*, 10 Cal.App.3d 582 (1970) at 593–597.

193 *Perkey v. Dept. of Motor Vehicles*, 42 Cal 3d 185 (1986) at 187–188.

194 Civ. Code, § 1798 et Seq.

195 Id. Section 1 of Article 1.

196 *Perkey v. Dept. of Motor Vehicles*, 42 Cal 3d 185(1986) at 193.

197 Id.

198 Id. at 202.

199 *Thom v. New York Stock Exchange*, 306 F Supp 1002 (S.D.N.Y. 1969) at 1011.

200 *Messing v. Bank of America N.D.*, 373 Md. 672, 821 A.2d 22 (2003).
201 Ibid. It provides: '(2) Upon demand of the person to whom presentment is made, the person making presentment must (i) exhibit the instrument, (ii) give reasonable identification and, if presentment is made on behalf of another person, reasonable evidence of authority to do so, and (iii) sign a receipt on the instrument for any payment made or surrender the instrument if full payment is made'.
202 See, for example, *Perkey v. Department of Motor vehicles*, 42 cal 3d 185(1986), *Brown v. Brannon* 399 F. supp.133 (1975). *Thom v. New York Stock Exchange*, 306 F. Supp. 1002 (1970).
203 Chief Judge Robert M. Bell and Judge John C. Eldridge signed a separate opinion, dissenting as to the reasonableness of the identification requirement. *Messing v. Bank of America N.D.*, 373 Md. 672, 821 A.2d 22 (2003) at 704.
204 Emphasis added.
205 Id.
206 Maryland Uniform Commercial Code (UCC)§1–201(39).
207 *Thom v. New York Stock Exchange*, 306 F Supp 1002 (S.D.N.Y.1969).
208 *People v. Stuller*, 10 Cal.App.3d 582 (1970).
209 See, for example, *Perkey v. Department of Motor vehicles*, 42 cal 3d 185(1986), *Brown v. Brannon* 399 F. supp.133 (1975), *Thom v. New York Stock Exchange*, 306 F Supp 1002 (S.D.N.Y. 1969).
210 *Perkey v. Dept. of Motor Vehicles*, 42 Cal. 3d 185 (1986).
211 95/46/EC, art 8(2)
212 In 1998 and 1999, eight states introduced bills to prevent banks from requiring fingerprints prior to cashing a cheque: California (1998, AB2265), Georgia (1999, HB 481 and HB50), Missouri (1998, AB 1828), Pennsylvania (1998, HB15; 1999, HB22), Connecticut (1999, SB 697), Maryland (1999. SB270), Delaware (1999, SB697), Nebraska (1999, LB306); there are also many examples of regulating biometrics, e.g. Colorado Senate Bill 174 prohibits the sale of biometric information; Washington Senate bill 6399 (1998) prohibits one-to-many facial recognition searches of DMV database; Missouri House Bill 659 prohibits businesses from requesting fingerprints.
213 Besides the bills mentioned here, there are also bills proposed specifically for facial recognition: California SB 169, 2001; Massachusetts Bill (H.B. 44833, 181st Gen. Ct., 1999 Reg. Sess (mass.1999). The Massachusetts legislation was stopped, but was amended into a 'study bill' for interim work. Five states have introduced legislation to regulate the use of biometrics in the private sector: California (1999 – SB 71 and SB 129); Colorado (1999 – HB 1293 and SB 174); Massachusetts (1999 – HB 4483); Missouri (1999 – HB 659); and Virginia (1999 – HB 507). The Missouri bill and California's HB 71 have been fully stopped. California's HB 129 is still moving through the legislative process. The Massachusetts legislation was stopped, but was amended into a 'study bill' for interim work. The Virginia bill was enacted and now requires that businesses (but not banks), return or destroy biometric records within 21 days, unless the parties otherwise agree. And in Colorado, the Department of Revenue is now specifically prohibited from selling biometric and other personal information.
214 Biometric Identifier Privacy Act. Assembly A 1194. (latest version in 2006) State of New Jersey. 210th Legislature. Introduced June 13, 2002 by Joan M. Quigley, Introduced Pending Technical Review by Legislative Counsel.
215 California Senate, SB71, as amended 05.11.1999.
216 Biometric Identifier Privacy Act. Assembly No. 2448. State of New Jersey. 210th Legislature.
217 California Senate, SB71, as amended 05.11.1999.
218 *Smith v. Maryland*, 442 U.S. 735 (1979).
219 Id. At 743–744.
220 *United States v. Miller*, 425 U.S. 435 (1976).
221 Id. At 443.

References

Benoit, M., and Lovoy, E. (2003). Update on consumer financial privacy legislation and regulation, *Business Lawyer*, 58: 1163–79.

Bolle, R.M., Connell, J.H., Pankanti, S., Ratha, N., and Senior, A.W. (2003) *Guide to Biometrics*, New York: Springer.

Brogan, J.D. (2002) Facing the music: the dubious constitutionality of facial recognition technology, *Hasting Communications and Entertainment Law Journal*, 25: 65–81.

Business, W. (2002) *No More Passwords or Debit Cards – City-Money's New Atm – Iteller – Requires Only Thumb Print or Eye Scan.* Available at http://www.findarticles.com/p/articles/mi_m0EIN/is_2002_March_14/ai_83758283.

Cochran, M.L. and Chavarria, L.R. (2000) Privacy regulations under the Gramm–Leach–Bliley Act: new strategies for financial institutions, in *The E-Commerce Economy 2000. PLI Corporate Law & Practice Course Handbook Series No.B-1206*, New York: Practicing Lawyer Institute, 1017–1022

Electronic Privacy Information Center (2003) *US-VISIT.* Available at http://www.epic.org/privacy/us-visit/.

Electronic Privacy Information Center (2005) *Comments of the electronic privacy information center.* Available at http://www.epic.org/privacy/us-visit/comments080405.html.

Evers, J. and McCullagh, D. (2006). E-passports pose security risk, *ZENT News.* Available at http://www.news.com/Researchers-E-passports-pose-security-risk/2100-7349_3-6102608.html.

Fischer, R.L. and Camper, C.D. (1999) Reform law and privacy: a road map, *American Banker*, 19: 6.

Gorman, C.E. (2004) Comment, publicity and privacy rights: everything out the playing field for celebrities and private citizens in the modern game of mass media, *DePaul Law Review*, 53: 1247–1268.

Higgins, P.T. (2003) Fingerprint and hand geometry, in Brownlow J. K. (ed.), *Biometrics*, New York: McGraw-Hill, 25–41.

Huddart, M. (2001) *Hearing on Biometric Identifiers and the Modern Face Of Terror: New Technologies in the Global War on Terrorism, Senate Hearing 107–657.* Available at http://www.judiciary.senate.gov/resources/transcripts/107transcripts.cfm.

Keaton, W., Dobbs, D., Keeton, R., and Owen, D.W. (eds.).(1984). *Prosser and Keaton on the Law of Torts* (*5th ed.*), St Paul, MN: West Publishing.

Kent, S.T. and Millett, L. eds. (2003) *Who Goes There? Authentication Technologies Through the Lens of Privacy Committee on Authentication Technologies and Their Privacy Implications*, Washington: National Academies of Science Press.

Kempler, C. and Woody, R. (2000) *Living with the Gramm–Leach–Bliley Act.* Available at http:/www.insurelegal.com/livingwith031500.html.

Liu, Y. (2011) Biometrics and privacy protection in USA's constitution, *International Journal of Private Law*, 4(1): 54–68.

Ludington, S. (2006). Reining in the data traders: a tort for the misuse of personal information, *Maryland Law Review*, 66: 140–93.

McCarthy, J. T. (2007). *The Rights of Publicity and Privacy*, (2nd ed.), New York: West Group.

McCoy, S. (2002) Comment: O' big brother where art thou? The constitutional use of facial-recognition technology, *John Marshall Journal of Computer and Information Law*, XX (3): 471–85.

Nguyen, A.T. (2002) Here's looking at you, kid: has face recognition technology completely outflanked the Fourth Amendment? *Virginia Journal of Law and Technology*, 7(2). Available at http://www.vjolt.net/vol7/issue1/v7i1_a02-Nguyen.PDF.

Open Congress (2009). *Official summary of the PDPSA*. Available at http://www.opencongress.org/bill/111-s1490/show.

Prosser, W. L. (1960). Privacy. *California Law Review,* 48: 383–9.

Schwartz, P.M. (1995). Privacy and participation: Personal information and public sector regulation in the United States, *Iowa Law Review*, 80(553): 584–92.

Schwartz, P.M. and Reidenberg, J.R. (1996) *Data Privacy Law*, Charlottesville, VA: Michie Law Publishers.

Schiller, J.C. (2003) Comment, information privacy *v.* the commercial speech doctrine: Can the Gramm–Leach–Bliley Act provide adequate privacy protection? *Comm Law Conspectus*, 11: 349–363.

Slobogin, C. (2002) Public privacy: camera surveillance of public places and the right to anonymity, *Mississippi Law Journal*, 72: 213–316.

Tomkovicz, J.J. (2003) Technology and the threshold of the Fourth Amendment: a tale of two futures, *Mississippi Law Journal*, 72: 317–433.

Woodward, J.D. (1997) Biometrics: Privacy's foe or privacy's friend? *Proceedings of the Institute of Electrical and Electronics Engineers*, 85(9): 1480–1492.

Woodward, J.D. (2001) Super Bowl Surveillance: Facing Up To Biometrics, Rand Issue Paper 209. Available at http://www.rand.org/pubs/issue_papers/2005/IP209.pdf.

Warren, S.D. and Brandeis, L.D. (1890). The right to privacy, *Harvard Law Review*, 4: 193–211.

6 Privacy regulations on biometrics in Australia*

6.1 Introduction

Australia, like most other industrialised countries, is characterised by its high level of technological development. Smart cards and national identity schemes, biometric data including biometric passports, genetic testing and discrimination, microchip implants and radio frequency identification technology are important issues that currently face the Australian privacy law system (Australia Parliament, 2006). There are many laws available to help resolve these issues, but they are based on decades-old technology. Hence, the challenge for the current legislative regime is adapting to the reality of emerging technologies and at the same time providing adequate privacy protection for Australian citizens.

Concerns about biometrics are of particular importance in light of the Australian government's adoption of the e-passport[1] to be in compliance with the US requirement for visa-waiver countries, as well as by Australian customs authorities to expand the SmartGate scheme,[2] along with the rollout of biometric databases by immigration authorities in connection with processing visa applications.[3] The government's adoption of biometric technology is leading the way to greater biometric adoption in the private sector, including physical access for employees.[4] Biometric systems are being increasingly used or contemplated by organisations such as taxi-booking services, banking, and entrance control (Australian Government Office of Privacy Commissioner, 2006). The growing importance of identity management and the spreading use of biometric technology have precipitated vigorous debate in Australia about the legal regulation of personal information and reasonable limits of privacy. Some of that debate focuses on plans to roll out extensive facial recognition systems at Australian airports (Clarke, 2003a) and the introduction of biometric information in smart cards (Australian Privacy Foundation, 2004). Discussion from a legal perspective is manifest largely in a paper by the previous federal privacy commissioner, Mr. Crompton, in documentation regarding the drafting of the *Privacy Code* of the Biometrics Institute,[5] and, more recently, in the Australian Law Reform Commission's report, *For Your Information*, in which biometric issues are tackled fairly extensively. Otherwise, discussion on biometrics from a legal perspective is quite limited.

This chapter aims to provide an analysis of the current regulatory environment, at the federal level, of privacy protection concerning biometrics in Australia. As Australia is a federation, the study only focuses on the federal Privacy Act 1988 (Cth) and the Biometrics Institute Privacy Code. These two instruments provide the central privacy and data protection rules for regulation of biometrics in the private sector and the federal government sector. There are a small number of state laws on privacy and data protection but these are not analysed as they are of relatively minor significance for the biometrics-related concerns addressed in this book.

Section 6.2 of this chapter briefly describes the relevant sources of privacy rights in Australia, with special focus on the Privacy Act 1988 and the Biometrics Institute Privacy Code. Section 6.3 details the legal concerns of the use of biometrics, and an analysis is then made concerning the implications of privacy protection sources. The chapter concludes in Section 6.4 with suggestions for potential legislation necessary to protect privacy in the biometric context.

6.2 Sources of privacy protection

Australia is a federation of former British colonies. The federal (Commonwealth) Parliament has specific legislative powers according to Section 51 of the Constitution. In Australia, there has until recently been no recognition of a general tort of breach of privacy.[6] Actions for breach of confidence,[7] defamation, trespass or nuisance are occasionally used in support of privacy rights (Privacy International, 2004). The International Covenant on Civil and Political Rights (ICCPR) Article 17 can also provide privacy protection to Australians.[8] Australian privacy laws are contained in a variety of Commonwealth, State and Territory Acts. The Commonwealth legislation, according to Section 109 of the Constitution, prevails over State or Territory privacy legislation to the extent that these laws are inconsistent. At the federal level, the Privacy Act 1988 (Cth) is the main source of protection that is relevant to biometrics.[9] In addition, the industry self-regulated Biometrics Institute Privacy Code also has special value as it is one of the few existing biometric-specific privacy codes.

6.2.1 The Privacy Act

The Federal Privacy Act incorporates the amendments made to it by the Privacy Amendment (Private Sector) Act 2000 (Cth). Before this amendment, the Act mainly covered only Commonwealth and ACT government agencies, but the scope is now wider and encompasses some private sector actors. The Act covers the collection, use and disclosure, as well as the quality and security of personal information. It provides various basic privacy principles depending on what area it is regulating. There are 11 Information Privacy Principles (IPPs) governing the handling of personal information by Commonwealth and ACT government agencies as set out in Section 14.[10] They specify matters such as (1) the purpose and manner of collecting personal information (IPPs 1–3); (2) the way the personal information should be stored (IPP 4); (3) the way the records are to be kept (IPPs 5, 8); (4) access to and revision of the personal information in records (IPPs 6–7); (5) the reuse of the personal information kept in records (IPPs 9–10); and (6) limitations on the disclosure of the personal information (IPP 11).

In addition, the Act applies to the handling of personal information by health-service providers and private-sector organisations that earn >$A3 million annually. It establishes minimum privacy standards for the Australian private sector called National Privacy Principles (NPPs).[11] Organisations are required to comply with the NPPs unless they are exempt. The Principles are 10 in number and deal with (1) collection and use of personal information (NPPs 1–2); (2) the quality and security of personal data stored (NPPs 3–4); (3) the level of openness, access and revision of personal information by data subjects (NPPs 5–6); (4) the adoption of identifiers (NPP 7); (5) anonymity (NPP 8); (6) trans-border data flows (NPP 9) and (7) sensitive information (NPP 10). The legislation does not generally apply to small businesses (i.e. businesses with an annual turnover of <$A3 million) and it does not apply to political parties, employee records held by current or former employers of the employees, or to acts and practices of the media in the course of journalism.[12]

The two sets of privacy principles share many similarities but are, in places, inconsistent and unclear. A unified privacy protection system has recently been proposed by the Australian Law Reform Commission (ALRC) in order to resolve much of the complexity in the current provisions (Australian Law Reform Commission, 2008). At the time of writing, it is not clear exactly how the ALRC proposals will be implemented.

6.2.2 Biometrics Institute Privacy Code

The independent Biometrics Institute representing the Australasian biometrics industry and users received approval from the federal Privacy Commissioner on 19 July 2006 for a binding Privacy Code. The Code is an industry self-regulatory code. It came into operation on 1 September 2006 and is intended to be ratified by the organisations in the biometrics industry in Australia on a voluntary basis. Once the Code is ratified by a certain organisation, it will be bound by the Code instead of the NPPs.

This regime is based on the provisions of the Privacy Act 1988 which give the option for organisations to develop their own privacy codes which, when approved by the federal Privacy Commissioner, replace compliance with the NPPs. Section 18BA provides that 'an organisation may apply in writing to the Commissioner for approval of a privacy code'.[13] The Privacy Commissioner must approve each such code in accordance with the Act. Where an organisation consents to be bound by an approved code, the code operates in place of the NPPs until the organisation ceases to be bound by the Code.

The Biometrics Institute Privacy Code largely replicates the wording of each of the NPPs (from NPP1 to NPP10) but incorporates higher standards of privacy protection in relation to certain acts and practices concerning employee records that otherwise would be exempted from coverage by the Privacy Act. It seeks, inter alia, to give guidance on how privacy principles shall be applied to the collection and use of biometrics. At the same time, the Code includes three new supplementary Privacy Principles (11–13) specifically concerning biometric technology.

Principle 11 prescribes an array of security measures (including encryption) for protecting biometric information either while in storage or transmission. It supplements the data security obligations in NPP 4.

Principle 12 prescribes a set of measures that are aimed at enhancing data subjects' control over biometric information collected on them. The measures involve requirements of notice, consent, and withdrawal of consent. Particularly significant is a prohibition on 'secondary analysis or function creep of biometric information' without the data subjects' 'express free and informed consent'. The measures supplement NPPs 1.3, 1.5, 2 and 4.

Principle 13 deals with accountability matters. It requires, inter alia, disclosure of the purposes of biometric systems, third-party auditing of such systems, together with holistic privacy management (including the use of privacy impact assessments) of such systems. These requirements augment NPPs 1, 4 and 5.1.

6.3 Coverage of Privacy Act and Privacy Code

6.3.1 Definition of biometrics

According to the Biometrics Institute Privacy Code,[14] biometrics refers to 'automated methods for identifying and/or verifying an individual on the basis of some biological or behavioural unique characteristic of the individual' (Section E). Concomitantly, a 'biometric' is 'a biological or behavioural unique characteristic of an individual which is captured for the purpose of identification and/or verification of the individual' (Section E). As for 'biometric information', this is defined as follows:

> Biometric information is any data that can be used to biometrically identify an individual. This data includes, but is not limited to, images, sounds, chemical or geometric properties. It also includes any information encrypted or unencrypted that is derived from these raw acquired biometrics, such as biometric templates or filtered or pre-processed data.
>
> (Section E)

If we consider the definitions of 'biometrics' and 'biometric' first, they emphasise the 'unique' characteristic of an individual. This is somewhat surprising as not all biometrics are unique (as discussed in Chapter 3, Section 3.2.1.2). Furthermore, verification usage is mentioned as one purpose in parallel with identification, yet the definition of biometric information omits reference to verification or authentication.

The definition of biometric information is nonetheless to be lauded for its relative degree of clarity. A clear definition of biometric information or biometric data has not been found in any legal documents in Europe concerning privacy and biometrics, but it is generally indicated there that biometric information shall include both the raw biometric image and the biometric template.[15] By contrast, the Biometric Identifier Privacy Act enacted by the State of New Jersey in the United States provides only an unclear and imprecise definition of what can be considered a biometric identifier: 'retina or iris scan, fingerprint, voiceprint or record of a hand or face geometry'.[16] It fails to give an accurate description of the main characteristics of biometric information in general, and only provides a list of examples instead. That list seems to be exhaustive, but it actually excludes many other aspects of biometric information in the general sense, and it remains nebulous as to whether it refers to the raw biometric image or the biometric template. Compared with both the European understanding of biometric data and the definition contained in the New Jersey legislation, the Privacy Code's definition includes more than raw biometric images and biometric templates, as it states '. . . such as biometric templates or filtered or pre-processed data', which tends to include more relevant data that may appear during the biometric processing.

Although the attempt to give a legal definition of biometric information in the Privacy Code can be seen as a leap forward compared with the other discussed definitions, the current definition of *biometrics* and *biometric information* in this code needs to be delineated more clearly and precisely. The code defines biometric information 'as any data that *can be* used to biometrically *identify* an individual' (emphasis added). First of all, the word 'biometrically' is obscure, although the meaning might be deduced from the definition of 'biometrics' (i.e. 'automated methods for identifying and/or verifying an individual on the basis of some biological or behavioural unique characteristic of the individual'). Also, instead of using the typical wording 'for the purpose of identification and verification', the specific use of the words 'can be' is important. Does this indicate that the Code accepts that biometric information is generally 'identifiable', since by its nature it is created to identify people? Or does it mean to exclude some part of biometric information that is not identifiable – if there is any?[17] Since biometric information can be used for authentication as well as for identification purposes, does this specification of the words 'can be' indicate that no matter what purposes the biometrics are intended to be used for, biometric data in general are able to identify people?

6.3.2 Personal information and identifier

The Privacy Act (Cth) section 6 defines personal information to be:

> Information or an opinion (including information or an opinion forming part of a database), whether true or not, and whether recorded in a material form or not, about an individual whose identity is apparent, or can reasonably be ascertained, from the information or opinion.

This definition is replicated in the Biometrics Institute Privacy Code section E. At the same time, the Code also operates with the concept of 'biometric information', which is defined such that it appears to be a category of 'personal information'. This approach conforms with an earlier analysis by Crompton of when biometric information may be personal information (Crompton, 2002). Crompton concluded that the Privacy Act will apply to biometric information when it is clearly information about 'an individual', and when biometric information is created to identify people (Crompton, 2002). Biometric data, by their nature, are intimately and specifically linked to a certain person. Even if sometimes collected only for authentication, there is the possibility that these data *can be* (as stated in both the Privacy Act and the Privacy Code) related to an identifiable individual. In this respect, biometric information falls within the scope of 'personal information', and so the Privacy Act should be applicable.

The Privacy Act also contains a principle dealing explicitly with identifiers; NPP 7 defines an identifier as including 'a number assigned by an organisation to an individual to identify uniquely the individual for the purposes of the organisation's operations'. However, an individual's name or ABN (as defined in A New Tax System (Australian Business Number) Act 1999 (Cth)) is explicitly excluded as an identifier. A similar provision does not appear in the IPPs, which indicates that it only applies to private-sector actors.

Biometric information as code that relates to physical or behavioural characteristics of individuals might be included in the current definition of an identifier. According to the Revised Explanatory Memorandum to the Privacy Amendment (Private Sector) Bill 2000, the identifiers are 'not limited to letters and numbers', although an identifier 'will often contain either, or both'.[18] However, NPP 7 also requires that the identifier 'identify uniquely the individual for the purpose of the organisation's operations'. As discussed in Chapter 3, while biometric information is relatively unique to an individual, a number of factors may affect the accuracy of a match and the quality of a certain biometric template, which as a result may affect the 'uniqueness' of the biometric identifier. As a result, it has been suggested that an amendment to the definition may be required to ensure that biometric information is in fact captured by the 'identifier' principle (Australian Law Reform Commission, 2008).

6.3.3 Exemptions

The Privacy Act exempts certain specified actions, practices and organisations from its coverage. Section 7B exempts individuals in a non-business capacity, organisations acting under commonwealth contract, employee records, journalism and organisations acting under state contract, and s. 7C exempts political acts and practices. Of these various exemptions, those for employee records and small businesses may appear to be most critical where biometrics is concerned. If these two exemptions continue to pertain, a significant number of biometric applications would be able to escape from the data privacy legislation in Australia, which would render all the principles and considerations therein meaningless for those applications. However, the Biometrics Institute Privacy Code includes employee records, which is an improvement on the Privacy Act. Unfortunately, the Code still keeps the small business exemption, even though that exemption is actually more problematic than that for employee records – as discussed below.

Using biometric technology as an authentication method for computer network access and access to office buildings is becoming more and more common. For example, as already noted, in the Australian Woolworths supermarkets, finger-imaging technology is used to monitor time and attendance for about 100,000 employees. If these records are regarded as employee records, it will mean none of the employees' biometric data will be protected by the Privacy Act. The vacuum created by the exemption is supposed to be filled by state government legislative initiatives on workplace privacy, but the legislation so far enacted would seem to be far from adequate, at least in the context of biometrics. For instance, the main legislation on workplace privacy in Victoria is the Surveillance Devices Act 1999, which covers certain uses of devices for 'data surveillance', 'optical surveillance' or 'tracking'. Arguably, biometric equipment might be regarded as a 'tracking device' under the Act. Such a device is defined as 'an electronic device the primary purpose of which is to determine the geographical location of a person or an object' (s. 3(1)). But if the biometric equipment's primary purpose is to authenticate or identify people, it would seem difficult to argue that such equipment is covered by this definition. As for other states, most of them (with the exception of New South Wales) have yet to enact workplace privacy legislation, which means that in most if not all Australian jurisdictions, the use of biometric data as employee records is left inadequately regulated.

It is even more difficult to find a justification for the small business exemption in the Privacy Act. Small businesses comprised up to 94% of Australian business in 2000 (Australian Department of Employment, 2000) which indicates that many consumers in Australia may enjoy comparatively weak protection of their privacy interests when interacting with small businesses. It can be reasonably assumed that a significant (and increasing) portion of these enterprises are adopting biometric technology as authentication or identification methods for various commercial purposes. The deficiency of the small business exemption is not only restricted to the biometric scenarios; rather, it is a deficiency of the Privacy Act in general. A number of criticisms of the exemption have been expressed. Considering the relevance to the protection of privacy in biometrics, they can be summarised as follows:

- Privacy rights do not disappear just because a consumer happens to be dealing with a small company (Electronic Frontiers Australia, 2005).
- The $3 million annual turnover threshold for the exemption is not usually apparent to the consumers (Privacy Commissioner New South Wales, 2005).
- By the expedient splitting of any constituent businesses into sub-businesses before they reach the $3 million threshold (s. 6D(4)(a)), big businesses can escape completely from the operation of the Act (Privacy Commissioner New South Wales, 2005).

6.3.4 Sensitive information

As previously discussed in Chapter 4 (Section 4.3.2.1), biometric information does not fit neatly into the category of 'sensitive information', besides its health and genetic-related nature. However, biometric information may carry greater risks than other forms of information due to its linking and tracking ability, its relative uniqueness and permanence, and its intimate link to physical characteristics. For these reasons, it is pertinent to discuss whether and how biometric information can be categorised as 'sensitive information' under the Privacy Act 1988 (Cth).

The concept of 'sensitive information' is defined in s. 6(1) of the Privacy Act. Sensitive information is a subset of personal information. It includes (a) information or an opinion about an individual's (i) racial or ethnic origin; or (ii) political opinions; or (iii) membership of a political association; or (iv) religious beliefs or affiliations; or (v) philosophical beliefs; or (vi) membership of a professional or trade association; or (vii) membership of a trade union; or (viii) sexual preferences or practices; or (ix) criminal records that also contain personal information; or (b) health information about an individual; or (c) genetic information about an individual that is not otherwise health information.

This definition is broader than the definition of 'special categories of personal information' in the EU Data Protection Directive, as discussed in Chapter 4, in that it clearly specifies that genetic information is sensitive information. By contrast, the Directive does not explicitly mention genetic information, though the latter is usually regarded as relevant to health information in most cases.

The NPPs provide additional restrictions on the collection of sensitive information; it is subject to a higher level of privacy protection than other 'personal information' handled by organisations (NPP 10). However, the IPPs which apply to government agencies contain no equivalent restrictions, which is problematic. The Biometrics Institute Privacy Code includes an additional principle (Principle 12.3) similar to the scope of protection provided for sensitive information by NPP2.1 (a):[19]

> Secondary analysis or function creep of biometric information collected for purposes such as authentication or identification is not permitted without express free and informed consent. For example, biometric information collected for the purposes of authentication and identification shall not be used to examine that information in search of genetic patterns or disease identification without express free and informed consent.

This may indicate that biometric information is classified as sensitive information in this Code, and thus is provided with a higher level of protection.

The ALRC has proposed that biometric information should be given the same level of protection as other information that is currently treated as sensitive information; concomitantly, the definition of sensitive information should be amended to include biometric information (Australian Law Reform Commission, 2008). The ALRC intends to limit the type of biometric information to be included in the definition of sensitive information. It proposes that the sensitive information should only include 'biometric information collected for use in automated biometric authentication and identification systems and biometric template information' (Australian Law Reform Commission, 2008, p. 452). This opinion is in accordance with the view of the Office of the Privacy Commissioner (OPC), which states it may be impractical and undesirable for all biometric samples to be included under the definition of sensitive information, especially where there is no intention to use the sample for biometric matching or identification (Australian Government Office of Privacy Commissioner, 2006). An example is given that taking a photograph of a person shall be excluded. This opinion is reasonable to the extent that it takes into account the practicability of enforcement. But it may also create a loophole for easily circumventing the protection of biometric information as sensitive information. Because 'intention' is a subjective term, suppose, in practice, biometric information may be collected from an individual without consent since it is classified as ordinary 'personal information' by a private agency, as there is no *intention* to use it for identification and authentication, and thus no biometric template is created. 'Intention' could change from time to time, once the collection is made; therefore, it becomes difficult to control the biometric information at the level of its individual source. All that can be done is to trust the collector.

6.4 Information privacy challenges of biometrics

6.4.1 Unauthorised collections: notice and consent

To realise the maximum control by a data subject over personal information, their free and informed *consent* is indispensable. However, as a basic means of 'controlling' personal information during the primary collection, consent is not clearly required by the Privacy Act 1988 (Cth). With respect to the federal government sector, IPP 1(1) requires that the collection of personal information be for lawful *purposes* and by lawful *means*. Inconsistently, NPP 1, for the private sector only, requires that the *means* of collection be lawful. In addition, IPP 2 requires that, where *practicable*, the individual concerned is to be made generally aware of the purpose and to whom the information will be disclosed. Solicitation is required for the collection of private information from the individual concerned, though it does not mention that *consent* is required under any circumstances. In NPP 1.3, the requirement of 'practicable' is replaced using the wording 'must take reasonable steps to ensure'.[20] The Biometrics Institute Privacy Code generally applies the same wording as in NPP 1, but its supplementary Principle 12.1 emphasises that 'enrolments in biometric systems shall be voluntary, unless required by law'. This indicates that the Privacy Code prohibits compulsory collection of biometric information in the private sector, unless the law regulates otherwise. However, it is not clear when it should be regarded as 'required by law'. If 'law' includes the Privacy Act, this would be very problematic as it would undermine the consent requirements of the Code. As noted above, the Act does not require consent for primary collection. Thus, the collection of biometric data can be compulsory under the Privacy Act. Individuals have no choice or control whatsoever concerning collection, use and disclosure of their personal information for the primary purpose of collection.

However, both the NPPs and IPPs forbid the covert collection of biometric data, as they both require that information be collected by lawful and fair means (IPP 1, NPP 1.2). The notice requirements form an important foundation for prohibiting covert collection. In the Guidelines to the National Privacy Principles and in the Plain English Guidelines to the Information Privacy Principles, the Office of the Federal Privacy Commissioner has interpreted this to mean that information must not be collected covertly, although exceptions have been made for investigations of criminal offences (Office of the Federal Privacy Commissioner, 2001a, 2001b). However, even if notice is given when collecting biometric information, it is not guaranteed, in practice, that the notice clearly includes all of the required information or provides links to such information (Australian Privacy Foundation, 2005). In the context of biometrics, the 'notice' becomes even more problematic. The understanding of biometric technology is quite limited if not non-existent among lay persons; moreover, most people are not aware of the privacy and security risks they are exposed to by providing their biometric data. Science fiction movies and misleading advertisements appear to be the most widespread ways of getting to know about the technology. Therefore, how and to what extent the amount of information should be presented to the data subjects become crucial. Notice consisting of a short and concise statement with links to further resources would be useful and desirable.

NPP2.1 and IPP11 (b) both require consent for secondary collection whenever *practicable*. However, this principle can be greatly abused. As mentioned above, lacking a clear and concise notice, many individuals are likely to be irritated by the length of tedious statements, and might tend to simply choose to sign the consent without being clearly informed. In addition, it is likely that few individuals will challenge the request due to the imbalance of power between individuals and organisations. When they have no better alternatives, they might simply give their consent because of the convenience of doing so. This might pose an even more serious problem in using the biometric technology. An obvious advantage of biometric technology is that it offers convenience in contrast to the old authentication/identification methods, and it is arguable that people generally do not understand the danger of secondary disclosure of their biometric information. How to ensure the *free and fully informed choice* to the individual concerned about biometrics is another question that needs to be considered. It has been suggested that the ability to revoke consent and withdraw from participation is a good indicator of consent being truly free (Australian Privacy Foundation, 2005). This is shown in Article 12.4 of the Privacy Code, which states that individuals who have enrolled in a biometric system should be given the opportunity to have their biometric information removed from the system upon request where *possible*. The 'possible' criterion might, in practice, become an obstacle for withdrawal of consent. Furthermore, as far as the author is concerned, it may be more meaningful to provide the individual concerned with 'truly equal' alternatives when they decide not to provide biometric data, instead of having to give up the service or not enter into a transaction. These alternatives would adhere to the 'no disadvantage principle' in the Australian Privacy Charter which states: 'People should not have to pay in order to exercise their rights of privacy described in this Charter (subject to any justifiable exceptions), nor be denied goods or services or offered them on a less preferential basis. The provision of reasonable facilities for the exercise of privacy rights should be a normal operating cost'.[21]

Besides the consent and notice problem, the covert collection of biometric data can also be facilitated by the use of radio frequency identification chips (RFID). RFID chips can make the unauthorised collection of various biometric data stored on the card much easier.[22] RFID chips can store and broadcast a great deal of biometric information, which could be used from a distance to identify and track individuals without their knowledge and consent (American Civil Liberties Union, 2004). However, the Australian Passport Act 2005 Section 47 gives the Minister broad powers to force individuals to carry RFID surveillance devices containing unspecified biometric data.[23] As long as this technology is used it will be very difficult to control the unauthorised collection in practice. The privacy protection afforded by the Privacy Act 1988 will be undermined, as any determination made by the Minister permitting the use and disclosure of biometric information will be likely to fall within the scope of 'authorised or required by law' (IPP2.d) or 'any law that requires the particular information to be collected' (NPP 1.3(e)) and therefore be exempted from privacy restrictions (Electronic Frontiers Australia, 2005).

6.4.2 Unnecessary collection

Biometric data may be unnecessarily collected under the following circumstances:

- when biometric technology is adopted in various circumstances where no strong verification or identification is needed;
- when biometric technology is used for identification when only authentication is needed.

In the Privacy Act, both IPP 1(b) and NPP 1 require that the collection of personal information should be *necessary to the function and/or activity* of the collectors. The OPC has interpreted 'necessary' as when 'an organisation cannot, in practice, effectively pursue a legitimate *function or activity* without collecting personal information . . .' (NPP 1.1). The Biometrics Institute Privacy Code Principle 1.1 largely repeats this same wording. However, by giving the individual the right to have their biometric information removed from the system upon request where *possible* (Principle 12.4), the Code tends to give more control to data subjects over the unnecessary collection of their biometric information, although the wording 'necessary' and its explication are a bit controversial.

The necessity provisions in the Privacy Act are fine on their face, but, in practice, they tend to be easily ignored. There are many examples of biometric applications being used in Australia without any apparent significant censure.[24] Is it reasonable to let the organisations themselves be the sole judge of whether biometric information needs to be collected?

Moreover, to determine what is considered to be *necessary* in the biometric context, the criteria of 'cannot, in practice, *effectively* pursue a legitimate *function or activity*' call for more clarification. The nebulous wording 'effectively' is quite likely to be a good excuse for adopting the putatively strong identification[25] offered by biometrics, and there is no clear standard to decide what should be regarded as 'effective'. If the 'effectively' requirement is interpreted as 'quick' or 'convenient', biometric technology may be likely to fulfil the requirement, especially in many commercial applications. If the interpretation is 'security' and 'accuracy', it is still quite likely to fulfil the requirement, especially considering that the organisations themselves are the judges, and given the general misconceptions of what biometric technology offers. However, it is quite doubtful if this is what biometrics can *really* guarantee. In many commercial applications, the accuracy rate and the threshold of acceptance is quite low; hence, the false acceptance rate (FAR) is quite high (Mansfield, 2001). Therefore, it might be dangerous to include the 'effectively' criterion when determining the *necessity* of biometric use, unless 'effectively' is interpreted as denoting a high degree of actual accuracy and reliability.

Since the Privacy Act is intended to provide privacy protection, it is at the least questionable if the effectiveness of the organisations' *functions and activity* ought to be the main standard in deciding what is necessary. Ought not the concern about using biometric technology to be whether or not it is disproportionate to the purpose of collection, not whether it is effective enough to the organisation's functions and activities? Although it is true that the 'function' concern is also an evaluation factor of 'proportionality', it is important to restrict the concern about 'function' and 'effectiveness' so that it is not excessive. Take, for instance, a scenario where only authentication is needed and biometric information is collected without any de-identifying measures, which in fact makes identification possible. This would be *unnecessary* to the collection purpose while it could be argued as necessary for *effective functions and activities* of the organisations. Consideration ought to be given to including in the Privacy Act the core principle of minimality as found in the EU Directive Article 6(1)(c), which explicitly require that the information being collected should be 'non-excessive' in addition to an abstract requirement of 'necessary'. This helps us to ensure that the collection of biometric data is proportionate to the primary purpose and not just for effective functions and activity. When no strong verification and/or identification are/is needed, other traditional means of verification or identification should be adopted instead of using biometrics.

As mentioned at the beginning of this section, there are generally two practices that result in the unnecessary collection of biometric information. To the extent that necessity should be based on the decision of whether it is necessary for the *purpose* of specific collection, the interpretation might be reformulated as follows.

If an organisation cannot, in practice, effectively pursue a legitimate identification or authentication purpose without collecting the biometric information (e.g. when strong identification is needed and there are no other alternative identification or authentication methods available that can achieve the same purpose), then the information would ordinarily be considered as necessary for the collection purpose. It would not ordinarily be acceptable for an organisation to collect biometric information on the off-chance that the information could become necessary for one of its functions or activities in the future. When biometric information is used only for authentication, privacy-enhancing measures shall be adopted to de-identify the biometric information and render it anonymous or pseudonymous.[26]

If strict control of adopting this controversial technology is ensured from the outset, the risks of using biometrics might be eliminated more effectively.

6.4.3 Function creep: secondary use and disclosure

Because of its special nature as relatively unique and health related,[27] biometric information has great potential for being used and collected elsewhere either as a unique identifier or to disclose private health information. This concern of function creep is at the centre of the debate surrounding the privacy risks of biometric technology.

As previously mentioned, NPP 7 deals with identifiers. It provides that an organisation must not adopt as its own identifier one that has been assigned by an agency. Accordingly, NPP 7.1 'prevents an organisation from acquiring a particular government assigned identifier from all the individuals with which it deals, and using that identifier to organise personal information it holds and match it with other personal information organised by reference to the same identifier'.[28] If biometrics falls within the scope of the definition of 'identifier', this principle will restrict the adoption of a biometric identifier unless for specified purposes;[29] thus, to some extent it will restrict the linking and matching ability of biometrics. Yet as pointed out by the ALRC this regulation still leaves much uncertainty (Australian Law Reform Commission, 2008). NPP 7.1 currently only prevents an organisation from adopting as its own identifier one that has been assigned by an Australian federal government agency; an agent of that agency acting in the capacity of an agent; or a contracted service provider of an Australian federal government agent. If a biometric identifier is issued by state or territory agencies, it will not be covered by the current provision. Moreover, although NPP 7 stops the private sector from adopting, using or disclosing Commonwealth government identifiers, it does not address the possibility of the private sector (1) collecting the biometric identifier themselves and developing it as its unique identifier, and (2) collecting the biometric identifier from other private-sector actors and linking the collected data for other purposes or using them to track people.

On the other hand, IPPs 9, 10 and 11 place some restrictions on use or disclosure for purposes other than collection of data. The same can be said of NPP2 and the Privacy Code's Principle 2. However, the additional control principle of the Privacy Code stipulates that secondary analysis or function creep of biometric information is not permitted without express free and informed consent (Principle 12.3).

- Consent is gained from the data subjects (NPP 2.1b, IPP 10.1.a).
- The personal information should only be used for relevant purposes. In cases of sensitive information, it should be directly relevant (NPP 2.1.a.i.) and the individual would reasonably expect the organisation to use or disclose the information for the secondary purpose (NPP 2.1.a.ii). However, in IPP 10.1.e it is generally required that the secondary use or disclosure should only be for directly relevant purposes.

- If the information is health information and the use or disclosure is necessary for research, or the *compilation or analysis of statistics*, it can be disclosed (NPP 2(d)).
- The use and disclosure for that other purpose is *reasonably necessary* for enforcement of the criminal law or of a law imposing a pecuniary penalty, or for the protection of public revenue (IPP 10.1(d)); IPP 11.1(e) or required or *authorised by or under law* (NPP 2.1(g)).

The Privacy Act provides a long list of exceptions to the above-mentioned restrictions on secondary use or disclosure of personal information. The following exceptions are worth mentioning in detail.

6.4.3.1 Consent exception

As is argued in Chapter 4 (Section 4.3.2.3), consent is only meaningful when it is truly informed and free. When biometric information is involved, a clear notification of the risks for secondary use is of great importance. However, the OPC considers that 'consent to the use or disclosure can be expressed or implied' and that '. . . [i]t may be possible to infer consent from the individual's failure to opt out . . .'[30] This passive implied consent may weaken the effectiveness of restricting function creep. Neglect of the risks and the irritation of unclear and tedious notices may become the main factors that contribute to 'implied' consent. A distinct characteristic of biometric technology is that its application is often accompanied by various potential security and privacy risks and wide misconceptions of its accuracy and security-enhancing capabilities. It is therefore important to consider the extent to which consent for secondary use should be by positive consent – as is stipulated in the Privacy Code's Principle 12.3 – rather than 'notice and opt-out'. Moreover, this also raises the question whether or not 'consent' could be an independent legal basis for any kind of use of personal information. Is it reasonable to allow the inappropriate or unnecessary use of biometric information as long as consent is gained?[31]

6.4.3.2 Relevant purpose exception

The NPP Guidelines from the OPC describe the term *relevant* as 'something that arises in the context of the primary purpose', and *directly relevant* is defined as 'a stronger connection between the use or disclosure and the primary purpose for collection'. Biometric information is generally not accepted as sensitive personal information, although it may be health and genetic related.[32] Thus, it followed that any secondary use and disclosure of biometric information for an indirectly relevant purpose is quite likely to be allowed. But what might be the indirectly relevant purpose of using biometric information? If the biometric information is collected for identification[33] on certain occasions, such as entering the workplace, logging onto a computer system in the office, or entering the workplace dining facilities, would using such information for surveillance be justified as *indirectly relevant?* It could be argued this arises from the primary purpose of security control. The ambiguity of the phrase 'the context of primary purpose' in the NPP Guidelines may be further increased if 'security' is being accepted as an excuse.

6.4.3.3 *Health information exception*

The possible health-related nature of biometric information is one thing that makes its use controversial.[34] There is concern that it might be possible to analyse and extract health information from biometric information. Although it might require significant development of the current biometric systems, the huge potential has raised much anxiety. Should some biometric information be regarded as health information?[35] Although at present biometric information is not treated as health information in general, the potential for doing so is possible. If this happens, will it be reasonable to disclose it for the *compilation or analysis of statistics*? Unlike other health data, which may be more easily anonymised, the relative uniqueness of biometric data and their inherent nature as identifiers make their use relatively problematic. Moreover, the purpose of such a *compilation or analysis of statistics* is not restricted in the Privacy Act. This could make function creep to a large extent unavoidable, since tracking and linking might also be justified as involving the *compilation or analysis of statistics.*

6.4.3.4 *Law enforcement exception*

Biometric information has a long history of being used in forensic research for purposes of law enforcement. It is, for example, trite that law enforcement agencies use fingerprints and DNA to identify criminals. Hence, other biometric information is quite likely to be required for law enforcement purposes. There is much concern that many innocent people's biometric information could be used for such purposes without there being reasonable grounds for suspicion. The long-standing legal doctrine of 'innocent until proven guilty' may to some extent be undermined by this use of an individual's biometric information. The nebulous words *law, under, authorised* may massively expand the scope of the law enforcement exception, as 'law' can be understood as federal law, state law, common law, and/or international conventions. The words 'under' and 'authorises' are also highly subjective and open to debate (Australian Privacy Foundation, 2005). The Australian Privacy Foundation pointed out that 'it is Exception 2.1 (g) more than any other that undermines the "honesty" of privacy statements and policies which seek to reassure individuals about confidentiality' (Australian Privacy Foundation, 2005, p. 20). A wide range of public agencies will gain enormous power to require disclosure of personal information. In the context of biometrics, function creep in this field becomes almost inevitable.

6.4.3.5 *Summary*

The potential of function creep is not likely to be effectively restricted by the Privacy Act or Privacy Code, despite principle 12.3 in the Code. It is reasonable to predict that function creep of biometric information will be inevitable, although it may take some time before it becomes a widespread unique identifier. There is no natural barrier against function creep of this kind: each step can be more or less convincingly justified, and the race to a state of ubiquitous surveillance can run unimpeded (Clarke, 2002). This poses the question of whether it would be necessary to adopt additional legislative measures around the use of biometrics to address the function creep issues.

6.4.4 Anonymity

Anonymity is an important aspect of privacy. Of special concern is that the application of biometric technologies generally undermines anonymity and pseudonymity (Clarke, 2002). While this concern is not clearly addressed in the IPPs, the NPPs include an anonymity principle (NPP 8) that requires organisations to give individuals the option of not identifying themselves when 'entering transactions with an organisation' where it is 'lawful and practicable' to have that option. In the Biometrics Institute Privacy Code, this principle is essentially repeated (Principle 8), but the supplementary Principle 11.2 suggests that, where practicable, biometric information shall be de-identified by removing the name, thus making it difficult to match with personal information. Privacy impact assessment and auditing are also stipulated in the Code's additional accountability principle (13.2), which could be seen as necessary when adopting privacy-enhancing measures to satisfy NPP 8. In addition, it is likely that the anonymity principle may also be applied to the public sector in the near future, if the ALRC's proposal to that effect is implemented (Australian Law Reform Commission, 2008).

The anonymity principle is regarded as a unique principle of privacy protection that does not occur in the OECD guidelines or the EU Directive. Although Hert's report mentions the anonymity approach as a good tool for enhancing the opacity of biometrics (Hert, 2005), it does not provide specific legal suggestions regarding how anonymity can be guaranteed in the biometric context.

It has been argued that the anonymity principle in Australian law has failed to live up to its potential as a significant protection device (Australian Privacy Foundation, 2005). To comply with this principle in the biometric context, it is necessary to note the following: biometric technology is generally inconsistent with anonymity (Crompton, 2002), though it is possible to use biometric technology for anonymous authentication (Grijpink, 2001; Impagliazzo and More, 2003). It has been stated that if you really want to be anonymous, biometrics is not an appropriate technology of choice (T. Dunstone, personal communication Aug17, 2006). For informatics systems generally, unless the anonymising technology is built in at the design stage, it can very likely be claimed as 'impractical' to assimilate it in a later stage. The OFPC advises that 'NPP 8 has substantial implications for the design of information technology systems' (Office of the Federal Privacy Commissioner, 2001a), and 'additional cost, inconvenience, or administrative inefficiency will not be sufficient grounds for refusing an anonymous transaction' (Office of the Federal Privacy Commissioner, 2001a, (3)). However, in practice, since a Privacy

Impact Assessment is not required under the Privacy Act, and there is no compulsory auditing, it is unlikely that any party will intervene at the design stage, and, as a result, the collector might always be able to claim 'impracticability' as an excuse for not offering anonymous use options.

The use of privacy-enhancing technologies to achieve privacy protection through anonymity has been of major interest (Hes and Borking, 1998). This is only mentioned in the draft NPP Guidelines as a possible means to assure anonymity (Office of the Federal Privacy Commissioner, 2001a,). The Privacy Code clearly encourages it in Section B of the Objective and Principle 11.1 by suggesting the encryption of biometric information immediately after collection. However, the encryption technology is more likely to serve the security purpose better than the anonymity purpose.[36] The listed means in the Privacy Code are somewhat abstract, and given that in many cases the object of biometrics is identification, the possibility of anonymity might seem impossible. It would also not seem to be *practicable* to store the biometric data separately from other personal information. In cases where biometrics can be used for anonymous authentication, it might fail to pass the 'necessity' test for collecting such identifiable personal data. However, it is still feasible for biometrics to be applied in a manner that protects identity, if 'pseudonymous' options are mandated as an alternative. This has also been proposed by the ALRC, which believes that if there is an option for an individual to interact pseudonymously, 'it will provide a more flexible application of the principle, by covering the situation where it would be impracticable or unlawful for an individual to transact anonymously but where these barriers would be overcome if the individual were to transact pseudonymously with an agency or organisation' (Australian Law Reform Commission, 2008, p. 696). Bygrave has expressed a similar idea when referring to regulation of profiling by data protection laws generally. He points out that the appropriate rules should stipulate anonymity as the primary goal, with pseudonymity as the first fall-back option when anonymity cannot be achieved for legal or technical reasons (Bygrave, 2002). In the scenarios of biometric applications, pseudonymity may be a practical alternative when real anonymity is either impractical or unlawful to implement.

Another weakness of NPP 8 is that it fails to clearly regulate who is supposed to take the responsibility of ensuring anonymity (Clarke, 2005). If the responsibility is in the hands of the collector alone, then the data subject of personal information will likely fall out of any considerations taken when making the decision. Once the disclosing organisations fail to comply with the principle, the indirect collector may be able to claim that since the possibility for anonymity is not built in when the information is disclosed, it would be 'impracticable' for it to be built in at a later stage (Electronic Frontiers Australia, 2005). Without clearly sharing the responsibility of being the disclosing party, this 'impracticable' claim is even more likely to remain in the biometrics scenarios.

Concerning the 'Privacy Impact Assessment' stipulated by Principles 13.3 and 13.4 of the Privacy Code, it may introduce a process under which due considerations can be given to how the anonymity principle can be respected and made effective. This is a very good process that should be encouraged in general, and not just for the practice of this principle alone. If the Privacy Impact Assessment reveals significant privacy risks, further privacy-enhancing measures should be adopted. Nevertheless, it is worth noting that there is some uncertainty about who is to conduct a Privacy Impact Assessment under Principles 13.3 and 13.4 of the Privacy Code. Principle 13.3 stipulates:

> A code subscriber must consider 'end-to-end' privacy management issues when providing a product or service to an information technology system . . . this also includes privacy audits, [and] privacy impact assessments . . . Principle 13.3

It is not clear here whether an independent assessor is to conduct the Privacy Impact Assessment as it is suggested for auditing in Principle 13.2, or whether it will be made public what is actually expected to be done in the 'Privacy Impact Assessments'. The Privacy Impact Assessment guidelines for public sector agencies developed by the OPC might be a good reference for the implementation (Australian Government Office of Privacy Commissioner, 2006). It could also be applicable in the private sector.

6.4.5 Information security risks and biometrics

Biometric technology has long been utilised as a weapon for combating fraud and providing security; however, as it has been pointed out, biometrics is no more than an alternative identification or authentication measure.[37] It has its own advantages and disadvantages, and can both solve and cause security problems. The insecurity of storage and transmission, permanent identity theft, and inaccuracy problems are the main security risks raised by biometrics (UK Government Biometrics Working Group, 2003).

6.4.5.1 Storage

As has been illustrated previously, the vast amount of biometric information stored in databases increases the risks of potential identity theft and fraud. Central storage and dissemination in various regions all over the world makes stored personal information vulnerable. In order to reduce the fears of fraud and misuse of personal information, IPP 4 and NPP 4 both require organisations to take reasonable steps to protect the information they hold from misuse and loss, and from unauthorised access, modification, or disclosure. In addition, IPP 4 requires that the personal information be kept in the same manner when it is necessary to give it to a third party. While NPP 4 does not mention the security requirement concerning secondary disclosures, it requires that organisations take reasonable steps to destroy or permanently de-identify personal information if it is no longer needed for any purpose for which the information may be used or disclosed under NPP 2. The Privacy Code repeats almost the same wording as NPP 4. Most of the additional provisions under the new 'protection' Principle 11 of the Code are in effect supplementary guidance on compliance with this security principle. Specifically, it requires encryption of biometric data immediately after collection; holding biometric data separately from other personal data; keeping the biometric data in storage only as long as necessary; transmitting the biometric data with 'due care'; and limiting access to the biometric data. A similar idea is reemphasised in the ALRC's report suggesting that the best practice for organisations is to hold certain information in an encrypted form, and mentioning biometric information as an example of such information (Australian Law Reform Commission, 2008).

Both the NPP and IPP guidelines elaborate the definition of *security* as being physical security, computer and network security, communications security and personnel security (Office of the Federal Privacy Commissioner, 2001a, 2001b). To make this principle practical and effective, the storage of biometric information may need more clarification. For example, it is clear from the legal provisions including the Code that organisations need to take security measures to protect the biometric information stored in their databases. However, it is not clear whether similar requirements should be extended to the biometric information stored in portable tokens such as in the new biometric passport, or any other kind of biometric smart card issued by government agencies or private-sector actors. It has been reported that the smart card used to store facial recognition information in the new Australian passport is encrypted by BAC technology.[38] However, the nebulous provisions in law may become a loophole for other biometric applications when biometric information is stored in the issued computer chips, and thus the data recorder does not actually hold them.

The Privacy Code has taken a further step by specifying some security measures. Encryption becomes a mandatory measure to protect the stored biometric data, but it is not clear what strength of encryption is appropriate, and whether there are alternative suggestions. 'Separation' storage is suggested for fulfilling the de-identification principle, which is also relevant to NPP 8. However, it is uncertain whether this measure is adequate enough to protect the data, as 'destroy or de-identification must be permanent, which means that an organisation is not able to match the de-identified information with other records to re-establish the identity of people' (Office of the Federal Privacy Commissioner, 2001a). The OFPC has suggested that they 'should provide guidance about the responsibilities that agencies and organisations have under the "data security" principle. This should include guidance on the manner in which personal information should be destroyed or rendered non-identifiable. This guidance should address both paper-based records and electronic media. It also may be useful for this guidance to refer to relevant standards for information destruction'[39] (Australian Law Reform Commission, 2008, p. 960). This is particularly complicated for digital records such as biometric data, because some systems only allow for simple deletion, which may only change the tag in the system, but still keep the data. This actually makes it possible to get back the *deleted* data. 'De-identification' also calls for more clarification, because when it comes to biometric data, simply removing information such as names, addresses and ID numbers does not necessarily 'de-identify' the biometric data. It has been suggested that an accepted industry standard for secure document destruction or de-identification should be established. Otherwise, biometric data may be easily matched with other information and therefore be identifiable again.

The Privacy Code keeps silent as to whether storage of biometric data should be centralised or not. Central storage of biometric data could pose many security risks such as function creep and identity theft. Hence, it would be useful to address where the master biometric templates should be stored. Should a portable token be suggested instead of central storage? If central storage is unavoidable, should it be strictly limited and made more secure with more security measures? Another related issue is this: since it has been proved that a raw biometric image may be closely linked to some health information, and since the biometric template is adequate enough for usage, should the raw biometric image be deleted right after the biometric template is created? As has been shown, it is possible to reconstruct the raw biometric image from the biometric template. If some relevant data are available, should it be reasonable to adopt additional security measures to prevent the reconstruction?

In addition, the Privacy Code is somewhat vague with respect to the transmission of biometric data; it mandates that biometric data can only be transmitted in a 'prescribed manner, giving due regard to the security of the medium involved' (Principle 11.5) This actually leaves much to the discretionary judgment of the transmitting entity to determine what the 'prescribed manner' will be.

6.4.5.2 Accuracy

Both IPP 8 and NPP 3 require that an organisation must take reasonable steps to ensure that the personal information it collects, uses or discloses is accurate, complete and up-to-date. The Privacy Code again simply repeats NPP 3. However, in its additional accountability principle, various measures such as auditing and privacy management are to be adopted.

As previously discussed in Chapter 3, the accuracy of biometric data has raised much concern. Given that there is no perfect security system and there is always room for errors, it is very important for organisations to try to minimise the error rate and choose an appropriate threshold to keep the right balance between the two rates. The access and correction principle in IPP 3.1, which provides individuals with a general right to access personal information about them, is a necessary measure to ensure the accuracy of the information. Concerning machine-readable biometric information, the ALRC has emphasised: 'An expert with the ability to interpret the results of a machine's analysis of biometric information should be made available to an individual seeking to exercise his or her right of access to this type of personal information' (Australian Law Reform Commission, 2008, p. 436).

Another major security concern is that biometrics adds a frightening new dimension to identity theft. A widely accepted biometric identifier may facilitate identify theft at one place for all.[40] If this happens, it may have a major impact on an individual's life. NPP 6 and IPP 6 suggest that it is an individual's right to get access to the personal information an agency or organisation holds about him or her, and to challenge or alter the result if it is wrong. However, this will often be impractical in the biometrics scenarios. Without special expertise, it would be very difficult to prove if the biometric data stored are wrong or inaccurate; neither can an ordinary person easily prove whether a match is wrong or inaccurate. To combat this problem, additional security measures still need to be explored by the technicians, but first the misconceptions on the accuracy of biometric technology should be avoided, and manual back-up solutions should always be provided. Finally, accuracy-enhancing technology such as 'liveness' detection may be recommended.

6.5 Physical privacy and biometrics

Compared with the information privacy concerns, biometrics' physical privacy intrusion is never at the centre of debate. However, there are still many cultural and religious concerns that make the use of biometric technology problematic, especially when the collection of biometric data happens in an unreasonably intrusive way.[41] The Privacy Act and the Privacy Code do not address this problem directly; however, both the IPPs and NPPs require that information be collected in a way that is not unreasonably intrusive. As discussed previously, generally the method of collecting biometric information may not involve means which are deemed too intrusive. Therefore, this might be an adequate protection when individuals are willing to accept the use of biometric technology, but to some individuals who have reservations about the technology on an ethical basis, or have certain cultural backgrounds, the very action of collecting biometric information may be considered as intrusive or unacceptable, even if it only requires them to put their fingertip on a scanner for a few seconds.

This brings us back to the 'necessity' issue of using biometric technology.[42] The Privacy Code suggested that the use of biometrics should be voluntary in the private sector, which would be a good solution to this issue provided that truly free and informed consent is guaranteed. However, the issue of choice is likely to arise concerning governments' use of biometrics. Mandatory collection of biometric data such as at customs and law enforcement agencies may appear to be difficult to avoid. It would still be advisable, however, to encourage systematic considerations. First, collection and use of biometric data should pass the necessity test as suggested: Are there any alternatives? Is a strong identification method really needed? Is it proportional to the purpose of collection? What safeguards may be needed? It has also been suggested that specific laws should be adopted for extremely intrusive circumstances, such as DNA collection, in order to provide more safeguards to protect individuals' right to privacy (Crompton, 2002).

6.6 Conclusions

With the current focus on privacy protection in Australia and its special provisions about identifiers, the anonymity principle may provide a useful legal basis of privacy protection for the use of biometric technology. However, the protection available is insufficient to deal with all the legal challenges posed by the use of biometric information. Several provisions are inconsistent, confusing and/or unclear.

The approach of the Biometric Institute to give guidance on how privacy principles can apply to the collection and use of biometrics is of considerable value. It is of the utmost importance to provide a higher standard in privacy protection in this sensitive area. The Privacy Code has included some good suggestions that need to be further developed in order to be effectively implemented. However, it is somewhat distressing to see that the Privacy Code cannot yet be seen as an effective and adequate measure for better protection of privacy in the biometric context due to its close linkage to the Privacy Act, where adequate legal support is lacking for the clear interpretation and effective implementation of the Code as was pointed out above. The utility of the large-scale repetition of the wording of the NPPs is at the very least questionable. This to some extent weakens the specialities of the Code, and introduces in to the Code many of the weaknesses of the NPPs. In spite of the additional principles, the Code is not obviously technology specific and seems to be too abstract and nebulous to be a totally effective privacy code intended to deal with a specific technology application. Its voluntary basis and self-regulated nature may, in practice, render many good improvements ineffective.

Accordingly, changes to the Code need to be made in response to the above inadequacies. The definition of biometrics needs to be further clarified concerning the verification and identification purposes of the biometric technology and the 'identifiability' of biometrics in general. The proposal of the ALRC about creating a unified set of privacy principles is also to be supported75.

1. The exemptions for small business and employee records should be deleted so as to provide privacy protection to the customers of small businesses and to employees who have been entered into a biometric system.
2. When biometric information is collected, the notice provided by the data collectors should include clear bullet points concerning privacy risks and security risks for adopting biometric technology.
3. Consent should be required not only for secondary collection, but also for direct collection unless there are exemptions. Objective criteria need to be specified to exclude certain biometric information from the protection of sensitive information.
4. The use of RFID technology, especially when combined with stored biometric data on the RFID chip, should be strictly controlled in legislation.
5. Strict legal control of the necessity of collection of biometric information and the use of biometric technology is encouraged. Positive consent should be required instead of passive consent.
6. Interpretation is needed for the 'indirectly linked usage' with regard to the use of biometric information in particular and the 'compilation and analysis of statistics'. Exceptions seem to be very problematic concerning the use of biometric information, calling for further explanations. The exceptions of 'authorised by law' need to be reformulated and clarified with special concern

given to the use of biometric information in forensic research. Amendments should be made to Principle 7 where a loophole exists for private-sector actors to use the same unique biometric identifier from other agencies for other identification purposes.

7. The anonymity principles need to be amended to clarify that the anonymity obligation should be considered at the beginning of the adoption of a system, especially when biometrics is being utilised.
8. Privacy-enhancing technology and Privacy Impact Assessments by an eligible third party should be required.
9. It is highly recommended to avoid central storage of biometric information, and where it is unavoidable, additional security measures should be adopted to increase the storage and transmission security.
10. Clear requirements for implementing security measures to protect biometric information stored in portable tokens are also indispensable. Liveness detection of biometric information should be obligatory when the biometric identification or authentication would have a significant detrimental impact on an individual's interests.

Notes

* This chapter is based on the author's article, Liu 2010.

1 The next generation of Australian passport—the 'e-passport'—was introduced on 24 October 2005. For further research, see Department of Foreign Affairs and Trade (2009).

2 It has been reported that the self-service facilities are installed at Sydney, Adelaide, and Brisbane, Cairns, Melbourne and Perth international airports. Smart Gate is an automated border processing system. It is a secure and simple system that performs the customs and immigration checks normally made by a customs officer when a person first arrives in Australia. It uses the data in the e-passport and face recognition technology to perform the customs and immigration checks that are usually conducted by a Customs and Border Protection officer.

Smart Gate can be used by Australian and New Zealand e-passport holders, aged 18 or over. It will be gradually opened to other nationalities that have e-passports. For more information on the Australian e-passport please go to: http://www.passports.gov.au/, Last visited 11.01.2010

3 The Department of Immigration and Citisenship is reportedly building a database of facial, fingerprint, and iris scans that will be linked to a global processing system and intelligence and security databases. The biometric data of New Zealanders and other foreign nationals entering Australia could be permanently stored in a central repository for identity verification and cross-checking between federal government departments, national and international anti-identity fraud bodies, and border control systems. This information will be stored in the department's central Identity Services Repository, see Pauli (2007).

4 Woolworths supermarkets in Australia operate a time and attendance system featuring biometrics. Finger-imaging technology is used to monitor time and attendance for about 100,000 employees. See *People Key* (2007).

5 The Biometrics Institute Privacy Code was approved by Australian Privacy Commissioner Karen Curtis 19.07.2006, and it came into operation 01.09.2006.

6 See *Australian Broadcasting Corporation v. Lenah Game Meats Pty Ltd* (2001) 185 ALR 1.

7 See, for example, Richardson (2002).

8 For example, in *Toonen v. Australia* (1994) 1 PLPR 50, the UN Human Rights Committee

found that Australia was in breach of the provision in the ICCPR because of legislation in an Australian state (Tasmania) which criminalised homosexual conduct in private. The Australian Commonwealth Government then legislated to nullify the effect of the Tasmanian legislation. No cases involving Australia have been brought before the Human Rights Committee with respect to data protection issues, let alone biometrics.

9 There are also some other laws and regulations that are relevant to privacy and data protection, such as The Spam Act 2003 (Cth), Medical care and Pharmaceutical Benefits Programs Privacy Guidelines, Data-Matching Program (Assistance and Tax) Act 1990 (Cth), National Health Act 1953 (Cth) and Telecommunications Act 1997 (Cth). These are, though, of relatively marginal significance for biometrics-related concerns addressed in this book.

10 See Office of the Federal Privacy Commissioner (2001b).

11 Office of the Federal Privacy Commissioner (2001a).

12 For more information on exemptions and coverage see Office of the Federal Privacy Commissioner (2001c).

13 'Organisation' is defined in s. 6C as any entity that is not a small business operator, a registered political party, an agency, a State or Territory authority, or a prescribed instrumentality of a State or Territory. A 'small business operator' is defined in s. 6D as a business that has an annual turnover of $3 million or less and, subject to some exceptions, is exempt from the legislation. A business may be exempt from the Privacy Act because it does not come within the definition in s. 6C; it may also choose to be treated as an organisation (see s. 6EA). The Biometrics Institute has chosen to be treated as if it were an organisation.

14 Hereinafter also referred to simply as the 'Privacy Code'.

15 See The Consultative Committee of Convention 108 (2005) 'the biometric data, either the picture or the template . . . .'

16 For further discussion, see Chapter 5 (Section 5.4.2).

17 There are a number of controversial discussions on whether biometric data can be made to be not identifiable. See Liu (2008a).

18 Revised Explanatory Memorandum, *Privacy Amendment (Private Sector) Bill 2000* (Cth), para 361.

19 NPP 2.1(a) states: 'An organisation must not use or disclose personal information about an individual for a purpose (meaning the secondary purpose) other than the primary purpose of collection unless: (a) both of the following apply:
(i) The secondary purpose is related to the primary purpose of collection and, if the personal information is sensitive information, directly related to the primary purpose of collection; (ii) the individual would reasonably expect the organisation to use or disclose the information for the secondary purpose.'

20 The nebulous words 'practical' and 'reasonable steps' are interpreted by balancing a number of possible factors. See Office of the Federal Privacy Commissioner (2001a).

21 The Australian Privacy Charter was launched in December 1994. It was developed by a specially formed group that styled itself the Australian Privacy Charter Council (APCC). The Council was established in 1992, under the Chairmanship of Justice Michael Kirby. It aimed to develop a Privacy Charter comprising principles that would encompass and apply to all forms of privacy and surveillance (i.e. not just information privacy); and to both private and public sector organisations and their clients. Note that the Charter as such is not legally binding. See further the Australian Privacy Charter Council (APCC) (1994).

22 It has been reported that some RFID-enhanced passports are easily cracked, allowing the unauthorised reading of the stored information. See Masnick (2006).

23 Sec 47 of the Australian Passport Act 2005 states: 'Minister may determine particular methods and technologies'.

24 No cases have come before the OPC or the courts concerning the necessity of using biometric technology.

25 Strong identification here refers to very accurate and reliable identification measure.
26 This is in accordance with NPP 8, and I return to this discussion further below.
27 See Liu (2008b).
28 Revised Explanatory Memorandum, Privacy Amendment (Private Sector) Bill 2000 (Cth), p. 380.
29 NPP 7 provides four exceptions permitting the adoption of such identifiers.
7.1A 'An organisation must not adopt as its own identifier of an individual an identifier of the individual that has been assigned by (a) an agency; or (b) an agent of an agency acting in its capacity as agent; or
(c) a contracted service provider for a Commonwealth contract acting in its capacity as contracted service provider for that contract. However, sub clause 7.1 does not apply to the adoption by a prescribed organisation of a prescribed identifier in prescribed circumstances.' 7.2 An organisation must not use or disclose an identifier assigned to an individual by an agency, or by an agent or contracted service provider mentioned in subclause 7.1, unless:
(a) the use or disclosure is necessary for the organisation to fulfill its obligations to the agency; or
(b) one or more of paragraphs 2.1(e) to 2.1(h) (inclusive) apply to the use or disclosure; or
(c) the use or disclosure is by a prescribed organisation of a prescribed identifier in prescribed circumstances.
30 Office of the Federal Privacy Commissioner (2001a, 2001b), (1), 2.1(b).
31 This is also a question I dealt with in Chapter 4 when examining the application of the proportionality principle for biometrics.
32 See, for example, Crompton (2002) For a discussion on the controversial issue of whether biometric data should be regarded as sensitive personal data , see Liu (2008a).
33 It is believed that without special privacy enhancing features installed from the outset, biometric information is generally identifiable.
34 See discussion in Chapter 3.
35 This question is relevant to the discussion of the sensitive nature of biometric information; see Chapter 3.
36 I will return to this in the following section.
37 See generally Chapter 3.
38 The International Civil Aviation Organisation, which created the international specifications for countries adopting RFID passports, designed specifications for a process called basic access control (BAC). Basic Access Control works as follows: The data on a passport would be stored on an RFID chip in the passport's back folder, but the data would be locked and unavailable to any scanner/reader that does not know a secret key or password to unlock the data. To obtain the key, a passport officer would need to physically scan the machine-readable text that is printed on the passport page beneath the photo (this usually includes date of birth, passport number, and expiration date). The reader would then hash the data to create a unique key that could be used to authenticate the reader and unlock the data on the RFID chip.
39 Office of Privacy Commissioner, Submission PR215, 28 Feb 2007.
40 See generally Chapter 3.
41 See more discussion on this in Chapter 3.
42 See section 4.3.2.2

References

American Civil Liberties Union (2004) *Naked Data: How the U.S. Ignored International Concerns and Pushed for Radio Chips in Passports Without Security*. Available at http://www.aclu.org/Privacy/Privacy.cfm?ID=17078&c=130//.

Australian Department of Employment (2000) *Workplace Relations and Small Business to the Standing Committee on Legal And Constitutional Affairs' Inquiry into the Provisions of The 2000 Bill.* Available at http://www.aph.gov.au/HOUSE/committee/laca/Privacybill/submiss.htm.

Australian Government (2004) *Overview of Smart Gate Trial.* Available at http://www.customs.gov.au/webdata/resources/files/medBackSmartgate0401.pdf.

Australian Government Office of Privacy Commissioner (2006) *Explanatory Statement Approval of Biometrics Institute Privacy Code.* Available at www.privacy.gov.au/business/codes/biometricses.pdf.

Australian Law Reform Commission (2008). *ALRC Report 108: For Your Information: Australian Privacy Law and Practice.* Available at http://www.austlii.edu.au/au/other/alrc/publications/reports/108/.

Australia Parliament (2006) *Senate Legal and Constitutional Committee, Inquiry into the Privacy Act 1988.* Available at http://www.aph.gov.au/Senate/Committee/legcon_ctte/privacy/report/index.htm.

Australian Privacy Charter Council (1994) *Australian Privacy Charter.* Available at http://www.privacy.org.au/apcc/>.

Australian Privacy Foundation (2004) *Submission to Senate Legal & Constitutional Committee Inquiry Into The Privacy Act 1988, Annex D, Biometrics Institute Draft Privacy Code Of Practice.* Available at http://www.aph.gov.au/senate/committee/legcon_ctte/privacy/submissions/sub32ann_d.pdf.

—— (2005) *Submission to Senate Legal & Constitutional Committee Inquiry into the Privacy Act 1988.* Available at http://www.privacy.org.au/Papers/SenateCteePrivacyAct0503.rtf.

Bygrave, L.A. (2002) *Data Protection Law: Approaching Its Rationale, Logic and Limits*, The Hague: Kluwer Law International.

Crompton, M. (2002) Biometrics and privacy, *Privacy Law and Policy Reporter,* 9(3): 53–8.

Clarke, R. (2002). *Biometrics' inadequacies and threats and the need for regulation.* Available at http://www.rogerclarke.com/DV/BiomThreats.html.

—— (2003a) *Smart Gate: A Face Recognition Trial at Sydney Airport.* Available at http://www.rogerclarke.com/DV/SmartGate.html

—— (2005). *Submission to the Senate Legal and Constitutional Committee Re Its Inquiry into the Privacy Act 1988.* Available at http://www.rogerclarke.com/DV/SenateReview0502.html.

Department of Foreign Affairs and Trade (2009). *The Australian e-passport.* Available at http://www.dfat.gov.au/dept/passports/.

Electronic Frontiers Australia (2005). *Inquiry into the Privacy Act 1988.* Available at http://www.aph.gov.au/senate/committee/legcon_ctte/completed_inquiries/2004-07/privacy/index.htm.

Grijpink, J. (2001) Privacy law: biometric and privacy, *Computer Law and Security Report,* 17(3): 154–60.

Hert, P.J.A. de (2005) 'Biometrics: Legal Issues and Implications', background paper for the Institute of Prospective Technological Studies, DG JRC – Sevilla. Available at http://www.statewatch.org/news/2005/apr/jrc-biometrics-paul-de-hert.pdf.

Hes, R., and Borking, J. (1998). Privacy Enhancing Technologies: the Path to Anonymity (Revised edn), *Background Studies & Investigations.* Available at http://www.dutchdpa.nl/downloads_av/AV11.PDF?refer=true&theme=purple.

Impagliazzo, R., & More, S. M. (2002). *Anonymous credentials with biometrically-enforced non-transferability.* Available at http://portal.acm.org/ft_gateway.cfm?id=1005150&type=pdf&coll=&dl=ACM&CFID=15151515&CFTOKEN=6184618.

Liu, Y. (2008a). Identifying legal concerns in the biometric context, *Journal of International Commercial Law and Technology*, 3(1): 45–54.

—— (2008b). Rational concerns about biometric technology: security and privacy, in R. Subramanian (ed.), *Computer Security, Privacy and Politics: Current Issues, Challenges and Solutions*, New York: IGI Global, 85–120.

Liu, Y. (2010) Privacy Regulation of Biometric Technology in Australia, *Computer Law and Security Review*, 2010, 26(4), 355–367.

Mansfield, T. (2001). *Biometric product testing final report* (Report X92A/4009309). Available at http://www.securimetrics.com/articles/gfx/cesg-trials-report.pdf.

Masnick, M. (2006). *UK RFID Passports Cracked Already*. Available at http://www.techdirt.com/articles/20061117/183947.shtml.

Office of the Federal Privacy Commissioner (2001a). *Guidelines to the National Privacy Principles, September 2001*. Available at http://www.privacy.gov.au/materials/types/guidelines/view/6482#c1.

Office of the Federal Privacy Commissioner (2001b). *Plain English Guidelines to Information Privacy Principles*. Available at www.privacy.gov.au/publications/page1.html#1.

Office of the Federal Privacy Commissioner (2001c). *Information Sheet (Private Sector) 12 - 2001 Coverage of and Exemptions from the Private Sector Provisions*. Available at http://www.privacy.gov.au/materials/types/infosheets/view/6544.

Pauli, D. (2007). *Aussies to Stockpile Kiwi Biometrics in Central Database*. http://computerworld.co.nz/news.nsf/news/4C244AE1ED29925CCC25731E000055D0

People. Key (2007). *Biometrics at Woolworth*. Available at http://www.peoplekey.com.au/about_us/c_woolworths.htm.

Privacy Commissioner New South Wales (2005). *Privacy New South Wales submission to the house of representatives standing committee on legal and constitutional affairs on the Privacy Amendment (Private Sector) Bill 2000*. Available at http://www.aph.gov.au/HOUSE/committee/laca/Privacybill/sub62.pdf.

Privacy International (2004). *PHR2004 – Commonwealth of Australia*. Available at http://www.privacyinternational.org/article.shtml.

Richardson, M. (2002). Whither Breach of Confidence: A Right of Privacy for Australia? (2002) *Melbourne University Law Review*, 26(20). Available at http://www.austlii.edu.au/au/journals/MULR/2002/20.html.

The Consultative Committee of Convention 108 (2005) *Progress report on the application of the principles of Convention 108 to the collection and processing of biometric Data*. Available at http://ec.europa.eu/research/science-society/pdf/3-t-pd-2005-biom_e_en.pdf.

UK Government Biometrics Working Group (2003). *Biometric Security Concerns*. Available at http://www.cesg.gov.uk/site/ast/biometrics/media/BiometricSecurityConcerns.pdf.

Part IV

7 Towards an appropriate regulatory approach for bio-privacy

7.1 Introduction

According to the analysis in the previous chapters, it can be maintained that the EU, the United States and Australia, have gradually extended the use of biometric technology in their information systems. However, they have not shown themselves to be adequately committed to strict rules on privacy and data protection.[1] Data protection rules will come under considerable pressure in the future as a result of the relatively open-ended approach to various privacy concerns when dealing with biometrics. Additionally, further problems may appear when the application of biometric systems is increased.

I have tried to explore whether the present data protection regimes are adequate and applied consistently to solve all the privacy problems raised by biometric technology. A question needing to be further addressed is: will the use of biometric data require a readjustment of the legal framework because the introduction of biometrics is an innovation that requires a new legal approach? Moreover, what kind of legal approach is likely to be appropriate to deal with the problems raised by biometrics? I have already touched upon some of the answers to these questions in the previous analysis, so, to avoid repetition, I focus in the following section only on the remaining issues needing further exploration.

7.2 Self-regulation

Industry self-regulation for bio-privacy issues has evolved in the United States and Australia. Most biometric industry groups, while expressing their appreciation of privacy concerns, have responded by saying that biometric products are 'privacy neutral', and that it is how they are used that leads either to privacy invasion or privacy protection (GAO, 2002). In the United States, the International Biometric Industry Association (IBIA) is advocating on the behalf of the industry to create responsible use guidelines and public policy. In 1999, the IBIA announced a set of principles to protect personal information collected by biometric systems. In this announcement, the IBIA stressed that it is very concerned with the issues of privacy and personal information use. The principles they propose as guidelines to manufacturers, integrators, customers and users are as follows.

- Biometric data is [*sic*] electronic code that is separate and distinct from personal information, and provides an effective, secure barrier against unauthorised access to personal information. There should be safeguards to ensure that biometric data is not misused to compromise any information or released without personal consent or the authority of law.
- In the private sector, the development of policies should clearly set forth how biometric data will be collected, stored, accessed and used, and the rights of individuals should be preserved to limit the distribution of the data beyond the stated purposes.
- In the public sector, clear legal standards should be developed to carefully define and limit the conditions under which agencies of national security and law enforcement may acquire, access, store and use biometric data (IBIA, 2000).
- Both the public and private sectors should adopt appropriate managerial and technical controls to protect the confidentiality and integrity of databases containing biometric data. The International Biometric Group (IBG) has also developed a framework called the IBG BioPrivacy Initiative. It encompasses: (i) the BioPrivacy Application Impact Framework, which is a tool for evaluating the potential risks of a specific biometric deployment; (ii) the BioPrivacy Technology Risk Ratings, which are a technology-by-technology assessment of potential privacy risks and (iii) the BioPrivacy Best Practices, which may be used to ensure that new or existing biometric deployments are consistent with generally accepted privacy principles in public and private sector institutions (IBG, 2005). The Privacy Code from the Biometrics Institute in Australia, as discussed in Part III, is also a typical example of self-regulation for biometric-related privacy issues although it features elements of 'hard' regulation as outlined in the next section.[2]

The advantage of self-regulation in any context is that industry groups are usually much more agile than legislatures and can act more quickly in response to changes in technology and business environments. This is regarded as being of particular importance for a relatively new technology and industry, such as is currently the case with biometrics. Moreover, this kind of regulatory approach can take into account the differences between different industries and technologies. For example, it is claimed that it may be appropriate to regulate differently for retinal information and voice prints respectively.[3]

However, the IBG, IBIA and the Biometrics Institute are biometric industry trade groups, not privacy advocates. Markets are erratic and the high value placed on data means the incentives for breaching privacy norms are high. Hence, it becomes very difficult for the industry, which always aims for profit, to regulate itself stringently and in a manner that enhances privacy interests. As Clarke aptly expresses in an aphorism, '[w]olves self-regulate for the good of themselves and the pack, not the deer' (Clarke, 2003). Although there are some valuable points for the protection of biometric information in the BioPrivacy Initiative and the Biometrics Institute's Privacy Code (such as the best practices and need for

Privacy Impact Assessment), self-regulation tends to suffer from a major drawback: a lack of application and enforcement. Sometimes, although, this problem is linked to underlying legislative weaknesses. For instance, in the latest Biometrics Institute Code Review Report (Biometrics Institute Independent Code Review Panel, 2009), it was reported that there has been a low take-up of the Biometrics Institute's Privacy Code. Several reasons for the small number of signatories were listed,[4] but the main reason is considered to be the 'Privacy Act itself with its exclusions and separations and continuing policy reviews' (Biometrics Institute Independent Code Review Panel, 2009). There was a strong view that the Privacy Code will not be fully effective until the relevant changes to privacy law are implemented. It was pointed out that the key requirements of the Code, such as mandatory Privacy Impact Assessment, and the inclusion of employee records are likely to be included in the revised Privacy Act. This means that the Code will still rely on the improvements and implementation of hard law to become fully effective. Yet if this happens, the Code will – somewhat paradoxically – become redundant, and the real value of self-regulation will be simply to convey concern.

As intimated above, many corporations are reluctant to be self-regulated because they are more concerned about profit, market share and share-holder interests than human rights. This is indicated by some of the corporate feedback received during the recent review of the Biometrics Institute's Privacy Code:

> . . . We would not be in a position to police compliance to a Code that our clients may choose not to sign up to; and in this respect it would be noted that the extra provisions in The Code may imply additional cost, time and otherwise vary the scope of initiatives.
>
> (Biometrics Institute Independent Code Review Panel, 2009)

To make the biometric self-regulation effective, it would have to apply to any/every business using the technology; however, many of these businesses do not subscribe to any of these regulations. Keeping track of the adherence to these self-regulatory initiatives may be made more difficult if companies do not audit how the personal information they hold is used and disclosed.[5]

Moreover, the rules in self-regulatory instruments sometimes have major flaws. For example, the substantive content of the Guidelines from the IBIA shows some significant inadequacies. Principle 1 of the Guidelines denies the fact that biometric information is personal information, which is the biggest flaw of this set of guidelines as it prima facie excludes the legal remedies for the protection of biometric information as personal information. In addition, none of these regulations take into account the difference between different industries and technologies; instead, the IBIA's Guidelines, for example, are quite abstract. The Guidelines make no mention of what sorts of technological solutions might be used to deal with privacy matters. Concomitantly, they give engineers and others involved in technology design very little concrete guidance on how best to create systemic mechanisms allowing, say, data subjects to have more control over their personal information[6].

7.3 Hard regulation

An alternative to self-regulation is 'hard' regulation in the form of binding legislation. By this is meant statutory requirements coupled with a regulatory agency wielding substantial powers to monitor and ensure compliance with the requirements.[7] In the biometric context, there are generally two categories of 'hard regulation'. One is the general privacy protection laws, which amount to the largest portion of legal remedies available to bio-privacy issues. The closest examples I have discussed are the European Data Protection Directive, Norway's Personal Data Act of 2000, Australia's Privacy Act of 1988 and the US Privacy Act of 1974. Furthermore, it could also include the privacy statutes at state level or special privacy legislation for specific sectors. Another category is binding biometric-specific regulations; the closest examples would be the Ontario Works Act of 1997 in Canada[8] and the Biometric Identifier Privacy Act (BIPA) in the State of New Jersey, United States. The chief advantage of having binding legislation is that the enforcement of the law is, in principle, guaranteed.

7.3.1 General privacy regulations

As demonstrated in the previous chapters, the European Data Protection Directive provides a solid point of departure for regulating processing of biometric information, although some special legal issues have arisen from the use of biometric technology which the Directive does not fully address. Despite the fact that the privacy principles in the Directive were drafted over 15 years ago – and, indeed, are largely based on principles originating from even longer ago (e.g. the principles in Convention No. 108) – they cope almost as well with the problems of biometric technology today as they do with ICT more generally. Conversely, their weaknesses in coping with biometrics are not radically different from their weaknesses in coping with other forms of data processing. This can also be demonstrated by the privacy regulations of the VIS as discussed in the previous chapters. The collection, use, storage and transmission of biometric information according to the existing privacy regulations are not entirely new privacy problems; instead many inadequacies for the protection of biometrics in these privacy regulations also exist for other personal information or sensitive personal information. For instance, the function creep issues which have raised much concern from privacy advocates over the use of biometric information are not really all that new. The issues also affect other personal information as well. Thus, the inadequacy of the US Privacy Act 1974 for the protection of biometric information in the context of function creep also exists for other personal information.

However, biometric information has a significance and special character that make these privacy issues more salient and noteworthy than is the case with many other types of personal information. The most important point is that biometric information, as I have argued before, while certainly ordinary personal information in the European and Australian legal context, does not fit exactly into the present category of 'sensitive personal information' (under Article 8 of the

Data Protection Directive).[9] However, in the Unites States it is regarded as a kind of 'sensitive personal information', although the name, scope and definition of this information category is not exactly the same as in Europe.[10] In my opinion, biometric information is a special category of information. The threat to privacy and related interests entailed by the abuse of biometric information is greater and involves more serious consequences than other ordinary personal information or, indeed, traditional categories of sensitive personal information (e.g. the categories specified in Article 8 of the Data Protection Directive). Moreover, although the use of biometric technology does not, on the whole, raise entirely new legal issues, it actually renders application of certain legal rules unclear and nebulous, which calls for further interpretations and clarifications. In addition, some special characteristics of biometric information require more restrictions on its use. Hence, it is necessary to insert some special guidelines or provisions into certain privacy laws concerning the use of biometric information.

Yet another weakness with the general privacy laws is the weakness that is mentioned above in relation to some of the self-regulatory initiatives – that is, that they are pitched at an abstract level and fail, concomitantly, to provide relatively concrete guidance on how to build in systemic privacy-enhancing mechanisms into biometric applications.

7.3.2 Biometric-specific regulations

An example is the Ontario Works Act, passed by the Ontario government in 1997, which deals mainly with the administering of benefits to unemployed persons. Section 75 of the Act contains a list of requirements specifically concerning use of biometrics in the provision of such benefits. These requirements, in summary, are:

- The encrypted biometric can only be used for the limited purposes listed in section 75(1), not as a unique identifier facilitating linkages to other biometric information or other databases (section 75(1));
- Biometric information collected under this Act must be encrypted (section 75(6));
- The administrator must ensure the original biometric is destroyed after the encryption process and the encrypted biometric information is stored or transmitted only in encrypted form and destroyed in the prescribed manner (section 75(6));
- Neither the director nor an administrator shall implement a system that can reconstruct or retain the original biometric sample from encrypted biometric information, or that can compare it to a copy or reproduction of biometric information not obtained directly from the individual (section 75(7));
- Biometric information can only be retained together with the individual's name, address, date of birth and sex (section 75(8)).

This is quite a strict limitation on the use of biometric information. The requirement of encryption of the biometric information makes reconstructing or recreating an original biometric pattern difficult. As a result, the database containing the encrypted biometrics of welfare recipients would be of little interest to the police and other organisations. The restriction of using the biometric as a unique identifier could be an important measure to prevent the detrimental consequences of function creep. As stated in the statute: 'Biometric information to be collected from the individual to whom it relates shall be collected openly and directly from the individual' (section 75(4)). The clearly stated purposes for the collection of biometric information also provide strict restrictions on the collection of biometric information. However, many controversial issues raised by the use of biometric information, as mentioned in the previous sections, are not adequately dealt with in these provisions. For instance, the Act does not require the destruction of useless biometric information and remains silent on the concerns about central storage of biometric data.

Furthermore, it is important to note that these provisions only apply to a very specific field concerning social assistance. Many provisions are not practical for other biometric applications. For instance, the requirement of encryption and not keeping original biometric information is unlikely to be adopted by the VIS or US-VISIT systems, where even raw biometric information is kept for security purposes. The relatively narrowly delineated list of purposes in the Works Act is also unlikely to be favoured by the national security sectors where relatively open-ended exceptions tend to be inserted for security reasons. Despite its value, the Ontario Works Act cannot provide a *general* practical model for other jurisdictions' regulation of biometric information. At most, it could be a model for similar social service applications. This leads to another question: should we regulate the use of biometric information in a sectoral way, that is, through sector-specific rules dealing with biometric applications in the particular sector only? One could envisage differential regulation for the public sector and private sector. Ought one go further down this track and also differentiate, for instance, the governmental agencies dealing with social services from the governmental agencies with the responsibility for national security? If the answer is yes, it may be inconsistent with the general privacy regulations which apply, as a point of departure, to both the public and private sectors. Then what would be the best solution?

Another example of specific biometric regulation is the Biometric Identifier Privacy Act (BIPA) of the State of New Jersey discussed previously. Besides the deficiency of the simplicity of the content of this piece of legislation, it is also only state legislation. Obviously, such state-applicable legislation cannot really deal with biometric systems that are implemented nationally. Considering the proliferation of interstate transmission of biometric data, state regulation would likely prove troublesome and potentially disastrous. The difficulty of trying to comply simultaneously with many different state laws would seriously affect the laws' enforceability with respect to biometric systems. These systems are usually connected via networks that cross state jurisdictional lines.

7.4 A hybrid approach

Due to the limitations of self-regulation, of general privacy legislative solutions, and of the existing biometric-specific regulations in tackling the problems of privacy in the biometric context, a hybrid approach may be more appropriate. Provided that the existing hard regulations are mostly general privacy legislation, the most practical way to regulate biometric technology is to enact some biometric-specific regulations or guidelines on the basis of the existing privacy legislation. The guidelines or regulations should have separate provisions for national security agencies and other public and private sectors. This is based on the consideration that national security agencies could be easily exempted from the existing privacy regulations, which may lead to a big loophole for privacy protection where biometric data is mostly used. Adopting open-ended exceptions for general 'national security or public interest reasons' is not desirable. On the other hand, it may be impractical to treat them exactly the same as other agencies, hence specific provisions are necessary for this group. Clearly formulated criteria for such exceptions are necessary to prevent abuse.

Australia has arguably come furthest in adopting a hybrid approach. However, its regime still is far from desirable. Its main problems are (i) deficiencies in the ambit of the federal Privacy Act 1988 with respect to the private sector and (ii) lack of industry take-up of the Biometrics Institute's Privacy Code.

There are four main players in the area of data protection supervision in Europe: national DPAs, the Article 29 Data Protection Working Party, the European Data Protection Supervisor (EDPS) and the European Commission.[11] The available European legislation is: Directive 95/46/EC, Directive 2002/58/EC (concerning the processing of personal data and protection of privacy in the electronic communication sector) and Regulation 45/2001 on the protection of individuals with regard to the processing of personal data by the community institutions and bodies and on the free movement of such data. It is suggested that the Article 29 Data Protection Working Party may be the best forum for producing guidelines, and the European Commission, as the guardian of the treaties has the obligation to supervise national implementation. The EDPS also contributes to the development of the guidelines by delivering opinions on new proposals for large-scale EU biometric systems. At present, there are already many initiatives, proposals and reports that attempt to put forward recommendations and guidance on the use of biometrics on the basis of the existing legal framework. As noted in Chapter 4, these initiatives are quite useful while remaining unclear or silent on some controversial issues (e.g. the application of the proportionality principle to biometrics). Moreover, these initiatives are issued separately, which makes them difficult to follow and co-ordinate. Cultural differences between member states lead to different interpretations of current legislation, which, in turn, leads to different approaches on some issues for handling biometric data. In my opinion, it will not be surprising if additional guidelines for dealing with biometric information are developed on a pan-European level by Article 29 Data Protection Working Party and/or the EDPS in the near future.

This approach is justified for the following reasons. First, as noted before, since biometric technology does not, on the whole, raise any absolutely new legal issues, there is no need to establish a brand new code for regulating biometrics. Second, as certain aspects of biometrics are not being covered by the current legislation and biometric data as a special category of data call for more specific protections, certain guidelines or additional provisions for the current privacy law framework are necessary. Third, the new guidelines or regulations will be in addition to the existing framework; hence the current supervisory and audit regimes will still be in place and could oversee the new biometric regulations. Moreover, any new interpretations of, or added provisions to, the existing privacy legislation will have a similar binding effect and could use a similar audit or supervision system to that of the current regime, which makes it practical to enforce. Fourth, the suggestion of enacting separate rules for ordinary public and private sectors, and the national security agencies, is based on the study of the current biometric applications. Treating the public sector and private sector equally in most cases is in accordance with the current privacy law framework in most regimes and is appropriate for the protection of biometric data too. However, under the current regime the exceptions of national security and public interest are likely to be abused; hence, more specific provisions about the use of biometric data in the national security agencies should be introduced.

Despite these reasons, it is also the case that different legal regimes may need different approaches. It would be impractical to treat them exactly the same, as their respective privacy protection frameworks and legal cultures differ. For instance, in the Unites States, this approach may not be appropriate because a basic unified and comprehensive legal framework for privacy protection is lacking. However, if such a framework is ever put in place – and, in my opinion, it ought to be – then it could be complemented by the biometric-specific rules that I call for, thus comprising the suggested hybrid approach.

7.5 Conclusion

We are just at the beginning of an expansion in the use of biometric technology made possible by technological development combined with security and commercial concerns. It is crucial that legislation keeps abreast of this development. It may be safe to conclude that the current privacy scheme in the EU could provide a solid legal basis for further guidelines; hence, a hybrid approach which uses the current privacy instruments as a legal basis, and issues special guidelines or recommendations about using biometrics may be an appropriate solution for dealing with the legal issues raised by the use of biometric data. A similar approach has already gained foothold in Australia but needs to be built out further.

However, in the Unites States, the situation becomes more complicated. The current privacy protection scheme consists of a 'potpourri' of sectoral legislation, federal legislation for public sectors, state laws, the federal Constitution, tort law and industry self-regulation, all of which makes the protection of privacy difficult to monitor and implement and leaves many privacy issues unclearly addressed or not addressed at all. To deal with the legal issues raised by biometrics more effectively, enacting federal regulation that encompasses both the public and private sectors would be an appropriate choice. For those sectors not subject to a specific regulatory authority, oversight and enforcement functions will remain with the federal law. This regulation could be based on the existing fair information practices, supplemented by more specific provisions and restrictions on the use of biometric information.

Notes

1 This general observation also applies to other biometric-related security systems, such as EURODAC, SIS, European biometric passports, and the US biometric passport.

2 Most notably, the Code is legally binding on its signatories (just as ordinary legislation would be); it includes and is largely based upon legislative rules (i.e. the NPPs in the Privacy Act 1988 (Cth); and complaints under it may be brought before a legislatively established body (i.e. the OPC; see section J (3)). Nevertheless, it is an example of self-regulation in that the initiative for its creation and adoption comes from the biometrics industry sector itself, which has also formulated the rules. See Chapter 6 for more detail.

3 Id.

4 '(a) The deficiencies of the Privacy Act, such as the separation between public and private sectors, and the lack of unified national coverage with the resultant complications for those who wish to sign up; (b) The fact that there are areas of the Privacy Act which will be changed with the proposed changes to the Act; potential subscribers are waiting to see what will be required; (c) The fact that Government agencies are not covered by the Code; (d) The complications for global companies that already have a significant number of jurisdictional requirements to meet; this makes it more difficult for Australian staff to gain head office approval to sign up to the Biometrics Institute's Privacy Code'.

5 Note, although, that subscribers to the Biometrics Institute's Privacy Code are under a duty to have a third party audit their biometric systems (Principle 13.2).

6 This flaw applies also to hard regulation – particularly the general privacy regulations.

7 Id.

8 Ontario Works Act, 1997, S.O. 1997, c. 25, Schedule A. Strictly speaking, this is not a biometric-specific privacy regulation but a statute primarily regulating the provision of unemployment benefits. Nevertheless, it includes several biometric-specific provisions in its section 75.

9 See Chapter 4. Cf. the latest government consultation paper for the review of the Personal Data (Privacy) Ordinance (PDPO) of Hong Kong has suggested that biometric data be classified as sensitive personal data: The Constitutional and Mainland Affairs Bureau and Privacy Commissioner for Personal Data (2009)

10 See Chapter 4.

11 These are not the only players – one could also mention the European Parliament and the so-called Article 31 Committee (set up under Article 31 of the Data Protection Directive and composed of representatives of Member State governments) – but they are the most important in terms of policy development.

References

Clarke, R. (2003) *Dataveillance*. Available at http://www.rogerclarke.com/DV/PPSwamp.ppt.

The Constitutional and Mainland Affairs Bureau and Privacy Commissioner for Personal Data (2009) *Consultation document on review of the Personal Data (privacy) Ordinance.* Available at http://tinyurl.com/lmdpce.

GAO (2002) *Technology Assessment: Using Biometrics for Border Security (GAO-03-174).* November 14. Available at http://www.gao.gov/products/GAO-03-174.

IBIA (2000) *IBIA Privacy Principles.* Available at http://www.ibia.org/privacy.htm.

IBG (2005, 2007) *Bioprivacy Initiative.* Available at http://www.bioprivacy.org/index.htm.

8 Concluding remarks and future vision

8.1 Introduction

This chapter aims to bring together the main findings in the book and propose how they could be used to address the legal issues and problems posed by the use of biometric technology. Overall, the book has examined the biometric-related privacy issues from the perspective of technical, social and legal realities. The development of biometric technology and its interaction with legal and policy changes, as discussed in the preceding chapters, indicate that the current legal environment in the United States, Europe and Australia needs to be improved. Currently, privacy and data protection laws in these jurisdictions (and, no doubt, others) are far from adequate to deal with the problems appearing or that will appear due to the increased use of biometric technology. This is due to both general weaknesses in the laws themselves and a lack of normative determination and/or guidance regarding biometrics specifically.

Throughout the discussion on the peculiarities of and challenges posed by biometric technology, this book has challenged the rationale behind the argument of trading off privacy for security. It has established justification for strict regulatory management of biometrics that meets the legitimate privacy and security expectations of people. Whilst the study has demonstrated the presence of inadequacies in the existing regulatory framework of several jurisdictions, the existing laws and policy also provide a road map for further improvements under the current legal framework. For example, the relevant case law and privacy protection principles and policy specifications could be developed as legal guidance for future revisions of law and policy.

8.2 Rational concerns of biometric technology

The first main question addressed the privacy risks of using biometric technology and its relevance to security. The book took as its point of departure, the realisation that biometric technology is special in the sense that it utilises the biometric data to identify/authenticate people. The rational concerns of biometric technology's special set of characteristics have been illustrated in Part II (Chapter 3).

The consequence of these characteristics is that biometrics is regarded as an effective technology to provide security on the one hand and a great threat to the individuals' privacy right on the other. If the challenges to privacy posed by this technology are not addressed properly, the use of the technology might introduce more significant security risks than it intends to protect against. It could make serious inroads into the basic human right of privacy in the democratic society.

The key issue regarding biometric technology is not choosing between security and privacy. If we allow ourselves to see through tunnel vision and balance solely the enhancement of security against the sacrifice of privacy, then the trade-offs are easily cast into doubt. The present developing biometric technology does not actually offer the gains in security that were anticipated. This is in spite of the invasion of privacy that occurs when it is implemented. If biometric technology is going to be adopted without strict restrictions, then it seriously risks undermining the very liberty and privacy that it is designed to protect. Even if the creation of a surveillance society may insulate us to some extent from some types of security threats, it may very well swap one inequality for another, and in the process, raise more problems than it solves.

Moreover, biometrics' current state of effectiveness still leaves much to be desired. The quality and accuracy of biometric technology as it now stands do not actually offer the gains in security as demonstrated by many vendors or advocates. Having addressed some of the misconceptions surrounding the technology, it is perhaps worth noting that biometric technology alone cannot provide an answer to security issues. The inherent nature of biometric technology provides enormous potential for undermining privacy. This is despite the fact that, as it stands now, such technology does not offer all the matching, tracking and linking possibilities that are commonly envisaged. The inaccuracies and the security risks posed by biometric technology have, rather ironically, added more security problems, something which is not generally known. As mentioned previously, biometric technology, for the near future at least, is more likely to function as a convenient alternative or a supplement to traditional authentication methods than as a much more effective security enhancement tool.

Nonetheless, it is also true that a lot depends on the details surrounding the various biometric technologies and their related applications. Biometric technology may evolve to be more privacy friendly and security enhancing than it is now. It is important to recognise that privacy and security are not necessarily two contradictory concepts where biometrics is concerned. The means and application of biometric technology are the key issues here. They need to be further studied by technical experts as well as by law and policy makers.

8.3 Adequacy of current legal framework

The analysis of the current legal remedies relating to the privacy issues raised by the use of biometrics is undertaken for several legal regimes. Different legal questions are raised regarding the use of biometrics. In terms of the existing respective legal frameworks in the EU, Norway, United States and Australia for regulating biometrics, this is deficient in significant aspects.

At the European level, the analysis of the European legal initiatives concerning the use of biometrics indicated that some legal issues remain quite controversial or insufficiently discussed. Despite the relatively comprehensive privacy regulations and case discussions surrounding the use of biometrics, the 'principle of proportionality' calls for further elaboration. Analysis of case law from Norway, Sweden and, indirectly, the UK (the latter emanating from the ECtHR) illustrates the complexities of interpreting the principle of proportionality in specific biometric contexts, and specific relevant factors to decide the 'proportionality' of a certain bio-privacy case have been elaborated.

Moreover, the book also argues that biometric data should be categorised as a special kind of personal data for stricter protection. The provisions permitting derogation from privacy protection in the public interest should be accompanied with specific criteria, especially when biometrics is concerned. Further discussion was undertaken concerning the understanding of 'consent'. It was concluded that it might be possible to introduce a ranking system for real informed consent. Consent cannot be an independent legal basis for processing biometric data especially when the proportionality test is failed. Based on the study of ECtHR case law, it is also found that the collection of biometric data in public places may still undermine individuals' right to privacy.

In the US legal regime, the constitutionality of the collection of biometric information raises much debate. The issue of a constitutional right to informational privacy has been touched upon in the Supreme Court decision regarding the *Whalen* case, and there is the possibility of applying it to biometric information. At present, most discussion seems to focus on whether the collection of biometric information in a public space constitutes a breach of the Fourth Amendment. The circumstances under which such collection could amount to violation of the Fourth Amendment need to be addressed by the legislature and the courts. Moreover, the collection and use of biometric information are also relevant to citizens' right to be free from self-incrimination under the Fifth Amendment and to their rights to anonymity and free movement under the First Amendment.

Besides the general insufficiencies of the federal Privacy Act of 1974, the collection of biometric information also poses higher requirements for 'notice and consent' than ordinary personal data. '[S]haring the information with government agencies' and the problem of considering the 'matching' as 'routine use' of personal data may undermine the effectiveness of the Act with regard to biometric data. The security problems relating to biometric data are not adequately addressed in this Act either. In the proposed draft of the Personal Data Privacy and Security Act of 2009 use of biometric data also poses questions for the implementation of notice obligations when 'compromise and harm happens'; there is no clear answer as to how the individual can control the personal identifiable information flows after collection; 'unauthorised' become a very nebulous term in the legislation. Besides the disappointing 'opt out' regulation in the Gramm–Leach–Bliley Act (GLBA), individuals may also be concerned about (i) the undisclosed sharing of their biometric information among affiliates and non affiliates; (ii) the exception for information disclosed to a non-affiliated third party in connection with a joint-marketing agreement and (iii) the distinction between 'consumers' and 'customers' for purpose of the notice requirement under the Act.

Further, the use of biometric data also calls for more clarification of tort law. Through the extensive analysis of the case law, it was concluded that tort law in its current form may not be able to sufficiently address the privacy problems raised by biometric technology. Specific problems and suggestions were put forward for the four torts applicable when biometric data is applied.

The bio-privacy case law in the United States concerning both the public and private domains shows that the collection of physically non-intrusive biometric data for a specific legislated agency purpose may be accepted by the US courts. Simultaneously, the courts admit that biometric information is a kind of sensitive information which needs to be protected as sensitive personal information such as health information. The lack of sufficient legal solutions to provide protection for biometric information after its collection may become a loophole for effective privacy protection.

Regarding Australia, the book has focused on the Privacy Code of the Biometrics Institute and the federal Privacy Act 1988. Through the risk allocation analysis, it is concluded that the Privacy Act fails to adequately protect and enforce individual privacy with regards to the processing of biometric data. This is due to the general deficiencies in the legislation, for instance the extensive exemptions for small businesses and employee records. This is accompanied by unclear and, at times, inconsistent provisions for processing of personal information and security measures. However, a raft of recommendations for changes have been put forward by the Australian Law Reform Commission (ALRC), some of which are particularly relevant to the use of biometrics, such as listing biometric data as a new category of sensitive personal data. In addition, the relatively unique anonymity principle and no disadvantage principle could be of special meaning to the use of biometric technology when clearer interpretations are provided. The prohibition upon organisations adopting as their own identifier an identifier that has been assigned by an agency is also of particular relevance for restricting function creep of biometric data. Specific suggestions have been provided for the improvement of the current privacy framework.

The study of all the above jurisdictions indicates there is a paucity of biometric-specific regulatory initiatives and those that do exist suffer from various weaknesses. Through the investigation of sample legal proposals and legislation of bio-privacy issues in the United States, it is clear that the existing legal proposals still need improving. The regulation about 'consent' is flawed and unclear; explanation of important concepts is lacking; there is still the possibility of 'function creep' even if the law is adopted; the important term 'verify identity' is inaccurate. Although encryption has been required in the transmission, it is far from adequate to protect a 'biometric identifier' as such. Australia has also put in place a code of conduct concerning biometrics; however, that code cannot be regarded as an effective and adequate measure for protection of privacy in the biometric context. Although some of the provisions added may provide more protection for biometric data, the code is weakened in part by its close links to the Privacy Act, nebulous wording and its voluntary basis.

In the light of the above findings, the use of biometric technology and biometric data raises a few innovative problems that stem from its special characteristics, which are generally not addressed in current privacy laws. They are not so relevant to other personal information but have raised attention in many bio-privacy initiatives or court decisions:

- the justification and regulation of differing biometric identification and authentication when privacy issues are concerned;
- the difficulty of dealing with compromised biometric data;
- the choice of exempting certain groups from certain biometric systems and relevant alternatives;
- privacy concerns about interoperability and international data linkability;
- types of biometrics used in applications;
- Potential health data extraction in the future.

Moreover, as highlighted in the book, there also exists a regulatory gap between current privacy regulations and the requirements of biometrics. Among the many inadequacies and controversies that are caused by the application of biometric technology, the following issues are of common importance in all the jurisdictions discussed:

- the understanding of the legal term 'reasonableness' or 'necessity' in a biometric context (proportionality);
- restrictions on the collection of biometric information for certain purposes. The specific requirement of effective notice and consent in the biometric context;
- legal status of biometric information as a special category of personal information (or sensitive personal information);
- legal provisions concerning the location, length and security measures taken to protect biometric data stored and when in transmission;
- legal measures to deal with not so accurate biometric information, quality assurance of the biometric technology;
- the effective regulations to restrict the unauthorised secondary use and disclosure of biometric information;
- how to interpret the exception (derogation) of governmental and law enforcement use;
- anonymous issue of biometric information.

All of the above issues have been discussed extensively; specific recommendations have been made respectively in different jurisdictions concerning the legal issues peculiar to those jurisdictions. The general recommendations about the above common legal issues are highlighted and summarised later in this chapter.

8.4 Appropriate regulatory approach

As the book has attempted to show, in different jurisdictions (Part II and Part III) the use of biometric technology has encountered different legal issues. In evaluating the various options for regulating biometric technology, the book has utilised several practical examples of regulatory models: industry self-regulations such as the International Biometric Industry Association (IBIA) Guidelines and the Privacy Code of the Biometrics Institute; the general privacy regulations such as the Data Protection Directive 95/46/EC or Privacy Act(s) at national level; biometric-specific regulations such as the BIPA in the state of New Jersey in United States and the biometric specific provisions in the Ontario Works Act in Canada. The book concludes that the most promising solution is a hybrid approach: a comprehensive privacy act in general, combined with special guidelines and interpretations in particular. The government should base the legislation on the existing privacy act and the numerous reports and initiatives, and academic studies about bio-privacy issues, as they provide many potential solutions to the problems at hand. Such 'patchwork' regulations may be practical and enforceable for the European and Australian regime, while impractical for the US legal regime, where a comprehensive privacy act that covers both public and private sectors is lacking. Hence, the book suggests that in the US domain, an independent federal biometric privacy regulation may finally enable the effective protection of the use of biometric data in the public and private sectors.

8.5 General recommendations and postscripts

In lieu of a conclusion, there is a need to postulate the direction of future research as the issues discussed in this book are still being implemented and developed. No doubt the use of biometric technology will steadily increase in both public and private sector applications. However, we should strive to restrict unlimited use of biometric technology and adopt it only under conditions where it is necessary, and then only in a privacy friendly manner. Some of the problems of the technology, such as quality and level of accuracy, may be improved as the technology advances. At the same time, some potential risks posed by the use of the technology such as the interoperability, automatic collection, linking tracking ability, health information indication, and so on, may also become reality as time goes by. In this respect, governmental thinking and associated strategy should be able to extend at least over 20–30 years. No implementation should go ahead without a systematic study of sustainability.

In this book, I have made some strong points and offered quite a few specific suggestions around the use of biometric technology in relation to the corresponding legislations, policies and regulations in each chapter. Many inadequacies and controversial legal problems raised by biometrics have been addressed in the book. However, to implement these measures effectively, the current legal safeguards should focus on two key aspects: control the adoption of biometrics and regulate the use of biometrics.

The first stage is of great significance to stop the abuse of the technology. It is much easier and more practical to prohibit the use of biometric technology where it is inappropriate than to try to control the use of biometric data after it is collected or built in a certain security function after the system is implemented. At present, most privacy cases about biometrics concern this stage. The evaluation of whether to adopt a biometric application is focused on the essential discussion about the reasonableness and necessity of adopting biometrics in certain applications. However, there exist two obstacles for effective privacy protection at this stage.

- Not all the jurisdictions require a licence to process biometric data as is the case in Norway. In many jurisdictions, biometric systems are adopted as the entities wish unless a specific case is raised by the individual data subjects. This, to a large extent, renders the subsequent restriction of biometric technology in privacy legislations meaningless or ineffective.
- Besides the concern about the purpose, proportionality or necessity, concerns about how the biometric data are processed afterwards are also very important. In some of the biometric-related cases in Europe such as in France,[1] Italy (Andenas and Zleptnig, 2003) and Greece,[2] the DPAs have already expressed concern about how the system will be used, the storage of the biometric data, the type of biometrics being chosen and the security measures adopted. Unnecessary collection and use of biometric data cannot be justified by the individual consent of the data subject. However, in the US courts, such discussions are generally lacking.

Accordingly, it is recommended that censuring and evaluation should be conducted at this stage in order to supervise the use of biometric technology more effectively. Both unnecessary purposes and inappropriate use of the biometric technology shall be prohibited. The target of the biometric system should not only be commercial use but also public use, especially the large-scale mandatory systems. More field trials with sample populations are indispensable for large-scale applications.

The common legal issues that are relevant at this stage should be addressed as follows:

1. To evaluate the necessity/reasonableness/proportionality of certain biometric applications specific factors need to be taken into consideration: the benefits verses the risks; existing alternatives to such an application; the necessity; types of biometrics chosen and its risk indications; the method of data storage (scale, length, and so on).
2. When consent is sought, besides the notice about the purpose of collection, in the biometric context there should be informed notice about the privacy and security risks of using biometrics. There should be real alternatives for people who don't consent, as well as an open policy for the security and privacy protections afforded to the biometric information. (It was suggested that an official assessment ranking the privacy and security risks may be published by the data protection authorities, and the ranking should be notified to the

individuals when they make decisions.) Consent should not be presumed to have been freely given in most instances. The justification of biometric information collection should not rely solely on consent, but should take into account transparent procedures and be bound by strict regulations.

3. Biometric information should be regarded as a special category of personal information and receive stricter privacy protection than ordinary personal information. The strict protection measures should include most of all: (i) an evaluation of the necessity of using the biometric system by data protection authorities in advance; on the other hand, this should not be considered as the only essential factor when deciding whether to allow the use of biometrics. The existing privacy safeguards of the system may need to be adapted in order to guarantee the legal requirements maintain effective force. It is therefore of utmost importance that in this phase, potential negative side effects of individual measures based on biometrics are identified and addressed. (ii) Consent shall be required on most occasions, only in specific cases, and precise legally circumscribed situations can compulsory collection be justified (see point 2). (iii) Security and accuracy evaluation: to be specific, the evaluation and concerns at this stage should be targeted at the privacy policy of the system; the quality of the adopted system; mandatory security measures (both technical security and human factors) (see points 4–6).
4. Biometric technology at present is generally used for identification and authentication. The privacy impacts of these two applications to a large extent depend on the existence of a central database, and the scope of other personal data collected. Generally speaking, biometric identification is more likely to entail privacy risks than biometric authentication. In authentication processes, only one-to-one matches are required and central data storage is not necessary. Accordingly, when only authentication of biometrics is needed, central storage of biometric data should be prohibited, and biometric identification should not be adopted. However, the purpose of a certain biometric application is just one of the many factors that determine the privacy protection level of the system; hence the practical value of having different regulation for biometric identification and authentication is minimal.
5. Different biometric applications may differ in the accuracy levels and privacy impacts. When deciding which kind of biometric system should be chosen, for an evaluation of all relevant factors: accuracy level, social acceptance, convenience of usage, privacy impacts, it should be considered according to the peculiar requirement of certain applications.
6. In addition, certain categories of people may have difficulty in enrolling in certain biometric applications. It is suggested that multiple biometric applications should be utilised in circumstances that may encompass such categories of people in order to avoid de-facto discrimination and enhance the accuracy level.

Once the system is allowed, the control of the biometric data becomes more difficult in practice. In the second stage, to regulate the function of the biometric systems, it is suggested that:

1. Centralised collection of biometric data should be strictly limited and needs to be legally justified. Smart card storage should be used on most occasions when only authentication is used. The length of storing biometric data should be strictly limited by its usage, biometric data should not be kept longer than necessary. Explicit regulation of the storage length for different situations should be provided for in biometric guidelines. The requirement of encryption, or use of other privacy enhancing technology, should not only concern the transmission process, but also the storage. Extra security measures should be built into the system at the design stage.
2. To deal with FRR/FAR problems, there should be specific requirements regarding the quality of biometric systems, especially for large-scale compulsory applications. For security reasons, biometric technology should never be used alone for important security controls. Use of multiple biometrics and a combination of smart card, password and other identification/authentication measures should be suggested. On the other hand, fall-back procedures should also be provided for false recognition/rejection and compromise of biometric data, or other occasions such as exclusion of certain group of people, when the biometric technology is not applicable. Possible solutions and protective measures should be required to be integrated into the biometric system design and the system management policies.
3. Transparency rules are important to ensure that the system is implemented as it is allowed to be. There should be measures to allow individual data subjects and the privacy authorities to trace how the data is processed after collection. Both individual data subjects and the privacy authorities should be given notice regarding the functioning of the system in various ways. However, as noted before, the individual participation principle is difficult to implement. It is difficult for data subjects to check the accuracy of biometric data even if the access right is given. Hence, there should be other measures provided by the data controllers to guarantee that the data subject can get technical assistance for checking the usage, completeness and correctness of the biometric information. Furthermore, economic benefit to the individuals for commercial exploitation or any secondary use (unless exempted by law) should be promoted in addition to the informed consent requirement.
4. Biometric technology is not totally security enhancing, and it should not be regarded as an effective tool to fight against terrorism. Derogations/exceptions from privacy protection under relevant privacy legislation should be applied strictly. In Europe, there should be comprehensive regulation of the activities that can result in interference of privacy, and prohibition of the abuse of biometric data. Serious privacy violations should in general be prohibited; derogation is necessary when no other less invasive alternatives are available.
5. Biometrics presents a novel way to identify humans, but one that puts at risk the ability to remain anonymous. All three jurisdictions considered anonymity as useful if not essential to privacy (e.g., the First Amendment in the US Constitution, the anonymity principle in Australia's federal Privacy Act and particular provisions of the Data Protection Directive). Biometrics inherently

eliminates anonymity, especially when identification happens or a central database is built. Hence, when anonymity is essential to a particular application, biometrics should not be adopted.

6. As highlighted, linking and tracking of biometric information through sharing and interoperable systems within the public sector or between public and private sectors and among the private sector, are a great threat to people's privacy. This kind of sharing and linking of biometric information should be strictly restricted; bio-privacy regulations and public policies should prohibit such sharing and linking in general with limited specific exceptions. On specific occasions where such linking and sharing is necessary, there should be monitoring by data protection authorities, and relevant formal criteria and procedures to be followed.
7. Given the body-related nature of biometric data, the potential of extracting health information from biometric information should not be underestimated. A simple general prohibition is unlikely to be a practical solution; further research is needed to deal with such risks.
8. Interoperability encompasses several ethical concerns revolving around biometrics and personal detection technologies. These concerns include, but are not limited to, function creep, data centralisation, standardisation and system combinations. Hence, it is important to include delineation of boundaries and clarifications of principles when interoperability is required.

Notes

1 See, for example, Deliberation 01-006 of 25 January 2001 (Louvre); Deliberation 02-045 of 18 June 2002 (URSSAF); Deliberation 02-070 of 15 October 2002 (School Refectory).

2 See, for example, Decision 510/17/15-05-2000 4. and 245/9-20-03.2000.

Appendix 1: Personal Data Act, Norway[1]

Act of 14 April 2000 No. 31 relating to the processing of personal data
Chapter I Purpose and scope of the Act
Chapter II General rules for the processing of personal data

Section 8 Conditions for the processing of personal data
Personal data (cf. section 2, no. 1) may only be processed if the data subject has consented thereto, or there is statutory authority for such processing, or the processing is necessary in order

a) to fulfil a contract to which the data subject is party, or to take steps at the request of the data subject prior to entering into such a contract,
b) to enable the controller to fulfil a legal obligation,
c) to protect the vital interests of the data subject,
d) to perform a task in the public interest,
e) to exercise official authority, or
f) to enable the controller or third parties to whom the data are disclosed to protect a legitimate interest, except where such interest is overridden by the interests of the data subject.

Section 9 Processing of sensitive personal data
Sensitive personal data (cf. section 2, no. 8) may only be processed if the processing satisfies one of the conditions set out in section 8 and

a) the data subject consents to the processing,
b) there is statutory authority for such processing,
c) the processing is necessary to protect the vital interests of a person, and the data subject is incapable of giving his or her consent,
d) the processing relates exclusively to data which the data subject has voluntarily and manifestly made public,
e) the processing is necessary for the establishment, exercise or defence of a legal claim,
f) the processing is necessary to enable the controller to fulfil his obligations or exercise his rights in the field of employment law,

g) the processing is necessary for the purposes of preventive medicine, medical-diagnosis, the provision of care or treatment or the management of health care services, and where the data are processed by health professionals subject to the obligation of professional secrecy, or
h) the processing is necessary for historical, statistical or scientific purposes, and the public interest in such processing being carried out clearly exceeds the disadvantages it might entail for the natural person.

Non-profit associations and foundations may process sensitive personal data in the course of their activities even if such processing does not satisfy one of the conditions laid down in the first paragraph, litra a-h. Such processing may apply solely to data relating to members or to persons who, on account of the purposes of the association or foundation, voluntarily have regular contact with it, and solely to data which are collected through such contact. The personal data may not be disclosed without the consent of the data subject.

The Data Inspectorate may decide that sensitive personal data may also be processed in other cases if this is warranted by important public interests and steps are taken to protect the interests of the data subject.

Section 10 Register of criminal convictions
A complete register of criminal convictions may only be kept under the control of official authority.

Section 11 Basic requirements for the processing of personal data
The controller shall ensure that personal data which are processed

a) are processed only when this is authorised pursuant to sections 8 and 9,
b) are used only for explicitly stated purposes that are objectively justified by the activities of the controller,
c) are not used subsequently for purposes that are incompatible with the original purpose of the collection, without the consent of the data subject,
d) are adequate, relevant and not excessive in relation to the purpose of the processing, and
e) are accurate and up-to-date, and are not stored longer than is necessary for the purpose of the processing, cf. sections 27 and 28.

Subsequent processing of personal data for historical, statistical or scientific purposes is not deemed to be incompatible with the original purposes of the collection of the data, cf. first paragraph, litra c, if the public interest in the processing being carried out clearly exceeds the disadvantages this may entail for natural persons.

Section 12 Use of national identity numbers, etc.
National identity numbers and other clear means of identification may only be used in the processing when there is a objective need for certain identification and the method is necessary to achieve such identification.

The Data Inspectorate may require a controller to use such means of identification as are mentioned in the first paragraph to ensure that the personal data are of adequate quality.

The King may by regulations prescribe further rules regarding the use of national identity numbers and other clear means of identification.

Section 13 Data security

The controller and the processor shall by means of planned, systematic measures ensure satisfactory data security with regard to confidentiality, integrity and accessibility in connection with the processing of personal data.

To achieve satisfactory data security, the controller and processor shall document the data system and the security measures. Such documentation shall be accessible to the employees of the controller and of the processor. The documentation shall also be accessible to the Data Inspectorate and the Privacy Appeals Board.

Any controller who allows other persons to have access to personal data, for example, a processor or other persons performing tasks in connection with the data system, shall ensure that the said persons fulfil the requirements set out in the first and second paragraphs.

The King may prescribe regulations regarding data security in connection with the processing of personal data, including further rules regarding organisational and technical security measures.

Section 14 Internal control

The controller shall establish and maintain such planned and systematic measures as are necessary to fulfil the requirements laid down in or pursuant to this Act, including measures to ensure the quality of personal data.

The controller shall document the measures. The documentation shall be accessible to the employees of the controller and of the processor. The documentation shall also be accessible to the Data Inspectorate and the Privacy Appeals Board.

The King may prescribe regulations containing further rules regarding internal control.

Section 15 The processor's right of disposition over personal data

No processor may process personal data in any way other than that which is agreed in writing with the controller. Nor may the data be turned over to another person for storage or manipulation without such agreement.

It shall also be stated in the agreement with the controller that the processor undertakes to carry out such security measures as ensue from section 13.

Section 16 Time limit for replying to inquiries regarding data, etc.

The controller shall reply to inquiries regarding access or other rights pursuant to sections 18, 22, 25, 26, 27 and 28 without undue delay and not later than 30 days from the date of receipt of the inquiry.

If special circumstances should make it impossible to reply to the inquiry within

30 days, implementation may be postponed until it is possible to reply. In such case, the controller shall give a provisional reply stating the reason for the delay and when a reply is likely to be given.

Section 17 Payment

The controller may not request compensation for providing data pursuant to Chapter III or for meeting demands of the data subject pursuant to Chapter IV.

. . .

Section 33 Obligation to obtain a licence

A licence from the Data Inspectorate is required for the processing of sensitive personal data. This does not apply, however, to the processing of sensitive personal data which have been volunteered by the data subject.

The Data Inspectorate may decide that the processing of data other than sensitive personal data shall also be subject to licensing, if such processing otherwise will clearly violate weighty interests relating to protection of privacy. In assessing whether a licence is necessary, the Data Inspectorate shall, *inter alia* take account of the nature and quantity of the personal data and the purpose of the processing.

The controller may demand that the Data Inspectorate decide whether processing will be subject to licensing.

The obligation to obtain a licence pursuant to the first and second paragraphs shall not apply to the processing of personal data in central government or municipal bodies when such processing is authorised by special statute.

The King may prescribe regulations to the effect that certain processing methods are not subject to licensing pursuant to the first paragraph. As regards processing methods which are exempt from licensing, regulations may be prescribed to limit the disadvantages which processing may otherwise entail for the data subject.

Note

1 Only the relevant chapter is included here

Appendix 2: Biometric Identifier Privacy Act

Status: Passed by the Assembly, February 5, 2004. Received in the Senate and referred to Senate Judiciary Committee, February 9, 2004.

ASSEMBLY, No. 1194
STATE OF NEW JERSEY
211th LEGISLATURE
PRE-FILED FOR INTRODUCTION IN THE 2004 SESSION
Sponsored by:
Assemblywoman JOAN M. QUIGLEY
District 32 (Bergen and Hudson)
Assemblyman NEIL M. COHEN
District 20 (Union)
Co-Sponsored by:
Assemblyman Chivukula, Assemblywomen Greenstein and Stender
SYNOPSIS Creates the

Current version of text

Introduced Pending Technical Review by Legislative Counsel.

An Act concerning biometric identifiers and supplementing Title 2A of the New Jersey Statutes.

Be It Enacted by the Senate and General Assembly of the State of New Jersey:

1. This act shall be known and may be cited as the 'Biometric Identifier Privacy Act.'
2. As used in this act:

   'Biometric identifier' means a retina or iris scan, fingerprint, voiceprint, or record of a hand or a face geometry.
   'Governmental entity' means the State, any agency, authority, or employee thereof, or any political subdivision of the State, including but not limited to any county, municipality, or school district, or any agency, authority, or employee thereof.

3. a. Notwithstanding any other provision of law to the contrary, no person shall obtain a biometric identifier of an individual, for the purpose of commercial advantage, without authorisation of the individual.
   b. A person who possesses a biometric identifier of an individual shall not sell, lease, or otherwise disclose the biometric identifier to another person unless:

      (1) the individual consents to the sale, lease or disclosure;
      (2) the sale, lease or disclosure completes a financial transaction requested or authorised by the individual;
      (3) the sale, lease or disclosure is required or permitted by federal or State law; or
      (4) the sale, lease or disclosure is made by or to a law enforcement agency for a law enforcement purpose.

   c. A person who possesses a biometric identifier of an individual shall store, transmit, and protect from disclosure the biometric identifier using reasonable care and in a manner that is the same as or more protective than the manner in which the person stores, transmits, and protects other confidential information.
   d. A person aggrieved by a violation of this section may bring an action in the Superior Court to enjoin further violation and to recover for the actual damage sustained by reasons of such violation, including costs and reasonable attorneys fees.
   e. Any person who violates any provision of this section shall be liable for a civil penalty of not more than $25,000 for each violation. Any such penalty shall be enforced and collected in accordance with 'The Penalty Enforcement Law of 1999', P.L.1999, c.274 (C.2A:58–10 et seq.).

4. a. A governmental entity that possesses a biometric identifier of an individual shall not sell, lease, or otherwise disclose the biometric identifier to another person unless:

      (1) the individual consents to the sale, lease or disclosure;
      (2) the sale, lease or disclosure is required or permitted by a federal or State law; or
      (3) the sale, lease or disclosure is made by or to a law enforcement agency for a law enforcement purpose.

   b. A governmental entity that possesses a biometric identifier of an individual shall store, transmit, and protect from disclosure the biometric identifier using reasonable care and in a manner that is the same as or more protective than the manner in which the governmental entity stores, transmits, and protects other confidential information.
   c. A governmental entity that possesses a biometric identifier of an individual shall establish a reasonable procedure under which an individual is entitled to have the governmental entity correct information about the individual

that is possessed by the governmental entity and that is incorrect. The procedure shall not unduly burden an individual using the procedure.

d. A person aggrieved by a violation of this section may bring an action in the Superior Court, to enjoin further violation and to recover the actual damage sustained by reasons of such violation, including costs and reasonable attorneys fees.
e. Information compiled pursuant to this section shall not be subject to disclosure pursuant to P.L.1963, c.73 (C.47:1A-1 et seq.) as amended and supplemented.

5. This act shall take effect immediately.

Statement

This bill, the 'Biometric Identifier Privacy Act', provides guidelines for the use and distribution of biometric identifiers and establishes civil penalties for the misuse of the information.

A biometric identifier is a retina or iris scan, fingerprint, voiceprint, or record of a hand or a face geometry. Biometrics technology is a noninvasive method of using computer technology to provide automatic identification or identity verification or authentication of individuals. The technology acquires an image of a physical feature which is then applied to the algorithm to produce a 'template.' This 'template' is then encrypted for data transmission and storage. This stored 'template' can then be stored and compared against the live 'template' when necessary. This technology is being used for criminal identification as well as in airport security systems, border clearances and for transaction verifications in internet businesses.

Under the provisions of the bill a person cannot obtain another individual's biometric identifier information, for the purpose of commercial advantage, without authorisation from that individual. The bill prohibits a person who possesses a biometric identifier of another individual from selling, leasing, or otherwise disclosing this information unless: the individual consents to the sale, lease or disclosure; the sale, lease or disclosure completes a financial transaction requested or authorised by the individual; the sale, lease or disclosure is required or permitted by federal or State law; or the sale, lease or disclosure is made by or to a law enforcement agency for a law enforcement purpose. A person who possesses a biometric identifier of an individual would be required to store, transmit, and protect from disclosure the biometric identifier using reasonable care and in a manner that is the same as or more protective than the manner in which that person stores, transmits, and protects other confidential information.

The bill provides that any person who has been aggrieved by a violation of the act may bring an action in the Superior Court, to enjoin further violation and to recover the actual damage sustained by reasons of such violation, including costs and reasonable attorneys fees. Furthermore, the bill establishes a civil penalty to be assessed against a violator of the act. The penalty would apply to each violation of the act, and would not be more than $25,000 for each violation.

In addition, the bill prohibits any governmental entity which possesses a biometric identifier of an individual from selling, leasing, or otherwise disclosing the biometric identifier to another person unless: the individual consents to the sale, lease or disclosure; the sale, lease or disclosure is required or permitted by a federal or State law; or the sale, lease or disclosure is made by or to a law enforcement agency for a law enforcement purpose. A governmental entity that possesses a biometric identifier of an individual would be required to store, transmit, and protect from disclosure the biometric identifier using reasonable care and in a manner that is the same as or more protective than the manner in which the governmental entity stores, transmits, and protects its other confidential information. The bill also requires the governmental entity to establish a reasonable procedure under which an individual is entitled to have the governmental entity correct information about the individual that is possessed by the governmental entity and that is incorrect. The bill states that this procedure cannot be unduly burdensome.

Index